T0275993

Equine Behavioral Medicine

Equine Behavioral Medicine

Bonnie V. Beaver

Academic Press is an imprint of Elsevier
125 London Wall, London EC2Y 5AS, United Kingdom
525 B Street, Suite 1650, San Diego, CA 92101, United States
50 Hampshire Street, 5th Floor, Cambridge, MA 02139, United States
The Boulevard, Langford Lane, Kidlington, Oxford OX5 1GB, United Kingdom

Library of Congress Cataloging-in-Publication Data
A catalog record for this book is available from the Library of Congress

British Library Cataloguing-in-Publication Data
A catalogue record for this book is available from the British Library

ISBN 978-0-12-812106-1

For information on all Academic Press publications
visit our website at https://www.elsevier.com/books-and-journals

Publisher: Charlotte Cockle
Acquisition Editor: Anna Valutkevich
Editorial Project Manager: Pat Gonzalez
Production Project Manager: Sreejith Viswanathan
Cover Designer: Christian Bilbow

Typeset by SPi Global, India

Horses reach into your soul, pull out the good and heal what isn't.

Sandy Collier

Contents

About the Author

Bonnie V. Beaver

Dr. Beaver is a veterinarian internationally recognized for her work in animal welfare and in the normal and abnormal behaviors of animals. She has given over 250 scientific presentations on these subjects and has discussed many areas of veterinary medicine for the public media. In addition, she has authored over 150 scientific articles and has nine published books, including *Feline Behavior*, *Canine Behavior*, *Efficient Livestock Handling* (with Don Höglund), *Your Horse's Health*, and *Horse Color* (with D. Phillip Sponenberg). Her expertise in behavior and welfare has been passionately shared with the students at Texas A&M University and veterinary colleagues worldwide.

Dr. Beaver is a member of numerous local, state, and national organizations. Professionally this includes having served as president of the American Veterinary Medical Association (AVMA) and Texas Veterinary Medical Association. She was also a founding member and first president of the American Veterinary Society of Animal Behavior (1975), the American College of Animal Behaviorists (AVMA recognized veterinary specialty 1993), and the American College of Animal Welfare (AVMA recognized veterinary specialty 2012). Additionally, Dr. Beaver is a long-time member of numerous horse organizations including the American Association of Equine Practitioners, American Horse Council, American Quarter Horse Association, Palomino Horse Breeders of America, and National Reining Horse Association.

About the Author

Bonnie V. Beaver

[illegible]

[illegible]

Acknowledgments

It may take a village to raise a child, but it takes a lot more people to create a book like this. Because of the curiosity of numerous researchers and their willingness to share results of their studies through publications, we are able to understand horse behavior in a scientific, rather than anthropomorphic, way. To these many colleagues, I express appreciation.

No book on horse behavior should ever be written by someone who has not worked with those splendid animals. Each teaches us something, but collective experiences change learning from individual anecdotal happenings to a more comprehensive understanding of this species' fascinating behavior. For this, I owe a lifetime of thanks to the collective group of horses I have owned, ridden, and/or worked with over the years. Some excelled in the show ring; some excelled at challenging their rider. Several are pictured throughout this book.

The author is very appreciative of the support and encouragement given by so many individuals. Some freely supplied photographs; others allowed access to their horses. A few people helped with some proofreading; still others just said "do it." With sincere apologies to those who I may have overlooked, I wish to express special appreciation to the following people:

Mark Bradford of Skyline Ranch
Alda Buresh
Carla Carleton
Pat Gonzalez
Don Höglund
Jon Levine
Tami Nelson
Nathan and Jean Piper of Nathan Piper Reining Horses
Nyla Rayburn
Stanley and Lauren Stephenson
John and Nancy Tague of Toyon Ranch
Maggie Gratny Young

Chapter 1

The History of Horses and Their Relationship to Humans

HORSES THROUGHOUT HISTORY

Throughout human history, humans have been fascinated by the horse, and this creature and its ancestors have held a fascination for scientists as well. The evolution of the horse is the best known of all animal species because of a rich archive of prehistoric fossil bones. How the equine species came to be is still somewhat up for debate because of the many subspecies that developed over time, but there is no debate about the major contributions this animal has made to the development of the world.

Evolution

The boney records of horse relatives are rich, and they show the skeletal and tooth adaptations that led to the modern horse. Because changes occurred slowly, scientists have a fairly good picture of how the modern horse came to be, at least until about 25 million years ago (mya). Then, the family tree started to change rapidly, developing many branches. This branching led to a variety of theories as to the exact lineages. While modern DNA analysis of ancient bones would be desirable, it is nearly impossible because of bone mineralization and soft tissue degradation or replacement. The oldest sample analyzed to date is approximately 750,000 years old, from a foot bone preserved in the Canadian permafrost.[1–3] While this date is 9–10 times older than analyses of archaic humans, it is nowhere near the age of the oldest horse relative or most of the branching. Comparisons, then, rely on physical features, and for ancient horses, those features are primarily in the skull, feet, and teeth. Modifications in these structures accompany adaptation to the environmental changes.[4–6] Tropical forests became dry plains, and tender plants growing on soft ground gave way to coarse grasses on hard earth. Feet changed from four or five digits to one. Legs grew longer to travel farther for food, and necks got longer as the animal got taller to allow it to reach grass as it changed from browser to grazer. Successive ancestors also became larger in body mass.[7,8] The general

Equine Behavioral Medicine. https://doi.org/10.1016/B978-0-12-812106-1.00001-2

description of body size suggests there was a relatively constant increase over time, although that description is a little oversimplified. The earliest ancestors gradually increased in weight, topping out around 110 lb (50 kg). This remained fairly constant from 57 mya until approximately 25 mya. After that, the increase tended to be steeper.[7] The cheek teeth of horse ancestors also changed over time. They became more adept for grinding the harsher grasses instead of chewing tender browse by developing flatter, grinding surfaces and taller crowns.[6,9,10] Internal casts from ancient horse skulls show that size and complexity of the brain also continued to develop.[6,11] Although physical proof is lacking, it is reasonable to assume that soft tissue structures would also change, particularly the gastrointestinal tract.[5]

For the most part, the ancestors of the horse evolved on the North American continent. Occasional branches crossed the Bering Strait land bridge when it was exposed, and some moved into South America after the Isthmus of Panama joined the two continents between 12 and 15 mya. It was not until several thousand years ago that the horse completely left North America.

Eocene (56–33.9 mya)

Eohippus (also known as *Hyracotherium*) is considered the "grandfather" in the horse's lineage, dating back to about 55 mya (Figure 1-1).[14,21] This was a tiny mammal typically described as the size of a large house cat but estimated to weigh about 50 lb (25 kg).[6,7] The head and neck were relatively short, but the legs were slightly longer than would be expected. The feet of the *Eohippus* had five toes on each forelimb, with digits 2–4 having small protohooves instead of claws, and the first digit, corresponding to the human thumb, was off the ground. On the hind limbs the second, third, and fourth digits had small hooves. The first and fifth digits were vestigial and did not touch the ground. Like the cat, the feet were padded. Teeth of the *Eohippus* were typical of those of today's omnivore, being similar in appearance to those of humans. There were 44 teeth, with three incisors, one canine, four premolars, and three molars on each side of the mouth of both upper and lower jaws.[5] While the *Eohippus* survived approximately 20 million years, the teeth started adapting from a diet of soft foliage and fruit to plants they could browse.

The *Eohippus* gradually transitioned into the *Orohippus* about 50 mya.[14] The size of this ancient horse was slightly larger than that of its predecessor, perhaps about that of today's Italian greyhound, but the weight did not change much.[7] The legs lost the vestigial toes, giving the animals four toes on the front and three behind. Pads were still present on the feet. The teeth had changed from the *Eohippus*, with the first premolar shrinking in size and the fourth premolar taking on the shape of a molar. This suggests the diet was made of tougher plant material.

About 47 mya, the *Epihippus* continued the changes initiated in its predecessors.[14] The prehistoric horse was now about 2 ft (0.6 m) tall and weighed about 65 lb (30 kg).[7] The premolars were becoming more molarlike. This part of the

Eohippus (*Hyracotherium*)
  Orohippus
    Epihippus
      Duchesnepippus
      Mesohippus
        Miohippus
          Kalobatippus (*Miohippus intermedius*)
          Anchitherium (some went to Asia, then Europe)
          Sinohippus (Eurasia)
          Hypohippus
          Megahippus (last three-toed horse)
          Parahippus
            Merychippus (may have come from *Miohippus*)
              Callipus
                Pliohippus
                  Astrohippus
              Hipparion (N. America, Asia, Europe)
                Cormohipparion
                Nannippus
                Neohipparion
                Pseudohipparion
              Protohippus
                Dinohippus
                  Plesippus (N. America, Asia—may be intermediate to *Equus*)
                  Equus
                    Subgenus *Equus*
                      E. ferus (wild horse)
                        E. andium (extinct)
                        E. algericus (extinct)
                        E. conversidens (Mexican horse—extinct)
                        E. lambei (extinct)
                        E. neogeus (extinct)
                        E. niobrarensis (extinct)
                        E. sanmeniensis (extinct)
                        E. scotti (extinct)
                        E. f. ferus (*tarpan*—extinct)
                        E. f. caballus (modern horse)
                        E. f. przewalski (Przewalski horse)
                      New World stilt-legged horses (extinct)
                      Hippidions (extinct)
                    Subgenus *Asinus*
                      E. africanus (*African wild ass*)
                      E. hemionus (*Asiatic ass*)
                      E. kiang (*Kiang*)
                      E. hydruntinus (*European ass—extinct*)
                    Subgenus *Dolichohippus*
                      E. grevyi (Grévy's zebra)
                    Subgenus *Hippotigris*
                      E. quagga (Plains zebra)
                      E. zebra (Mountain zebra)
                    Subgenus *Parastylidequus* (extinct)
                    Several extinct species of unknown relationships

FIGURE 1-1 The lineage of the horse from earliest times. Around 25 million years ago the number of branches increased significantly, making tracing of ancestors less exact. This has resulted in comparisons between numerous sources being different. There also remain questions as to the number of branches relative to *Equus*. Unless indicated otherwise, all the following lived in North America.[12–20]

ancestral tree lived for about 10 million years. While there remains a debate as to whether the *Epihippus* gave rise to the *Duchesnehippus* as a distinct genus or whether the latter was a subgenus, it is known that *Epihippus* did give rise to *Mesohippus*. This probably happened in response to climate changes occurring in North America.

Mesohippus came about rather quickly 40 mya as a result of the selective pressures of foliage changes in a drier environment.[14,21] Lush forests became grass and brush as land dried, so animals had to travel farther to eat, run faster to avoid predators, graze instead of browse, and adapt to coarser vegetation. The *Mesohippus* was the first ancient horse to have three toes on the front leg rather than four toes. As its limbs grew longer, the third digit was enlarging, although the second and third digits still supported some weight. Longer legs made the *Mesohippus* slightly taller than its *Epihippus* predecessor and also meant that the head and neck became somewhat longer so the animal could graze. Because eating behavior included both browsing and grazing, it was becoming harder to differentiate premolars from molars in the *Mesohippus*. With the exception of the rudimentary first premolar, all premolars and molars were grinding "cheek teeth," similar to what we associate with modern horses. Changes in appearance of these teeth and crown heights continued to evolve in relationship with the diet of the ancient horses. This animal weighed about 100 lb (45 kg).[21]

Oligocene (33.9–23 mya)

Miohippus evolved rather abruptly from the *Mesohippus* about 30 mya and then coexisted with it for at least 8 million years while gradually replacing it.[14,21] The *Miohippus* ultimately branched into forest dwelling and plains dwelling groups. Some scientists report that members of both groups probably weighed about 50 lb (23 kg), although others suggest they may have weighed as much as 120 lb (54 kg).[7] There were three complete toes on both front and back feet, with the third digit getting larger and the side toes bearing weight only on soft ground. Digital pads remained and continued to be weight bearing. The upper cheek teeth were beginning to develop a prominent crest, a characteristic of modern horse teeth.

It was during the Oligocene, about 25 mya, that the ancestral tree really started to branch at a fast rate and in multiple directions. This has resulted in considerable variation in opinions as to which ancient genus gave rise to the next, eventually leading to *Equus*.

The forest-dwelling group of *Miohippus* gave rise to *Kalobatipippus* (also known as *Miohippus intermedius*) 24 mya. The *Kalobatipippus* lived to about 19 mya and possibly gave rise to the *Anchitherium* group.[14] Some of the *Anchitherium* group crossed the Bering Strait land bridge into Asia and continued from there to Europe. In Eurasia, the *Anchitherium* gave rise to the *Sinohippus*, which would later become extinct. The *Anchitherium* that remained in North America gave rise to the *Megahippus* and *Hypohippus*, also dead-end branches.

Miocene (23–5.3 mya)

Between 20 and 25 mya, the plains-dwelling *Miohippus* that remained in North America gave rise to the *Parahippus*, an animal that was the size of a small pony or large dog, weighing about 200 lb (90 kg).[7,14] The third digit had continued to enlarge and bore all the weight. This increase in digit size was necessary to accommodate bone stresses of the increased body size.[22] The digital pads were present but did not carry weight, and side toes supported weight only in boggy ground. The teeth continued to evolve. All the premolars resembled molars except the small first ones, and the incisors remained shaped much like human incisors. Only the upper ones had a small indentation that marked the beginning of a cup.

The *Parahippus* gave rise to the *Merychippus*, the first genus of the proto-horse to be primarily grazers.[16,17,19] *Merychippus* stood about 3 ft tall (1 m) and may have weighed up to 300 lb (135 kg). From this point onward, size increased at a much greater rate than in previous eras.[6] Side toes still had tiny hooves that are thought to have only touched the ground when the animal was running, if at all. This genus evolved for approximately 15 mya and eventually developed into at least three lineages: *Callippus*, *Hipparion*, and *Protohippus*.[14]

The *Callippus* gave rise to the *Pliohippus* about 12 mya. Until recently *Pliohippus* was thought to be the continuation of the ancestral line of the modern horse. This has recently been questioned because the lateral side of its skull had a deep facial fossa that was not continued in other direct horse ancestors. Instead, it is now felt that the *Pliohippus* was the likely ancestor of the *Astrohippus* and both became dead-end groups.[14]

The *Hipparion* is also a dead-end group that came from the *Merychippus*. It proliferated into many kinds of horse-like animals, some of which migrated into Eurasia. These animals were a lot like today's antelopes with slim legs and one main toe flanked on each side by a smaller one.[21] Their weight was probably around 240 lb (110 kg).[7]

The *Protohippus* branch of the *Merychippus* is the likely one to have given rise to modern equids. This branch gave rise to the *Dinohippus* between 5 and 13 mya. It was a small horse-sized animal weighing 300–530 lb (135–240 kg).[7,14]

Pliocene (5.3–2.6 mya)

In North America, *Equus* arose about 5 mya. Some scientists feel the *Dinohippus* gave rise to the *Plesippus* and it to the *Equus* group. Others do not recognize the *Plesippus* as part of the family tree and attribute *Dinohippus* as the immediate ancestor of *Equus*.[1–3,20] It is the primitive *Equus* that ultimately gave rise to all caballine (true horses) and noncaballine species that are similar in appearance to horses (Figure 1-2). The first major branching that did not lead to a dead end occurred between 2.1 and 3.4 mya with the evolution of *Equus asinus*, the ancestor of today's wild asses, zebras, and donkeys. Asiatic and African asses split into new subspecies from their common ancestor about 1.7 mya, with zebra branches separating about 1.1 mya.[15]

Pleistocene (2.6–11,700 ya)

The ancient horse flourished in the Western Hemisphere, ultimately dividing into three main groups (Figure 1-2).[21,31,33] The *Hippidion* split from the Equus

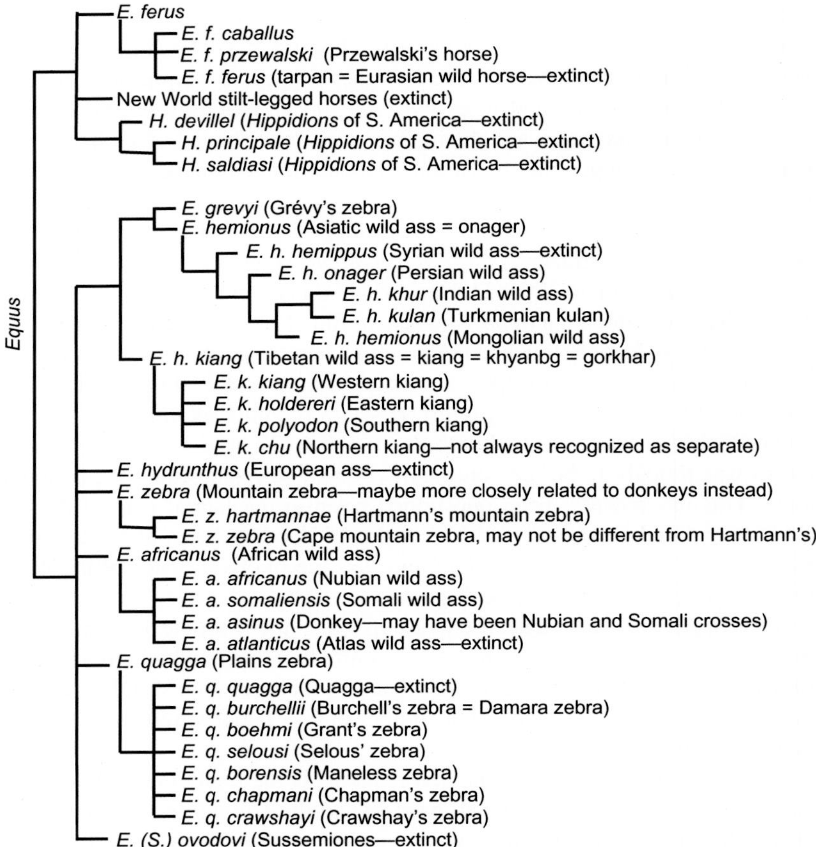

FIGURE 1-2 The relationship of modern equids. There remain controversies relative to several of the relationships, so this representation combines several anatomical and DNA analysis studies to estimate the most current thinking.[3, 15, 18, 20, 23–32]

genus between 5.6 and 6.5 mya.[34] These animals resembled modern horses except for their three toes.[35] Some members of the group moved into South America about 2.5 mya, shortly after the formation of the Isthmus of Panama, possibly because of the chilling climate. Here it coexisted with other *Equus* groups. Ultimately, the *Hippidions* went extinct.

The second group has been given the name of New World stilt-legged horses (NWSL). These animals ranged widely across North America and shared physical features with some modern Asian asses. Eventually this line also went extinct about 30,000 years ago.[35]

Equus ferus, the third group of the *Equus* genus, ultimately gave rise to three "true horses." Przewalski's horse separated from the *Equus ferus caballus* (modern horse) and *Equus ferus ferus* (tarpans) between 38,000 and 72,000 years ago. It has remained separate since then and is the last surviving wild horse.[3,36] Tarpans became extinct near the beginning of the 20th century.[5,37]

Holocene (11,700 ya to present)

Horses continued to exist in North America in diminishing numbers and size until 10,000–14,200 years ago, and then disappeared relatively suddenly for reasons that are not totally clear.[12,14,21,38,39] It is known that the vegetation in Alaska had changed significantly about 20,000–26,000 years ago. This probably forced some horses across the Bering Strait land bridge into Asia. This was also a time of the extinction of other megafauna species in both North and South America. Not until Christopher Columbus brought horses to the New World in 1493 did *Equus* return to its ancestral home.

Genetics

In the past, the knowledge about genetic differences between the various equids was limited to chromosome numbers. As an example, Przewalski's horse has 66 chromosomes (33 pairs), compared to the 64 of domestic horses, 62 in donkeys, 63 in mules, and between 32 and 46 in zebras.[15,35,40–42] Modern genomic analyses have provided a great deal more information. Specifically, the onset of mitochondrial deoxyribonucleic acid (mtDNA) and Y chromosome marker analyses have provided tremendous insights into the evolution of the modern horse—far more than morphologic comparisons alone. Not only are researchers starting to understand the relationships between current equids (Figure 1-2), but they can also now look for genetic diseases and traits that are associated with outstanding performance in modern horse breeds.

Studies of mtDNA, which is passed only from mares to their offspring, indicate that there were multiple ancient maternal origins for today's horse.[36,43,44] Over 100 distinct mtDNA sequences (haplotypes) exist in horses around the world today, and these are fitted into seven haplogroup clusters (A–F).[43] The clusters can be dated to 320,000–630,000 years ago, long before domestication, so they are shared by different breeds of horses, including Przewalski's horse.[45]

Y chromosomal analysis is used to follow the male line of progeny. For the modern horse, only six haplotypes are found, suggesting that at some time in the past selective breeding narrowed the gene pool to descendants of a small group of closely related stallions.[36,43,46] Recent sequencing of genomes from stallions in an ancient tomb suggests the original number of stallions following domestication was plentiful and included intermittent breeding with wild horses until at least 2300 years ago.[47,48] Some time after that, inbreeding practices narrowed the haplotype to six, and these are distinctly different from the two haplotypes identified by Y chromosomal analysis in Przewalski's horses.

With the advent of DNA sequencing, researchers have been able to determine the genome of the horse. Its chromosomes consist of approximately three billion DNA base pairs, most of which are part of an estimated 20,000 distinct protein-coding genes. These numbers are essentially the same for dogs and humans. The big difference is that the genes are carried on 78 chromosomes in dogs, 64 in horses, and 46 in people.[49–51]

Inbreeding and line breeding are common practices in modern horse breeds, done in an attempt to concentrate desirable genes for traits that make the offspring successful in the show ring or on the race track. Unfortunately, the heritability of desirable traits is not a sure thing, even with tight breeding practices. Heritability is also hard to calculate. The heritability of a specific trait is typically indicated by a number ranging from 0.0 when there is no genetic contribution to the trait, up to 1.0 when inheritance is the only reason for the trait. For racing Standardbreds, the heritability score for speed is 0.29.[52] This means approximately 29% of a Standardbred's speed may be attributed in some way to the individual's genes. It does not mean that 71% of the speed is due to the environment, but rather that several other things contribute to the trait. Conformation, motivation, temperament, training, and environment contribute a lot more to the horse's racing success than does inheritance. Genetics plays some role in temperament characteristics related to trainability and some behavior problems like cribbing in Finnhorses at a level of 0.10 and 0.68, respectively.[53,54]

How heritability is measured affects the results. As an example, heritability for thoroughbred racing speed is anywhere from 0.09 to 0.76 depending on the distance run and what measurement was used—race time, money won, or handicap value.[55–57] Even clones are not identical in abilities. Three mule clones of a highly successful racing mule have ended up being quite different in personality and performance ability.[58]

Modern techniques of genetic analysis are now being used to evaluate changes that affect performance and health.[59–61] The *DMRT3* mutation in several breeds, including the Standardbred, has surfaced as something to be considered relative to racing success. The AA genotype of *DMRT3* is better for racing success, while horses with CC or CA genotypes are better for general riding.[60,62] *MSTN*, the myostatin gene, is associated with muscle fiber proportions in the Quarter Horse and potentially other racing breeds.[47,62] Other genes associated with racing performance—*ACN9*, *CKM*, *COX4I1*, and *COX4I2*—have been present in horses since at least the mid-Holocene.[47] Loci on *ECA11* affect the extremes in size variations from the Miniature horse to the draft breeds.[62]

Behavioral genetics is more difficult to study due to the large variation in how traits are expressed. The dopamine D4 gene (*DRD4*) has been shown to relate to a horse's curiosity and vigilance.[59] In Tennessee Walking Horses, candidate loci have been identified that may be associated with anxious, tractable, and agonistic temperaments.[63] Also within that breed, mares that are black in color (aa on the *ASIP* locus) are more independent than bay (A-) mares.[64]

The practice of breeding for genetic closeness also concentrates undesirable recessive traits that can compromise the foal's welfare or even life.[65] Modern horses have considerably more recessive deleterious genes than did ancient horses.[36,47] Some of these genes tag along with desirable traits, such as happens with roan and tobiano coat colors. The first is related to embryonic loss and the second with reduced fertility.[36] As these deleterious genes are identified, owners can make informed decisions about their breeding programs. For specific medical conditions that are particularly devastating, breed organizations are beginning to require genetic testing in an attempt to educate breeders and buyers (Table 1-1). Relative to the heritability of negative behavior traits, there are certain ones that seem to be passed down through family lines, particularly fearfulness and reactivity.[66] But "seem to" is all that can be said at this time. Determining the heritability of behavior is even more difficult. Finite endpoints are less easily defined, particularly when that endpoint is really based on what the horse might be thinking. As an example, stallions and their offspring known for their aggression in the racing or show barn may not show that trait at all when pastured.[66]

Epigenetics

In a discussion of heritability, the topic of "nature vs. nurture" invariably comes up. As shown with the heritability of racing, genetics do not provide all that is needed for top performance. Research studies on human identical twins who were separated at birth have tried to tease out answers, only to show how complicated genetics and environmental interactions can be.[67] Recently, research has taken off in a genetics-related field of study—epigenetics. While the term has been around since the 1940s, research capabilities are just catching up to the concept that heritable changes in gene expression can occur that do not involve genotypic change. In other words, the phenotype changes but the genotype does not change.[68,69] Starvation or obesity of a paternal grandfather has been linked to increased risks for diabetes and cardiovascular disease.[70,71] Stress events can also have transgenerational effects.[68,72–76] While epigenetic effects on behavior have been studied mainly in laboratory rodents, the findings are remarkable enough that it is prudent to consider how horses might be affected.

Animals inherit many more genes than the ones that are active. Some have lost their evolutionary need. Others are simply not activated. Think of genes as having an on/off switch, and many genes are turned off all the time. Then, with the right situation, something might turn a gene on that changes a physical trait, health condition, or behavior. Research is suggesting that DNA methylation is that switch.[71] Epigenetics is basically the environmental flipping of the on/off switches of certain genes. What scientists are learning is that this altered expression of the gene is then passed on to the next generation, and even the one after that.[77–79] Retention of fear of specific cues, such as an odor, can be transmitted at least two generations from either the female or male side, as an example.

TABLE 1-1 Recessive or Suspected Recessive Conditions in Horses and Some of the Major Breeds With Which They are Associated

	American Quarter Horse	Arabian	Draft Horse	Warm-Blood	Appaloosa	Morgan	Thoroughbred	Standardbred	Haflinger	Akhal-Teke
HYPP	X									
HERDA	X									
GBED	X									
PSSM	X		X		X				X	
MH	X				X					
OLWS	X						X			
DSLD	X	X	X	X			X			
SCID		X								
CA		X								
LFS		X								
EMS		X								
JES		X								
GPT		X		S						
OAAM		X				S		S		
JEB			X							
NFS										X
IMM	S			S						
RU			S	S	S					
T							S			
OD							S	S		
RLN			S				S			

Known conditions include: hyperkalemic periodic paralysis (HYPP), hereditary regional dermal asthenia or hyperelastosis cutis (HERDA), glycogen branching enzyme deficiency (GBED), polysaccharide storage myopathy (PSSM), malignant hyperthermia (MH), overo lethal white syndrome (OLWS), degererative suspensory ligament desmitis (DSLD), severe combined immunodeficiency disorder (SCID), cerebellar abiotrophy (CA), lavender foal syndrome (LFS), equine metabolic syndrome (EMS), juvenile epilepsy syndrome (JES), guttural pouch tympany (GPT), occipitoatlantoaxial malformation (OAAM), junctional epidermolysis bullosa (JEB), and naked foal syndrome (NFS). Suspected genetic conditions include immune mediated myositis (IMM), recurrent uveitis (RU), tying-up (T), various developmental orthopedic diseases (OD), recurrent laryngeal neuropathy (RLN) (X = known genetic problem, S = suspected genetic problem).

There is a great deal of ongoing research in an effort to better understand epigenetics. While we do not know the implication of this in horses and are only beginning to grasp what it might mean in humans, epigenetics may someday be shown to play a role in better understanding individual horses. It certainly could have implications in understanding the environment's role in gene expression.

Variety of Equids

The wide variety of ancient horses was ultimately narrowed down to one genus, *Equus*, and from that evolved the multiple subspecies that exist today: horses (*ferus*), Grévy's zebra (*grevyi*), Asian wild asses (*hemionus* and *kiang*), mountain zebras (*zebra*), African wild asses and donkeys (*africanus*), and plains zebras (*quagga*) (Figure 1-2). Of these multiple subspecies, humans have domesticated members of only two—the true horse and the donkey. The evolutionary distance between true horses and their Asian and African relatives is part of the explanation of why donkeys are behaviorally different from horses and why their domestication took place in different parts of the world.

In the United States, horses under 14.2 hands (58 in., 147 cm) are generally called ponies. This is arbitrary, with some people calling any horse a pony. Some animals are regarded as horses even though they are under the 14.2 hand height. It is likely that the original *Equus ferus caballus* was somewhat smaller than the 14.2 hand horse size, probably limited because of harsh environments.[80] Most modern pony breeds come from areas having similar harsh conditions and perhaps where bred to continue that small size to meet a specific need, such as working in underground coal mines. The *HMGA2* gene on chromosome 6 has been linked to the short stature of ponies.[81] Taller horses developed where food became plentiful or there was a need for speed or distant travel. For purposes of this book, the size is irrelevant and all will be called horses.

Horses Around the World

Different breeds have been developed around the world to meet a specific need, to highlight a specific appearance or gait, or because of relative isolation. The Thoroughbred is noted for racing in the United States and England, as an example of a breed highlighted for a specific purpose. Paso Finos, Tennessee Walking Horses, and Missouri Fox Trotters have specialized gaits. The Thai Native Pony probably originated in Mongolia before Genghis Khan, but its isolation in Thailand has resulted in it becoming a genetically unique breed.[82] As gene sequencing becomes less expensive, it is being used to find similarities, differences, and relatedness between horse breeds.[83–85] The Food and Agriculture Organization of the United Nations (FAO) reports 618 local horse breeds worldwide (Table 1-2).[86] There are a number of sources that list or describe some or all of these breeds, but general behaviors are shared.[87–90]

TABLE 1-2 Location and Number of Local Breeds Reported to the Food and Agriculture Organization of the United Nations (FAO) in 2012[86]

Geographic Location	Number of Local Breeds
Africa	38
Asia	138
Europe & Caucasus	306
Latin America & Caribbean	76
Near & Middle East	14
North America	22
Southwest Pacific	24
Total	**618**

Population Statistics

Exact numbers of horses are imprecise because census counts are not done and free-ranging horses are difficult to count. Population statistics are also a moving target, since numbers will vary based on local economics and weather conditions. Data for the last several years indicate a worldwide horse population of approximately 59 million horses, 43 million donkeys, and 10 million mules (Tables 1-3 and 1-4).[86,89,91,92]

TABLE 1-3 Distribution of Horses Around the World in 2010 and 2014 as Reported to the Food and Agriculture Organization of the United Nations (FAO)[91]

Region	2010 Horse Population	% of Horse Population—2010	2014 Horse Population	% of Horse Population—2014
Americas	33,494,042	56.0%	32,528,035	55.2%
Asia	13,620,634	22.8%	14,348,443	24.4%
Africa	5,985,701	10.0%	6,076,168	10.3%
Europe	6,269,921	10.5%	5,556,545	9.4%
Oceania	404,350	0.7%	404,766	0.7%
Total	**59,774,648**		**58,913,957**	

TABLE 1-4 Top Five Countries With Horses in 2010 and 2014 as Reported to the Food and Agriculture Organization of the United Nations (FAO)[92]

Country	Number of Horses—2010	Number of Horses—2014
United States	10,000,000	10,260,000
Mexico	6,355,000	6,355,000
China (mainland)	6,785,000	6,027,400
Brazil	5,514,253	5,450,601
Argentina	3,600,000	3,600,000
Total	**32,254,253**	**31,693,001**
% of world's horses	**53.9%**	**53.8%**

The United States has the largest number of horses of any country in the world with an estimated 9–10 million animals.[91,93,94] Approximately 1 million of those are located in the state of Texas, with California second in horse population, and Florida third (Table 1-5). There is almost an equal division between horses used for recreation (3,906,923 horses) and those used for showing (2,718,954) and racing (844,531). An additional 1,752,439 horses are used in other ways.[93,95]

TABLE 1-5 The Top 10 States for Horses in the United States and Their Estimated Horse Population[93, 95]

State	Number of Horses
Texas	1,000,000
California	700,000
Florida	500,000
Oklahoma	326,000
Kentucky	320,000
Ohio	307,000
Missouri	281,000
North Carolina	256,000
Colorado	256,000
Pennsylvania	256,000

Free-Ranging Horses

In a discussion of free-ranging horses, the terms "wild" and "feral" often come up. As is described under domestication, both of these words may or may not accurately describe some of these horses. For that reason, "free-ranging" is a more accurate description of horses that live free or generally free from humans. These animals are often considered to embody the free-spirit nature and historical tie to an aggrandized past. The general public typically fails to appreciate how rapidly numbers can increase in free-ranging populations and what impact that can have on horse and environmental welfare. Without human interference or harsh environmental conditions, the number of free-ranging horses will increase by 15%–20% per year, with the herd size doubling every 4 years.[96,97] When herds are small, isolated, partially culled, or part of a contraceptive management program, the preservation of genetic diversity becomes particularly important.[98] Additional concerns have to do with habitat destruction—overgrazing, soil compaction, and water contamination. Debates continue about what is the appropriate action needed to preserve the environment and keep free-ranging horses. Expenses are real and resources are finite.

There are populations of free-ranging and semi-free-ranging horses and ponies throughout the world. In some areas, they represent unique breeds, and in others they are genetic hybrids (Table 1-6). The largest number of free-ranging horses is found in Australia, where numbers are estimated to be over 1 million.[96] The free-ranging horses of Australia, called brumbies, are creating the same problems as the mustangs are in the western United States.

Mustangs are managed by the United States Department of the Interior's Bureau of Land Management (BLM). This agency estimates that public grazing lands will support a maximum of 26,715 free-ranging equids before range conditions begin to deteriorate. As of March 1, 2016, there were an estimated 55,311 free-roaming horses, and an additional 11,716 free-roaming burros. In addition to on-range animals, the BLM rounds up a few thousand horses and a couple of hundred burros every year. Events are held to adopt out and sell as many animals as possible, but the supply outnumbers the demand. The BLM is currently managing >44,000 horses and 1200 burros in corrals, pastures, and ecosanctuaries off-range. Maintaining these captured animals is expensive and currently takes up over 60% of the Bureau's annual budget.[97]

Other countries have free-ranging or semi-free-ranging horses, but usually in small numbers. The exception is China, where the Mongolian people have a horse-related culture predating Genghis Kahn. Their horses are kept in a semi-free-ranging state, where they fend for themselves during the winter months and only a few are caught each year to ride.

THE HORSE AND THE HUMAN

The horse's association with humans probably started with it being hunted for food. Gradually, this gave way to the domesticated version. Why the horse?

TABLE 1-6 The Estimated Number of Free-Ranging and Semi-Free-Ranging Horses by Breed Around the World

Free-Ranging Horses/Ponies	Country	Estimated Number
Banker Horse	USA	120–130
Brumby	Australia	400,000–1,000,000+
Chincoteague Pony (Assateague Horse)	USA	300
Cumberland Island Horse	USA	150–200
Danube Delta Horse	Romania	4000
Elegesi Qayus Wild Horse (Brittany Triangle Horse)	Canada	886
Garrano	Portugal	2000
Kaimnawa Horse	New Zealand	300
Kundudo Horse	Ethiopia	11
Lavradeiros	Brazil	40
Marismeño	Spain	Near extinction
Misaki Horse	Japan	100
Mustang (on range)	USA	55,311
Namib Desert Horse	Nambia	90–150
Nakota horse	USA	70–110
Sorraia	Portugal	>200
Sable Island Horse (Sable Island Pony)	Canada	160–360
Welsh Pony (feral)	Wales	180
Semi-Free-Ranging Horses/Ponies		
Camargue Horses	France	varies
Dartmoor Pony	England	800
Exmoor Pony	England	800
Konik (Polish Primitive Horse)	Poland	550
Mongolian Horse	China	3,000,000
Mustang (off-range)	USA	44,219
New Forest Pony	England	Thousands
Pottok (Pottoka)	France, Spain	150

Several other countries, like China, have similar horse populations that are mixed breeds or not well documented, and thus hard to identify.[84, 96, 97, 99, 100]

Certain characteristics favor domestication, with the most important being some human need. Being able to keep many horses close to where humans lived would provide a ready source of food, as an example. While many other species could also meet that need for domestication, other characteristics also favored the human-horse connection. These animals live in large social groups having a hierarchical structure, and males are associated with the group. This allows several individuals to be kept in a smaller space, even in pens or fenced areas. Promiscuous mating is needed for selective breeding. A strong female-young bond allows easier care of the young and time for them to learn from their dam. Adaptability to a wide range of environments also favored equine domestication.[101]

In a discussion of animals and their relationship with humans, there are a number of related terms frequently used, often inappropriately. The most common of these are the terms "wild," "feral," "domestic," and "tame" (Table 1-7).

TABLE 1-7 Terms Used to Indicate Relationships Between Humans and Animals

Term	Meaning of the Term	Example
Wild	Animal species with little human interference in mating. They can be controlled and bred in captivity but are no different from free-roaming forms.	Muskrat; Lion; Zebra
Domesticated	Subset of a wild species that has undergone extensive selective breeding for one or more unique characteristics over several generations. Differ from wild forms morphologically, physiologically, and/or behaviorally.	Cat; Dog; Horse
Feral	Subset of domestic animals that has undergone reverse domestication over several generations such that they have become "wild" animals which retain some features of their domesticated ancestors.	Regional populations of dogs (i.e., Georgia) and some cats
Free-roaming	Animals that are not under human control. This can be on a part-time basis, like a dog let out of the house in the morning, or full time. It would include feral animals, most wild animals, and domestic animals reverting to a feral state.	Local populations of dogs and cats; Most mustangs
Tame	A wild or domestic animal that has learned to accept close proximity to a human.	Bart the bear; Secretariat the horse

There is a general understanding of what a wild animal is. We think of lions, deer, and skunks—animals living relatively free from human interventions, with the occasional exception of a few individuals in zoos or wildlife parks.

Domestication is a process of changing a fairly large subset of a wild population over many generations.[101] In order for domestication to occur, generations of selective breeding, whether at the conscious level or not, result in one or more of three types of changes. Appearance can change. An excellent example of this is seen in a comparison of Przewalski's horses with modern riding horses. The second change, and probably the most common one, affects the offspring's physiology. Wild animals live in a state of hypervigilance, which would prove to be extremely stressful for animals that live close to humans. While horses retain some of this vigilance, as evidenced by shying, they do not overreact to the presence of a dog unless they have had a bad experience with one. Without hypervigilance, a domesticated animal is easier to tame and acclimate to unique situations. The third change is in behavior. One example is the domestic horse allowing a rider on its back. Prior to domestication, the "something on the back" would have been a mountain lion or other predator, so wild horses evolved to not tolerate things on its back.

Ultimately, the degree of selection for a specific trait, the length of time to sexual maturity, and the heritability of that trait determine how fast this domestication process occurs. The trait chosen for intense selective breeding is also frequently connected to other traits that are modified at the same time. A well-published experiment with fur foxes in Russia was started several years ago in which foxes were bred for tameness or for aggression. In only 20 generations, the group selected only for tameness became dog-like in their behaviors. Additionally, physical changes that were not part of the selection process happened. The ears drooped and fur coloration changed to include white spotting.

In its strictest use, the term "feral" applies to a group of animals that have undergone reverse domestication.[101] For this, a closed breeding population starts as domesticated animals and over several generations changes genetically to be "other than domesticated." There is always controversy about which groups of animals are truly feral, as opposed to "feral-tending," and one should look for changes in the same three areas—appearance, physiology, and behavior—to help in that distinction. There are certain groups of dogs in the southeastern United States that have the distinction of being truly feral. Over many years without human control, group members have all taken on a similar appearance. Today's mustangs in the western United States still show appearances of their domestic ancestors in color and build. Arguments that these horses are feral point to changes such as smaller size, better hoof quality, higher level of alertness, and tendency to be leerier in strange situations compared to domestic horses.

It might be more appropriate to describe mustangs as "feral-tending" instead of "feral." In other words, their population is somewhere in the middle of the process of becoming feral, with some groups being closer than others.

"Free-roaming" and "free-ranging" are terms used to describe domestic-looking animals that are not under human control. This can be applied to any loose animal that is still similar to the domestic one but living primarily on its own. Free-roaming animals range from dog packs and cat groups to mustangs. Unfortunately, the terms also have been used to describe the domesticated dog while it is free to roam outdoors, even though it may spend part of its time indoors with the owner.

Another term to consider in a discussion of wild and domestic is "tame." In this case, discussion is about individual animals rather than a group of animals. By definition, a tame animal is an individual animal, domesticated, feral, or wild, that does not fear being close to a human because of a certain degree of trust having been developed toward the human. Examples of tamed wild animals include the Asian elephants used in logging and Bart the trained bear used in the movies. Domesticated animals are not automatically tame, as exhibited by free-roaming cats that run away when humans approach and actively resist capture. The process of taming an individual typically starts during the socialization period of very young animals, although it can be done with older animals to some extent. The process of taming is more difficult and results more restricted in wild species than in domesticated ones.

Human-Horse Bond

Domestication

Between 8000 and 12,000 years ago, most livestock species were domesticated in the Fertile Crescent area of what is now the Middle East—except the horse.[102,103] The ancient *Equus ferus* colonized the steppes of western Eurasia in an area that is now Kazakhstan and southern Ukraine, and it is likely that domestication first occurred there.[12,102–106] Exactly when is hard to pinpoint because it is so gradual. Evidence suggests domestication happened between 4000 and 6000 years ago, and potentially up to 9500 years ago, at a time when the diminishing horse population underwent a significant expansion.[36,102–105,107–114] Studies of mitochondrial DNA show genetic clusters around certain modern breeds or geographic locations within some, but not all, segments of the horse population. This suggests that the horse may actually have been domesticated multiple times, and it is likely that local wild horses also interacted with domesticated mares.[105,106,110,115,116] Very recent research suggests that Przewalski's horses may be feral descendants of early domestic horses from the ancient Botai culture rather than true wild early horses.[108]

Certain physical changes have also been associated with domestication. Two of the most obvious changes in the peri-domesticated horse were variability in body size and changes in coat color. Both suggest that selective breeding was part of the domestication process.[105,111,117] Even though leopard spotting was found to exist in wild type horses 25,000 years ago, the typical color of the wild horse was bay or bay-dun. About 6000 years ago, black horses appeared in what is now Spain and Portugal, comprising about 25% of the population. This is thought to be a natural mutation that was then actively bred for by humans. Chestnut coats appeared about 5000 years ago, and their numbers increased rapidly so that in another 1000 years chestnut horses made up just over one-fourth of the population.[111] Also during this 1000-year period in what is now Turkey, leopard spotted horses apparently became very popular because they made up an unusually high portion (60%) of horses.[118] Paint/pinto color patterns first appeared as the sabino pattern about 4000 years ago, followed by the tobiano pattern 500–1000 years later.[110] In the meantime, the leopard spotted horses apparently fell out of favor again and became relatively rare. Their numbers did not rise again for another 500–1000 years, probably by a reintroduction of the color from crosses with wild horses.[118] Diluted body colors such as buckskin, dun, and the body color now known as silver or silver dapple (which is common in today's Shetland ponies and often confused with chestnut) appeared about 2700 years ago.[111,119]

Humans quickly understood the value of the horse as a means of transportation, and genetic similarities spread more broadly than they did for other livestock species. This suggests horse travel was widespread. As a result, today there is a relatively homogenous genetic pattern in horses across the globe.[45,102,104]

Genomic sequencing is beginning to reveal interesting changes that have resulted because of domestication. By comparing the genes of ancient wild horses with those from Przewalski's horses, little difference is found.[114,120] But when these wild horses are compared to the domestic breeds, variations are found in two gene groups. The first is related to muscular and limb development, joints, and the cardiovascular system.[114] Perhaps these are associated with physiological adaptations relating to their uses by humans. Differences related to the second gene group affect cognitive functions, including social behavior, learning abilities, fear responses, and "agreeableness."[47,114] These changes may have related to the ability to tame individuals and ultimately domesticate them. Genetic evidence also shows that domestication was associated with inbreeding and the increase of "deleterious mutations," as has also been seen in other domesticated plant and animal species (Table 1-1).

Uses Throughout History

In the late Pleistocene, horses existed in the open tundra areas of what is now Europe, as the one type of animal that could tolerate cold during the last glaciation period. For the indigenous hunter-gatherer populations, these animals meant survival by providing much-needed meat.[109,121,122] Evidence now suggests that by 4500 years ago, peri-domesticated horses in the Eurasian Steppes of modern day Kazakhstan were not only a source of meat, they were being ridden and milked.[109,113,121,123] These horses had gene variants favoring mammary gland development but minimizing water loss.[47] Riding was useful for caring for herds of horses and other species. It increased productivity and efficiency by allowing one person to shepherd more animals over larger grazing areas than could be done on foot. As an example of this point, a man on foot and a dog can herd approximately 200 sheep, but if that man is on horseback, he can care for 500 animals.[107] This represents more food, clothing, and overall wealth.

Plaques from Mesopotamia and correspondence between kings 4000 years ago show that horses were being ridden with reins going to a nose ring and riders sitting near the rump—techniques sometimes used today for donkeys. Both techniques probably were first adopted from oxen and donkeys, but they were not efficient for steering horses or riding at speed. By 3500 years ago, Egyptians were riding a more traditional style with a saddle cloth and leather girth. Actual use of a saddle did not happen until approximately 2300 years ago with Scythian and Chinese nomads.[112] By then, riding had become more widespread throughout Europe, the Near East, Central Asia, and perhaps India too.[121]

As the traditional Chinese idiom goes, "When horses arrive, success follows." The value of using horses for rapid deployment was not lost on kings of ancient times either.[112] While donkeys were the original animal of transportation, horses quickly became favored by nobility because of their speed and maneuverability. This not only led to their popularity with the mighty, it led to the widespread distribution, especially throughout the Near East and southern Europe (Figure 1-3). Artifacts indicate horses had been trained to pull chariots about 4000 years ago in the southeastern Ural steppes.[107,112,124] In warfare, each

FIGURE 1-3 Nobility prized horses and used them to ride and pull chariots. This wall panel, now in the National Museum of Italy in Rome, was in or near the Basilica di Giunio Basso (Basilica of Junius Bassus) and dates back about 1600 years.

chariot was supported in battle by a small band of foot soldiers. Tribal raiding on horseback probably occurred soon after domestication, but the use of horses by a cavalry did not occur until about 3000 years ago.[107,124] Shortly thereafter, chariots accompanied the cavalry horses for large-scale military use, at least as long as the ground was navigable for the wheeled carts. Mountainous terrain limited fighting to foot soldiers and mounted warriors. Until 600 years ago, it was the horse that determined whether an army would face victory or defeat.[109]

Selective breeding, even in these most ancient of times, resulted in differences in physical size. Horse bones from ancient Egypt were refined, similar to today's Arabian, while those from Kazakhstan and Ukraine remained short and stocky. In the Middle East, horses used for war gradually became taller so they could carry more weight over longer distances at a lower "biologic cost."[112]

Approximately 3500 years ago, the Hittites had instructions for conditioning, interval training, and feeding their horses, and early written veterinary protocols existed. Four-horse chariot races were added to the Olympic games 2500 years ago, followed by mounted racing some 60 years later.[12] The Chinese emperor Qin Shi Huangdi had the famous terra cotta horses and warriors created 2300 years ago (Figure 1-4).[12]

FIGURE 1-4 The terra cotta horses and warriors are part of the burial site for China's first emperor, Qin Shi Huangdi.[12]

Uses of Modern Horses

Historically, Native Americans took great pride in their horses. They served as mounts for the warriors. Horses also took on religious meaning in some Native American cultures.[125] As in the past, the horse remains a measure of wealth, standard for trade, and symbol for status in many tribes.

The U.S. Army officially eliminated the horse cavalry during World War II, but individual horses were still associated with later wars. As an example, Sergeant Reckless was purchased by members of the 5th Marine Regiment and used as a pack horse during the Korean War.[126] She was later recognized as one of America's 100 all-time heroes by several national magazines. While the usefulness of the horse in military conflicts has ended, they now are proving helpful in the war on crime as police mounts.

Worldwide, today's horse is used in ways that generally fall into three broad categories: food, work, and pleasure. The first use of the horse probably was related to its being a source of food. That role continues in some parts of the world.[109] While the thought of eating horseflesh is repugnant to many in English-speaking countries, the meat continues to provide high-quality protein for people around the world, including Asia, Europe, and South America. The milk of mares is another food source, and it is particularly associated with Mongolia where the fermented form is a common drink.[109] There are a few horse facilities in other countries that market horse milk as a delicacy and as an alternative for people allergic to cow's milk.

Horses are used as work animals.[109] On horseback, a person can get to places that four-wheeled vehicles cannot and travel over terrain at speeds that humans cannot match. In many locations, they guide flocks of sheep and herds of cattle or horses better than mechanical devices. Draft equids pull carriages for tourists in many cities, and in the United States, the Amish people are common sights with their horse and buggy moving down a highway. Horses pull plows, produce-laden carts, and wagons with hay for livestock in winter. Mounted police are more visible in cities because of their height. Horses are public goodwill ambassadors and excel in crowd control when things start to go wrong. The horse remains a symbol of the Old West to many in the United States, and that harkening back to "the good old days" has put it on a pedestal. What many forget is that the cowboy of the Wild West used the horse for work, and much less for pleasure. That work use continues today as well.

In the developed world, many horses are considered an animal for recreation. Most horse owners enjoy their animals for trail riding or light training in an arena. That activity verifies a saying frequently, but inaccurately, attributed to Winston Churchill: "the outside of a horse is good for the inside of a man."[127] But the use of the horse for recreation and entertainment goes way beyond pleasure riding. Horse shows today have honed the skills of these four-legged athletes to the point that they excel in specific types of athletic prowess. Reining horses show acceleration of speed, fast turns on their rear limbs, and fast stops. Cutting horses quickly change directions, accelerate and decelerate to keep up with the cow trying to get back to the herd. Jumpers clear fences higher than their riders are tall. Several breeds have "plantation gaits" that make them seem to float. Thoroughbred and Standardbred horses are bred for racing, and some horses have been especially bred to buck as rodeo stock. Horses entertain humans in movies, television, circuses, pony rides, rodeos, and parades too. The animals are working, but the amount of pleasure the public receives from watching these beautiful animals is undeniable.

Horse "Whisperers" and "Naturalists"

There are a variety of reports about when and where "whispering" began. It is perhaps the secret nature of the successful horse trainers of old that popularized the theory that they "whispered" something to the rogue horse to make it behave.[128] What is not known is what was really happening to the animal in the secrecy of the barn or training arena.

The changing perception of the horse as a pet rather than livestock is accompanied by a change in attitudes about how the animal should be treated. The vision of the cowboy roping, saddling, and letting a wild bronc buck until it gives in to the will of the human is not one that was idealized. That image has given way to the trainer who gradually teaches the animal how it can be successful with its given talents using stress-free techniques. With these new expectations comes a change in the definition of what it means to be a "whisperer." The popularization of the term in the last 20 years has shifted the definition to that of gently communicating what is expected, at least in the mind of the general public.

When the term "whispering" caught on in the United States, trainers in Europe were commenting that the techniques had actually been in use there for at least 2000 years. The gypsies (Romany) of Europe were considered to be "whisperers "300 years ago.[129] Some horse trainers are now calling themselves "whisperers" regardless of the actual training methods they use. It is important to realize that the term "whisperer" is a colloquial one. Just because a person calls himself "a whisperer" does not mean that their training techniques are any more humane or that they are any better than someone not using the term.

"Natural horsemanship" has many of the same concepts of how training should be done as with the "whisperers." Both advocate kinder, gentler methods of training. Both imply that humans learn to communicate better with the horse, and both tend to use quasiscientific explanations of horse behavior. Those who follow "natural horsemanship" also tend toward anthropomorphism, considering the animal to be almost human.[130]

REFERENCES

1. Hayden EC. First horses arose 4 million years ago. *Nature* 2013; https://doi.org/10.1038/nature.2013.13261.
2. Millar CD, Lambert DM. Towards a million-year-old genome. *Nature* 2013;**499**(7456):34–5.

3. Orlando L, Ginolhac A, Zhang G, Froese D, Albrechtsen A, Stiller M, Schubert M, Cappellini E, Petersen B, Moltke I, Johnson PLF, Fumagalli M, Vilstrup JT, Raghavan M, Korneliussen T, Malaspinas A-S, Vogt J, Szklarczyk D, Kelstrup CD, Vinther J, Dolocan A, Stenderup J, Velazquez AMV, Cahill J, Rasmussen M, Wang X, Min J, Zazula GD, Seguin-Orlando A, Mortensen C, Magnussen K, Thompson JF, Weinstock J, Gregersen K, Røed KH, Eisenmann V, Rubin DJ, Miller DC, Antczak DF, Bertelsen MF, Brunak S, Al-Rasheid KAS, Ryder O, Andersson L, Mundy J, Krogh A, Gilbert MTP, Kjær K, Sicheritz-Ponten T, Jensen LJ, Olsen JV, Hofreiter M, Nielsen R, Shapiro B, Wang J, Willerslev E. Recalibrating *Equus* evolution using the genome sequence of an early middle Pleistocene horse. *Nature* 2013;**499**(7456):74–81.
4. Cantalapiedra JL, Prado JL, Hernández Fernández M, Alberdi MT. Decoupled ecomorphological evolution and diversification in Neogene-quaternary horses. *Science* 2017;**355**(6325):627–30.
5. Dixon PM. The evolution of horses and the evolution of equine dentistry. *Proc Am Assoc Equine Pract* 2017;**63**:79–116.
6. MacFadden BJ. *Fossil horses: systematics, paleobiology, and evolution of the family equidae*. New York: Cambridge University Press; 1992.p.369.
7. MacFadden BJ. Fossil horses from "eohippus" (*Hyracotherium*) to *Equus*: scaling, Cope's law, and the evolution of body size. *Paleobiology* 1986;**12**(4):355–69.
8. Shoemaker L, Clauset A. Body mass evolution and diversification within horses (family Equidae). *Ecol Lett* 2014;**17**(2):211–20.
9. Eronen JT, Evans AR, Fortelius M, Jernvall J. The impact of regional climate on the evolution of mammals: a case study using fossil horses. *Evolution* 2009;**64**(2):398–408.
10. Mihlbachler MC, Rivals F, Solounias N, Semprebon GM. Dietary change and evolution of horses in North America. *Science* 2011;**331**(6021):1178–81.
11. Radinsky L. Oldest horse brains: more advanced than previously realized. *Science* 1976;**194**(4265):626–7.
12. Davis B. Timeline of the development of the horse. In: *Sino-Platonic Papers No. 177*; 2007. p. 1–186.
13. Equus *(Genus)*. Wikipedia, https://en.wikipedia.org/wiki/Equus_(genus) [downloaded May 13, 2016].
14. *Evolution of the horse*. Wikipedia, https://en.wikipedia.org/wiki/Evolution_of_the_horse [downloaded Jan. 28, 2016].
15. Jónsson H, Schubert M, Seguin-Orlando A, Ginolhac A, Petersen L, Fumagalli M, Albrechtsen A, Petersen B, Korneliussen TS, Vilstrup JT, Lear T, Myka JL, Lundquist J, Miller DC, Alfarhan AH, Alquraishi SA, Al-Rasheid KAS, Stagegaard J, Strauss G, Bertelsen MF, Sicheritz-Ponten T, Antczak DF, Bailey E, Nielsen R, Willerslev E, Orlando L. Speciation with gene flow in equids despite extensive chromosomal plasticity. *Proc Natl Acad Sci USA* 2014;**111**(52):18655–60.
16. MacFadden BJ. Fossil horses—Evidence for evolution. *Science* 2005;**307**(5716):1728–30.
17. MacFadden BJ, Bryant JD, Mueller PA. Sr-isotopic, paleomagnetic, and biostratigraphic calibration of horse evolution: evidence from the Miocene of Florida. *Geology* 1991;**19**(3):242–5.
18. Oakenfull EA, Lim HN, Ryder OA. A survey of equid mitochondrial DNA: implications for the evolution, genetic diversity and conservation of *Equus*. *Conserv Genet* 2000;**1**(4):341–55.
19. Radinsky L. Ontogeny and phylogeny in horse skull evolution. *Evolution* 1984;**38**(1):1–15.
20. Weinstock J, Willerslev E, Sher A, Tong W, Ho SYW, Rubenstein D, Storer J, Burns J, Martin L, Bravi C, Prieto A, Froese D, Scott E, Xulong L, Cooper A. Evolution, systematics, and phylogeography of Pleistocene horses in the New World: a molecular perspective. *PLoS Biol* 2005;**3**(8).

21. Haemig PD. Evolution of horses. *Ecology Info* 2012;33.
22. McHorse BK, Biewener AA, Pierce SE. Mechanics of evolutionary digit reduction in fossil horses (Equidae). *Proc R Soc B* 2017;**284**.
23. Asinus. *Wikipedia*. https://en.wikipedia.org/wiki/Asinus [downloaded May 13, 2016].
24. Beja-Pereira A, England PR, Ferrand N, Jordan S, Bakhiet AO, Abdalla MA, Mashkour M, Jordana J, Taberlet P, Luikart G. African origins of the domestic donkey. *Science* 2004;**304** (5678):1781.
25. Eisenmann V, Sergej V. Unexpected finding of a new *Equus* species (Mammalia, Perissodactyla) belonging to a supposedly extinct subgenus in late Pleistocene deposits of Khakassia (southwestern Siberia). *Geodiversitas* 2011;**33**(3):519–30.
26. *Kiang*. Wikipedia, https://en.wikipedia.org/wiki/Kiang [downloaded May 13, 2016].
27. Moodley Y, Harley EH. Population structuring in mountain zebras (*Equus zebra*): the molecular consequences of divergent demographic histories. *Conserv Genet* 2005;**6**(6):953–68.
28. *Mountain zebra*. Wikipediahttps://en.wikipedia.org/wiki/Mountain_zebra [downloaded May 13, 2016].
29. *Onager*. Wikipedia, https://wikipedia.org/wiki/Onager [downloaded May 13, 2016].
30. Shaw N, St. Louis A, Qureshi Q. *Equus kiang*. http://www.iucnredlist.org/details/7953/0; 2015 [downloaded 5/18/16].
31. Vilstrup JT, Seguin-Orlando a, Stiller M, Ginolhac A, Raghavan M, Nielsen SCA, Weinstock J, Froese D, Vasiliev SK, Ovodov ND, Clary J, Helgen KM, Fleischer RC, Cooper a, Shapiro B, Orlando L. Mitochondrial phylogenomics of modern and ancient equids. *PLoS ONE* 2013;**8**(2)https://doi.org/10.1371/journal.pone.0055950. eff950.
32. *Zebra*. Wikipedia, https://en.wikipedia.org/wiki/Zebra [downloaded May 13, 2016].
33. Orlando L, Metcalf JL, Alberdi MT, Telles-Antunes M, Bonjean D, Otte M, Martin F, Eisenmann V, Mashkour M, Morello F, Prado JL, Salas-Gismondi R, Shockey BJ, Wrinn PJ, Vasil'ev SK, Ovodov ND, Cherry MI, Hopwood B, Male D, Austin JJ, Hänni C, Cooper A. Revising the recent evolutionary history of equids using ancient DNA. *Proc Natl Acad Sci* 2009;**106**(51):21754–9.
34. Sarkissian CD, Vilstrup JT, Schubert M, Seguine-Orlando A, Eme D, Weinstock J, Alberdi MT, Martin F, Lopez PM, Prado JL, Prieto A, Douady CJ, Stafford TW, Willerslev E, Orlando L. Mitochondrial genomes reveal the extinct Hippidion as an outgroup to all living equids. *Biol Lett* 2015;**11**(3):20141058.
35. Orlando L. Equids. *Curr Biol* 2015;**25**(10):R973–8.
36. Bailey E. Genetics after twilight. *J Equine Vet* 2015;**35**(5):361–6.
37. *Tarpan*. Wikipedia, https://en.wikipedia.org/wiki/Tarpan [downloaded May 19, 2016].
38. Buck CE, Bard E. A calendar chronology for Pleistocene mammoth and horse extinction in North America based on Bayesian radiocarbon calibration. *Quat Sci Rev* 2007;**26** (17–18):2031–5.
39. Guthrie RD. Rapid body size decline in Alaskan Pleistocene horses before extinction. *Nature* 2003;**426**(6963):169–71.
40. Benirschke K, Low RJ, Brownhill LE, de Venecia-Fernandez J. Chromosome studies of a donkey-Grévy zebra hybrid. *Chromosoma* 1964;**15**(1):1–13.
41. Benirschke K, Malouf N, Low RJ, Heck H. Chromosome complement: differences between Equus caballus and Equus przewalskii, Poliakoff. *Science* 1965;**148**(3668):382–3.
42. Breen M, Gill JJB. The chromosomes of two horse x zebra hybrids; *E. caballus* x *E. grevyi* and *E. burchelli*. *Hereditas* 1991;**115**(2):169–75.
43. Kavar T, Dovč P. Domestication of the horse: genetic relationships between domestic and wild horses. *Livest Sci* 2008;**116**(1–3):1–14.

44. Lei CZ, Su R, Bower MA, Edwards CJ, Wang XB, Weining S, Liu L, Xie WM, Li F, Liu RY, Zhang YS, Zhang CM, Chen H. Multiple maternal origins of native modern and ancient horse populations in China. *Anim Genet* 2009;**40**(6):933–44.
45. Vilà C, Leonard JA, Götherström A, Marklund S, Sandberg K, Lidén K, Wayne RK, Ellegren H. Widespread origins of domestic horse lineages. *Science* 2001;**291**(5503):474–7.
46. Wallner B, Vogl C, Shukla P, Burgstaller JP, Druml T, Brem G. Identification of genetic variation on the horse Y chromosome and the tracing of male founder lineages in modern breeds. *PLoS ONE* 2013;**8**(4). https://doi.org/10.1371/journal.pone.0060015.
47. Librado P, Gamba C, Gaunitz C, Der Sarkissian C, Pruvost M, Albrechtsen A, fages A, Khan N, Schubert M, Jagannathan V, Serres-Armero A, Kuderna LFK, Povolotskaya IS, Seguin-Orlando A, Lepetz S, Neuditschko M, Thèves C, Alquraishi S, Alfarhan AH, Al-Rasheid K, Rieder S, Samashev Z, Francfort H-P, Benecke N, Hofreiter M, Ludwig A, Keyser C, Marques-Bonet T, Ludes B, Crubézy E, Leeb T, Willerslev W, Orlando L. Ancient genomic changes associated with domestication of the horse. *Science* 2017;**356**(6336):442–5.
48. Scythian stallions offer heads-up on horse domestication. *Science* 2016;**352**(6288):875.
49. *Animal genome size database*. http://www.genomesize.com/results.php?page=1 [searched December 21, 2016].
50. *Horse genome*. Wikipedia, https://en.wikipedia.org/wiki/Horse_genome [downloaded December 21, 2016].
51. *Human genome*. Wikipedia, https://en.wikipedia.org/wiki/Human_genome [downloaded December 21, 2016].
52. Tolley EA, Notter DR, Marlowe TJ. Heritability and repeatability of speed for 2- and 3-year-old Standardbred racehorses. *J Anim Sci* 1983;**56**(6):1294–305.
53. Hemmann K, Raekallio M, Vainio O, Juga J. Crib-biting and its heritability in Finnhorses. *Appl Anim Behav Sci* 2014;**156**:37–43.
54. Houpt KA. Treatment of aggression in horses. *Equine Pract* 1984;**6**(6):8–10.
55. Bailey E. Heritability and the equine clinician. *Equine Vet J* 2014;**46**(1):12–4.
56. *Performance genetics: heritability of performance*. http://performancegenetics.com/heritability-of-thoroughbreds-performance/ [downloaded 4/7/14].
57. Velie BD, Hamilton NA, Wade CM. Heritability of racing durability traits in the Australian and Hong Kong Thoroughbred racing populations. *Equine Vet J* 2016;**48**(3):275–9.
58. Pereira R. *Q&A: the stubborn truth about cloned mules*. http://www.npr.org/templates/story/story.php?storyId=5504350; 2006 [downloaded June 24, 2016].
59. Deesing MJ, Grandin T. Behavior genetics of the horse (*Equus caballus*). In: Grandin T, Deesing MJ, editors. *Genetics and the behavior of domestic animals*. 2nd ed. New York: Academic Press; 2014. p. 237–90.
60. Fegraeus KJ, Johansson L, Mäenpää M, Mykkänen A, Andersson LS, Velie BD, Andersson L, Árnason T, Lindgren G. Different *DMRT3* genotypes are best adapted for harness racing and riding in Finnhorses. *J Hered* 2015;734–40.
61. Velie BD, Shrestha M, François L, Schurink A, Tesfayonas YG, Stinckens A, Blott S, Ducro BJ, Mikko S, Thomas R, Swinburne JE, Sundqvist M, Eriksson S, Buys N, Lindgren G. Using an inbred horse breed in a high density genome-wide scan for genetic risk factors of insect bite hypersensitivity (IBH). *PLoS ONE* 2016;**11**(4). https://doi.org/10.1371/journal.pone.0152966.
62. Petersen JL, Mickelson JR, Rendahl AK, Valberg SJ, Andersson LS, Axelsson J, Bailey E, Bannasch D, Binns MM, Borges AS, Brama P, da Câmara Machado A, Capomaccio S, Cappelli K, Cothran EG, Distl O, Fox-Clipsham L, Graves KT, Guérin G, Haase B, Hasegawa T, Hemmann K, Hill EW, Leeb T, Lindgren G, Lohi H, Lopes MS,

McGivney BA, Mikko S, Orr N, Penedo MCT, Piercy RJ, Raekallio M, Rieder S, Røed KH, Swinburne J, Tozaki T, Vaudin M, Wade CM, McCue ME. Genome-wide analysis reveals selection for important traits in domestic horse breeds. *PLoS Genet* 2013;**9**(1). https://doi.org/10.1371/journal.pgen.1003211.

63. Staiger EA, Albright JD, Brooks SA. Genome-wide association mapping of heritable temperament variation in the Tennessee Walking Horse. *Genes Brain Behav* 2016;**15**(5):514–26.
64. Jacobs LN, Staiger EA, Albright JD, Brooks SA. The MC1R and ASIP coat color loci may impact behavior in the horse. *J Hered* 2016;**107**(3):214–9.
65. Bellone RR. Pleiotropic effects of pigmentation genes in horses. *Anim Genet* 2010;**41**(Suppl. 2):100–10.
66. Sellnow L. Heritability of behavior. *The Horse* 2003;**XX**(5)49, 50 and 52.
67. Krueger RF, South S, Johnson W, Iacono W. The heritability of personality is not always 50%: gene-environment interactions and correlations between personality and parenting. *J Pers* 2008;**76**(6):1485–521.
68. Jensen P. Behaviour epigenetics—The connection between environment, stress and welfare. *Appl Anim Behav Sci* 2014;**157**:1–7.
69. Johannes F, Porcher E, Teixeira FK, Saliba-Colombani V, Simon M, Agier N, Bulski A, Albuisson J, Heredia F, Audigier P, Bouchez D, Dillmann C, Guerche P, Hospital F, Colot V. Assessing the impact of transgenerational epigenetic variation on complex traits. *PLoS Genet* 2009;**5**(6). https://doi.org/10.1371/journal.pgen.10000530.
70. Radford EJ, Ito M, shi H, Corish JA, Yamazawa K, Isganaitis E, Seisenberger S, Hore TA, Reik W, Erkek S, Peters AHFM, Patti M-E, Ferguson-Smith AC. In utero undernourishment perturbs the adult sperm methylome and intergenerational metabolism. *Science* 2014;**345** (6198):785.
71. Susiarjo ML, Bartolomei MS. You are what you eat, but what about your DNA? *Science* 2014;**345**(6198):733–4.
72. Franklin TB, Mansuy IM. Epigenetic inheritance in mammals: evidence for the impact of adverse environmental effects. *Neurobiol Dis* 2010;**39**(1):61–5.
73. Franklin TB, Russig H, Weiss IC, Gräff J, Linder N, Michalon A, Vizi S, Mansuy IM. Epigenetic transmission of the impact of early stress across generations. *Biol Psychiatry* 2010;**68**(5):408–15.
74. Leshem M, Schulkin J. Transgenerational effects of infantile adversity and enrichment in male and female rats. *Dev Psychobiol* 2011;**54**(2):169–86.
75. Miska EA, Ferguson-Smith AC. Transgenerational inheritance: models and mechanisms of non-DNA sequence-based inheritance. *Science* 2016;**354**(6308):59–63.
76. Rodgers AB, Morgan CP, Bronson SL, Revello S, Bale TL. Paternal stress exposure alters sperm microRNA content and reprograms offspring HPA stress axis regulation. *J Neurosci* 2013;**33**(21):9003–12.
77. Bauer T, Trump S, Ishaque N, Thürmann L, Gu L, Bauer M, Bieg M, Gu Z, Weichenhan D, Mallm J-P, Rüder S, Herberth G, Takada E, Mücke O, Winter M, Junge KM, Grützmann K, Rolle-Kampczyk U, Wang Q, Lawerenz C, Borte M, Polte T, Schlesner M, Schanne M, wiemann S, Geörg C, Stunnenberg HG, Plass C, Rippe K, Mizuguchi J, Hermann C, Eils R, Lehmann I. Environment-induced epigenetic reprogramming in genomic regulatory elements in smoking mothers and their children. *Mol Syst Biol* 2016;**12**(3):861.
78. Dias BG, Ressler KJ. Parental olfactory experience influences behavior and neural structure in subsequent generations. *Nat Neurosci* 2014;**17**(1):89–96.
79. Rassoulzadegan M, Grandjean V, Gounon P, Vincent S, Gillot I, Cuzin F. RNA-mediated non-Mendelian inheritance of an epigenetic change in the mouse. *Nature* 2006;**441**:469–74.

80. *Pony*. Wikipedia, https://en.wikipedia.org/wiki/Pony [downloaded June 21, 2016].
81. Frischknecht M, Jagannathan V, Plattet P, Neuditschhko M, Signer-Hasler H, Bachmann I, Pacholewska A, Drögemülleer C, Dietschi W, Flury C, Rieder S, Leeb T. A non-synonymous HMGA2 variant decreases height in Shetland ponies and other small horses. *PLoS ONE* 2015;**10**(10). https://doi.org/10.1371/journal.pone.0140749.
82. Carleton C. *Personal Communication*.
83. Cothran EG, Luis C. Genetic distance as a tool in the conservation of rare horse breeds. *Eaap Public* 2005;**2005**(116):55–71.
84. Cothran EG, McCrory WP. *A preliminary genetic study of the wild horse (*Equus caballus*) in the Brittany Triangle (Tachelach' ed) Region of the ?Elegesi Qayus (Nemiah) Wild Horse Preserve of British Columbia*. http://www.Irgaf.org/articles/Wild%20Horse%20DNA%20Report%202015.pdf; 2014. downloaded June 22, 2016.
85. Jun J, Cho YS, Hu H, Kim H-M, Jho S, Gadhvi P, Park KM, Lim J, Paek WK, Han K, Manica A, Edwards JS, Bhak J. Whole genome sequence and analysis of the Marwari horse breed and its genetic origin. *BMC Genomics* 2014;**15**(Supplement 9):54.
86. Commission on Genetic Resources for Food and Agriculture. *Status and trends of animal genetic resources—2012*. Rome: Food and Agriculture Organization of the United Nations; 20137.
87. Edwards EH. *Smithsonian handbooks: horses*. New York: Dorling Kindersley; 2002.p.255.
88. Edwards EH. *The new encyclopedia of the horse*. New York: Dorling Kindersley; 2008.p.464.
89. Khadka R. *Global horse population with respect to breeds and risk status*. Masters Thesis in European Masters in Animal Breeding and Genetics, Swedish University of Agricultural Scienceshttp://stud.epsilon.slu.se/7676/17/khadka_r_150305.pdf; 2010[downloaded June 21, 2016].
90. McBain S. *The illustrated encyclopedia of horse breeds*. Edison, NJ: Wellfleet Press; 2008. p.256.
91. FAOSTAT. *Food and Agriculture Organization of the United Nations, Statistics Division*. http://faostat3.fao.org/browse/Q/QA/E [downloaded June 22, 2016].
92. *Decline in world's horse population*. www.horsetalk.co.nz/2013/10/03/decline-worlds-horse-population/#axzz4CEFkKszh [downloaded June 21, 2016].
93. American Horse Council. *Economic impact of the United States Horse Industry*. http://www.horsecouncil.org/economics/; 2005[downloaded June 22, 2016].
94. Kilby ER. *The demographics of the U.S. equine population: chapter 10*. http://www.humanesociety.org/assets/pdfs/hsp/soaiv_07_ch10.pdf; 2007[downloaded June 21, 2016].
95. Kleine K. *Current trends in the horse industry*. http://www.animalagriculture.org/Resources/Documents/Conf%20-%20Symp/Conferences/2015%20Annual%20Conference/Speaker%20Presentations/Equine/Equine-Kleine-stateoftheindustry.pdf; 2015[downloaded June 22, 2016].
96. Burdon A. *Where the wild horses are*. Australian Geographichttp://www.australiangeographic.com.au/topics/wildlife/2016/03/where-the-wild-horses-are; 2016[downloaded June 22, 2016].
97. Bureau of Land Management. *Wild horse and burro quick facts*. http://www.blm.gov/wo/st/en/prog/whbprogram/history_and_facts/quick_facts.print.html [downloaded June 22, 2016].
98. Ballou JD, Traylor-Holzer K, Turner A, Malo AF, Powell D, Maldonado J, Eggert L. Simulation model for contraceptive management of the Assateague Island feral horse population using individual-based data. *Wildl Res* 2008;**35**(6):502–12.
99. *Feral horse*. Wikipedia, https://en.wikipedia.org/wiki/Feral_horse[downloaded June 22, 2016].
100. *The mighty Mongolian horse*. http://www.theadventurists.com/the-jibber/29/1/2014/the-horses-of-the-mongol-derby[downloaded 6/23/16].
101. Price EO. Behavioral aspects of animal domestication. *Q Rev Biol* 1984;**59**(1):1–32.

102. Bruford MW, Bradley DG, Luikart G. DNA markers reveal the complexity of livestock domestication. *Nat Rev Genet* 2003;**4**:900–10.
103. Driscoll CA, Macdonald DW, O'Brien SJ. From wild animals to domestic pets, an evolutionary view of domestication. *Proc Natl Acad Sci USA* 2009;**106**(suppl 1):9971–8.
104. Jansen T, Forster P, Levine MA, Oelke H, Hurles M, Renfrew C, Weber J, Olek K. Mitochondrial DNA and the origins of the domestic horse. *Proc Natl Acad Sci USA* 2002;**99**(16):10905–10.
105. Warmuth V. *On the origin and spread of horse domestication.* PhD Dissertation, Corpus Christi College; 20111–138.
106. Warmuth V, Eriksson A, Bower MA, Barker G, Barrett E, Hanks BK, Li S, Lomitashvili D, ochir-Goryaeva M, Sizonov GV, Soyonov V, Manica A. Reconstructing the origin and spread of horse domestication in the Eurasian steppe. *Proc Natl Acad Sci USA* 2012;**109**(21):8202–6.
107. Anthony DW, Brown DR. The secondary products revolution, horse-riding, and mounted warfare. *J World Prehist* 2011;**24**(2–3):131–60.
108. Gaunitz C, Fages A, Hanghøj K, Albrechtsen A, Khan N, Schubert M, Seguuuin-Orlando A, Owens IJ, Felkel S, Bignon-Lau O, de Barros Damgaard P, Mittnik A, Mohaseb aF, Davoudi H, Alquraishi S, Alfarhan AH, Al-Rasheid KAS, Crubézy E, Benecke N, Olsen S, Brown D, Anthony D, Massy K, Pitulko V, Kasparov A, Brem G, Hofreiter M, Mukhtarova G, Baimukhanov N, Lõugas L, Onar V, Stockhammer PW, Krause J, Boldgiv B, Undrakhbold S, Erdenebaatar D, Lepetz S, Mashkour M, Ludwig A, Wallner B, Merz V, Merz M, Zaibert V, Willerslev E, Librado P, Outram AK, Orlando L. Ancient genomes revisit the ancestry of domestic and Przewalski's horses. *Science* 2018;**360**(6384):111–4.
109. Hintz HF. Thoughts about the history of horses. *J Equine Vet* 1995;**15**(8):336–8.
110. Lippold S, Matzke NJ, Reissmann M, Hofreiter M. Whole mitochondrial genome sequencing of domestic horses reveals incorporation of extensive wild horse diversity during domestication. *BMC Evol Biol* 2011;**11**(1):328–37.
111. Ludwig A, Pruvost M, Reissmann M, Benecke N, Brockmann GA, Castaños P, Cieslak M, Lippold S, Llorente L, Malaspinas A-S, Slatkin M, Hofreiter M. Coat color variation at the beginning of horse domestication. *Science* 2009;**324**(5926):485.
112. McMiken DF. Ancient origins of horsemanship. *Equine Vet J* 1990;**22**(2):73–8.
113. Outram AK, Stear NA, Bendrey R, Olsen S, Kasparov A, Zaibert V, Thorpe N, Evershed RP. The earliest horse harnessing and milking. *Science* 2009;**323**(5919):1332–5.
114. Schubert M, Jónsson H, Chang D, Sarkissian CD, Ermini L, Ginolhac A, Albrechtsen A, Dupanloup I, Foucal A, Petersen B, Fumagalli M, Raghavan M, Seguin-Orlando A, Korneliussen TS, Velazquez AMV, Stenderup J, Hoover CA, Rubin C-J, Alfarhan AH, Alquraishi SA, Al-Rasheid KAS, MacHugh DE, Kalbfleisch T, MacLeod JN, Rubin EM, Sicheritz-Ponten T, Andersson L, Hofreiter M, Marques-Bonet T, Gilbert MTP, Nielsen R, Excoffier L, Willerslev E, Shapiro B, Orlando L. Prehistoric genomes reveal the genetic foundation and cost of horse domestication. *Proc Natl Acad Sci USA* 2014;**111**(52):E5661–9.
115. McGahern A, Bower MAM, Edwards CJ, Brophy PO, Sulimova G, Zakharov I, Vizuete-Forster M, Levine M, Li S, MacHugh DE, Hill EW. Evidence for biogeographic patterning of mitochondrial DNA sequences in eastern horse populations. *Anim Genet* 2006;**37**(5):494–7.
116. Yang Y, Zhu Q, Liu S, Zaho C, Wu C. The origin of Chinese domestic horses revealed with novel mtDNA variants. *Anim Sci J* 2017;**88**(1):19–26.
117. Lewin HA. It's a bull's market. *Science* 2009;**324**(5926):478–9.
118. Ludwig A, Reissmann M, Benecke N, Bellone R, Sandoval-Castellanos E, Cieslak M, Fortes GG, Morales-Muñiz a, Hofreiter M, Pruvost M. Twenty-five thousand years of

fluctuating selection on leopard complex spotting and congenital night blindness in horses. *Philos Trans R Soc Lond B* 2014;**370**(20130386).

119. Sponenberg DP, Beaver BV. *Horse color*. College Station: Texas A&M University Press; 1983.p.124.
120. Achilli A, Olivieri A, Soares P, Lancioni H, Kashani BH, Perego UA, Nergadze SG, Carossa V, Santagostino M, Capomaccio s, Felicetti M, Al-Achkar W, Penedo MCT, Verini-Supplizi A, Houshmand M, Woodward SR, Semino O, Silvestrelli M, Giulotto E, Pereira L, Bandelt H-J, Torroni A. Mitochondrial genomes from modern horses reveal the major haplogroups that underwent domestication. *Proc Natl Acad Sci USA* 2012;**109** (7):2449–54.
121. Barclay HB. Another look at the origins of horse riding. *Anthropos* 1982;**77**(1/2):244–9.
122. Bendrey R. From wild horses to domestic horses: a European perspective. *World Archaeol* 2012;**44**(1):135–57.
123. Levine MA. Botai and the origins of horse domestication. *J Anthropol Archaeol* 1999;**18** (1):29–78.
124. Schulman AR. Egyptian representations of horsemen and riding in the new kingdom. *J Near East Stud* 1957;**16**(4):263–71.
125. Lawrence EA. American Indians and their horses' health. *J Am Vet Med Assoc* 1989;**194** (12):1690–1.
126. Hoffman NLW. *Sgt Reckless: combat veteran*. https://www.mca-marines.org/leatherneck/sgt-reckless-combat-veteran; 1992[downloaded August 31, 2016].
127. Potter L. *Who said that? Probably not Winston Churchill*. http://www.horsechannel.com/media/the-near-side-blog/2013/0128-winston-churchill-horse-quotes.aspx.pdf; 2013[downloaded Sept. 6, 2016].
128. Beck A. *The origins of horse whispering*. http://www.equine-behavior.com/Origins_of_horse_whispering1.htm; 2003[downloaded 9/13/16].
129. Crank C. *Horse whispers part 1: origins, societies and secrets*. https://horsesandhistory.wordpress.com/2011/04/05/horse-whisperers-part-1-origins-societies-and-secrets/; 2011[downloaded Sept. 13, 2016].
130. Birke L. Talking about horses: control and freedom in the world of "natural horsemanship" *Soc Anim* 2008;**16**(2):107–26.

Chapter 2

Equine Behavior of Sensory and Neural Origin

There are many unique features of the neural and sensory systems that relate to behavior and behavioral problems. It is not the intent here to link every behavior to the nucleus and axons associated with it, but rather to create a better understanding of how the horse might view its world and how that relates to behavior.

SENSES

Successful escape from predators was the highest priority for ancient horses. To do that, the senses adapted over time to identify threats and allow the animal to avoid them. Humans primarily rely on vision, but the horse uses cross-modal perception involving multiple inputs. For example, a foal's call may first alert its dam to where it might be, and then vision becomes important to recognize that the calling foal is actually the correct one. Finally, touch and smell help reunite the bonded pair.[1–3] No single sense predominates. Horses use visual, auditory, or combined cues with equal proficiency.[4]

The senses represent a biological filter that determines whether the horse will even be aware of surrounding events. If stimuli pass through one of these filters, the brain becomes a second filter to determine the significance of the incoming message. Interpretation of this information will vary by "mood." This means things like previous experience, time of day, specific environment, and weather conditions can be important in the ultimate expression of behavior. Wind whirls a lot of scents in the air, making the locations and odor concentration hard to determine. As a result, horses go into high alert on windy days so that any message getting to the brain is more likely to be interpreted as potential danger.

Vision

In prey species like the horse, vision, hearing, and smell are complementary and about equal in importance. Because of this, the three sensory systems will differ from those in humans, particularly vision. Horses use the differences in

Equine Behavioral Medicine. https://doi.org/10.1016/B978-0-12-812106-1.00002-4

brightness, motion, distance, texture, and orientation to survey their environments. Color is of minor importance.

The physical structure of the horse eye suggests there are differences from human eyes. The most obvious is the horizontal rectangular shape of the pupil. This shape extends the area of visual perception, compared to that of round pupils.[5,6]

The changing shape of the lens facilitates the sharpest vision as the distance between the animal and the approaching image changes. For years it was thought that the horse's lens was less flexible than those of other species; however, that has been found not to be true. There are limits to lens accommodation with both distant and very close objects. When the approaching object is within approximately 1.5 ft (0.5 m), the lens is no longer able to sharply focus.[7] As a result, a horse might back away or move its head to try to see better.

Inside the eye, photoreceptor cells are not evenly distributed across the retina. Along the edges, there can be as few as 16–304 cells/mm^2, but near the optic disc, there is a concentrated band of nerve cells called the "visual streak." This band parallels the shape of the pupil and can have greater than 6500 cells/mm^2.[8–10] Images landing on the visual streak are perceived in greatest detail and are part of binocular vision.[5] For an image to land on the visual streak, it needs to enter near the bottom of the eye, much like the lower part of bifocal glasses.[11] Behaviorally, this explains why a running horse carries his head raised with nose slightly extended to see distant objects and one walking over obstacles bends the head and neck downward.[12] Previously, the behavior had been explained by statements that the back of the eye was angled like a ramp and where on that "ramp" the image landed was important for clear vision. Close objects needed to land on the lower part so the image entered the top of the eye and distant objects were best seen when their image landed on the upper "ramp" so the head was up and extended. Although still occasionally mentioned relative to horse vision, subsequent research has shown that the so-called "ramp effect" was an artifact of microscopic processing and did not occur in the live animal.[12,13]

For people, color vision is important, but it is not for domestic animals. Photoreceptor cells are one indicator of this. The ratio of rods to cones shows the importance of black/white compared to color vision, respectively. In humans, there are approximately 9 rods for every 1 cone.[5,14] In horses, the ratio is approximately 20:1.[5,9] The difference hints at two things: color vision is less important because of the low cone numbers, and nighttime vision is better since rods are adapted for low light.

The type of color vision experienced by horses is different than that for people. Humans are trichromats because cone sensitivity peaks at three different light wavelengths. The red peak is about 560–565 nm and is detected by long wavelength–sensitive (L) cones. Medium wavelength–sensitive (M) cones (530–535 nm) cover the yellow spectrum, and the short wavelength–sensitive (S) cones (430–440 nm) respond to the bluish colors.[15] Horses are dichromats

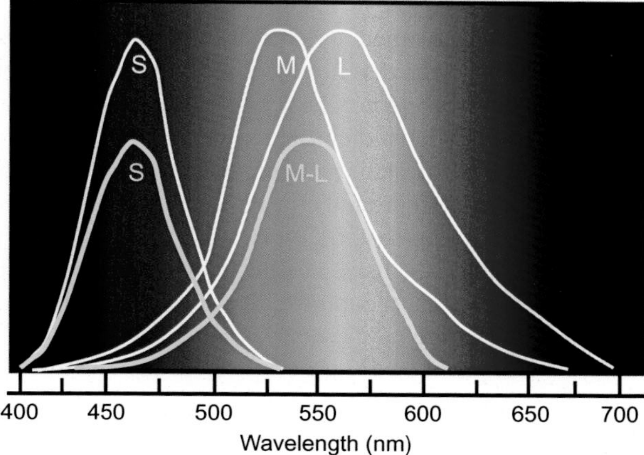

FIGURE 2-1 Equine dichromatic color spectrum (peak acuity shown as *yellow lines*) compared to the trichromatic spectrum peak acuity (shown as *white lines*).

View as seen by humans

View approximating that seen by horses

FIGURE 2-2 Color vision for horses is quite different than it is for humans. As a dichromat, the horse lives in a blue-green-yellow world. It has a much wider panorama because of the elongated pupil, but that panorama is slightly blurred. *(Based on images by Nickolay Lamm about cat vision in consultation with Drs. Kerry Ketring, D.J. Haeussler and the Veterinary Ophthalmology Group at the University of Pennsylvania. Used with permission.)*

with S cones that detect light of wavelengths of approximately 439–456 nm and middle-to-long wavelength–sensitive (M/L) cones for 537–557 nm wavelengths (Figure 2-1).[9,15–21] Thus horses see a continuous scale of colors ranging from bluish through yellowish colors, similar to the spectrum of a red-green color-blind person (Figure 2-2).[20]

Early studies of color vision in animals failed to take into account the amount of illumination of the object used in the test. As an example, the shine off a red object compared to the drabness of a dark green one might be the real signal the horse used as a cue and not the actual color. Research controlling for this factor still shows horses have dichromatic vision.[19,22] Instead of detecting things by color, the relatively high number of rods suggests that horses use contrasts of texture from plain backgrounds for safe movement and predator detection.[14] It also explains why horses and other livestock avoid sharp light/dark contrasts in pathways. This is the reason cattle guards work. Unless there is time to investigate the cause of a sharp contrast, horses will avoid having to go over or past such distinct differences.

The visual system of prey species is designed for scanning the environment rather than picking out sharp details. The larger the area they can scan at one time, the safer the animal is from a sneak attack. The visual field for the horse reflects that adaptation. Although the visual fields can vary somewhat by the shape of the head and set of the eyes, some generalities can be made. Each eye (monocular vision) sees across an arc of approximately 200–210 degrees around the body at one time. There is an overlapping area for the monocular fields straight in front of the horse's face such that the horse has a binocular field between 65 and 80 degrees (Figure 2-3).[9,23] The overlapping stereoscopic input in the binocular field is responsible for depth perception. The areas of monocular vision not associated with the overlap do allow a small degree of depth perception too, but it is about one-fifth of that occurring in the binocular field.[7,24] Even though the horse needs some amount of depth perception for fleeing danger or jumping obstacles, their capability is significantly less than for humans. Horses can detect an object that is 4.5 in (9 cm) tall from a distance of 6.5 ft (2 m), while a human is able to detect a 1/8-in. (a few millimeters) difference at that distance.[21]

The area directly behind the animal's head and body is called the blind area, and it is an arc of approximately 20 degrees. The more laterally placed the eyes are, the smaller is the blind area. Regardless, the blind area can easily be scanned by moving the head slightly side to side (Figure 2-4).[25] There are also blind areas located immediately in front of the forehead, directly under the head, and above and below the body. It is estimated that the vertical plane of vision from each eye is about 178 degrees, creating an almost complete sphere of vision.[9]

Head positioning affects vision too. Within reason, eye muscles maintain the horizontal position of the pupil regardless of head position.[26] The more extreme head positions, such as the Rollkur posture of hyperflexion, probably do affect vision, and even the ability to breathe. When the head is held in extreme flexion, horses will show behavioral signs of discomfort such as tail swishing, fighting the bit, and attempting to buck.[27] Limiting the field of vision can be stressful. Wearing blinkers is stressful enough to cause an increased heart rate for unfamiliar sounds.[28]

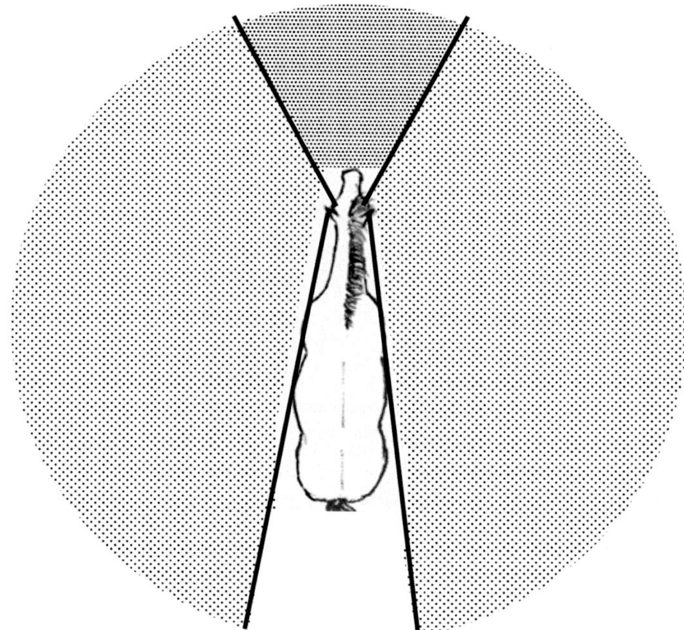

FIGURE 2-3 The visual field of the horse, as with other prey animals, is extensive. It includes a large monocular field on each side of the head which overlaps at the front to produce a relatively small binocular field. Directly behind the horse is a small blind area caused by the placement of the eyes and shape of the body.

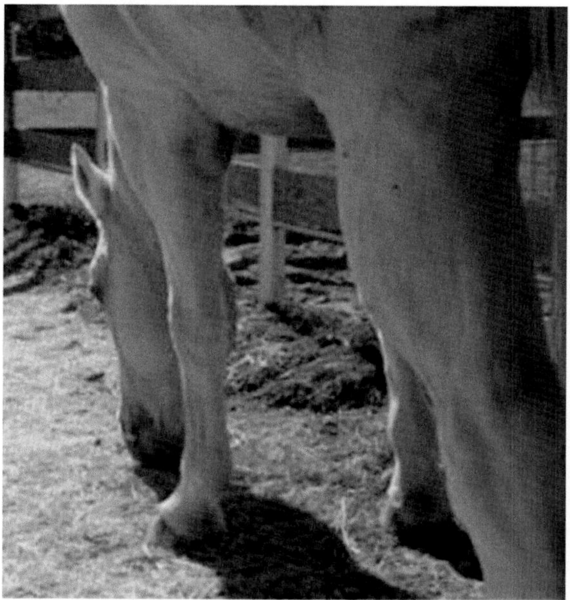

FIGURE 2-4 By lowering its head and moving it slightly side to side, a horse can easily scan a 360-degree horizontal periphery.

Experimentally, horses have relatively strong preferences for light in the morning hours between 6:00 and 10:00.[29] However, the real-world environment is different because they do not control when lights are on or off. Equids need visual capabilities to provide safety, particularly at dawn and dusk, because they eat throughout a 24-h period. That suggests low-light vision is important. Natural darkness comes on slowly and color vision gradually gives way to black and white at about the same visual light threshold as happens in people.[30] Although rods are very important in daytime vision for detecting movement, they become the only photoreceptors active in low light conditions.[31] Horses' eyes are not particularly adept at making a rapid transition between bright and dark locations. Accommodation from extreme brightness to darkness can take over 30 min.[14,21] This explains the reluctance to move between areas where illumination differences are great, such as loading into a dark trailer or going from outdoors into a dark, unfamiliar building.

The tapetum lucidum (or technically the choroidal tapetum fibrosum in horses) is an adaptation for animals that must function in low light environments. It enhances vision at night but complicates it somewhat during the day.[9,31,32] The tapetum acts much like a mirror, magnifying the amount of light that enters the eye. When light enters the eye, it triggers a photoreceptor on the retina, but the light is then reflected by the tapetum to hit another part of the retina and trigger additional receptors. As a result, the horse is able to see at lower light levels than can humans and other animals not having a tapetum. But the reflections also blur images by reducing the resolution (Figure 2-2).[10,16] Because light is magnified, the pupil must constrict more during sunlight to protect the eye. For horses, details and smaller objects are harder to identify, and moving ones are more likely to trigger the flight response.

As is well known by people who ride hunters and jumpers, horses more easily judge vertical objects than horizontal ones.[33] This is the evolutionary result of the horse needing to easily distinguish the vertical lines of a predator from the horizontal background when binocular vision is limited.

Visual acuity is another measure of what an animal sees, but it is difficult to determine in animals. For horses, test results suggest their acuity is approximately 20/30, which is a little worse than the human standard of 20/20.[34] It means horses see at 20 ft what "average" humans can see at 30 ft.

Different eyes, and thus different parts of the brain, are used when distinguishing familiar from potentially harmful objects. The preferential eye for novel stimuli used by social species, including the horse, is the left one, particularly in flightier animals.[35–37] The proposed explanation is that the right hemisphere of the brain, with input mainly from the left side of the body, developed to differentiate ongoing information from novel and unexpected events.[38,39] This left-viewing preference will not necessarily be shown by calm-natured horses except for interactions with people.[35,36,40]

Although horses rely on the combination of senses in their daily lives, genetic, congenital, and environmental factors can eliminate one or more of the senses. Blindness is one result. Affected horses learn the boundaries of stalls and paddocks so well that it is hard to identify them as blind. The giveaway is the way they carry their head—tipped to one side. Blindfolds create a temporary blindness and are useful in high stress situations. They facilitate handling of difficult horses and in dangerous events such as removing animals from burning barns. Blinkers are used to significantly restrict the visual field. Both increase heart rate, but it is much higher in blindfolded horses.[41]

Hearing

The curve of an audiogram is similar in shape between horses and humans but shifted slightly to the right (Figure 2-5).[42,43] Livestock generally do not hear the lowest sounds humans hear, but they can hear slightly higher ones. The audible range for humans is approximately 20 Hz to 20 kHz, with a peak sensitivity of 1–3 kHz.[21] The range for a horse is from 55 Hz to 33.5 kHz, with the best sensitivity between 1 and 16 kHz, which is within the normal range for most equine vocalizations.[43,44] Just as with humans, horses can experience a significant reduction in hearing as they age.[45]

The mobility of each ear suggests that hearing is an important sense. There is auditory laterality based on familiarity. Horses use their right ear and left cerebral hemisphere when listening to neighboring horses call.[46] Group member and strange horse whines do not elicit a preferential side, however. Localization of the source of other sounds is inversely proportional to the width of the field of best vision.[2,21] Large monocular visual fields mean less use of hearing for

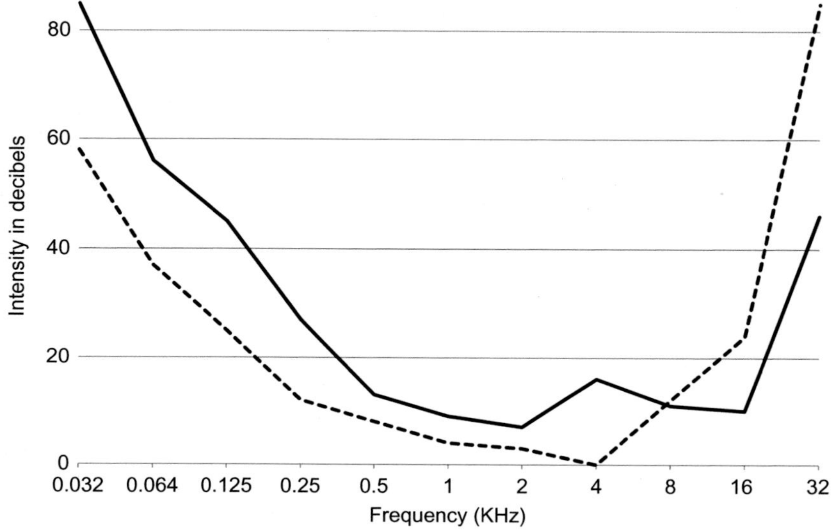

FIGURE 2-5 A comparative audiogram between a horse (solid line) and human (dashed line).[42, 43]

localization, meaning sound localization is poor in horses.[2,47] Practically, hearing is probably used to draw attention to something, with other senses taking over to define the actual source.[3]

A favorite high school science project is to study an animal's reaction to different types of music. Scientists have also done this in horses. There is no significant music preference or aversion between various types of music.[48] There are trends: an increase in time spent eating with country music and a decrease with jazz. Neighing tends to be reduced during the silent control periods and with rock music, but again, this was only a trend.

Deafness is common in horses. Affected animals are usually born deaf, and most deafness is usually related to an endothelin B receptor (*EDNBR*) gene mutation.[49] The problem usually relates to the presence of excessive white, occurring more often in horses with an extensive white blaze and blue eyes and with pintos/paints, particularly those having the splashed white or splashed white-frame blend coat patterns.[49,50] Affected horses have a very alert, ears-forward look, because they are more dependent on vision. Working with a deaf animal requires using its other senses for cues and rewards.

Smell

In horses, the size of the olfactory bulbs in the brain and behaviors shown suggest smell is an important sense. Conversely, it is hard for humans to appreciate an animal's sense of smell because of our poor capabilities. Human olfactory bulbs are microsomatic; the ones in the horse are macrosomatic. They are relatively large in size and include numerous folds that increase the surface area even more. The density of olfactory receptor cells over the olfactory bulb surface area is constant regardless of species, so the larger the surface area, the larger the number of receptors.[31]

Behaviors demonstrate that olfaction is an important sense in the social life of horses, and that domestication has changed its use. Wild Przewalski stallions sniff the genital regions of strangers, but modern horses are much more likely to sniff the noses or bodies.[51] Horse urine contains volatile components with chemically detectable differences based on the uniqueness of the individual, sex of the horse, and stage of the estrous cycle. This means it is possible that horses can detect these differences.[52–54] Feces also provides information. Horses can differentiate feces of competitors from that of nonthreatening herdmates through smell.[55] If horses are able to distinguish competitors from others, it is likely they can also distinguish individuals. Body odors may be significant too.[56] Stallions and geldings generally sniff an area prior to rolling, and they tend to roll in the same area previously used by another horse.[31]

Another role for olfaction in prey species is to identify approaching predators. Odors of a predator result in an increase in vigilance without fright. If another avoidance-inducing stimulus such as a sharp noise is also present, then

FIGURE 2-6 Flehmen is the extension of the upper lip, usually accompanied by an extension of the neck and upward pointing of the nose.

the heart rate will increase.[57] Windy days typically increase alertness in horses, probably because of the difficulty in examining incoming odors thoroughly.

The flehmen response, commonly called the "horse laugh," is associated with a second olfactory system. The behavior is expressed by the horse extending its neck, raising its nose, opening its mouth slightly, and lifting the upper lip (Figure 2-6). There are associated tongue and jaw movements, and the angles of the lips are slightly retracted.[58] In most species that flehmen, the head posture causes the paired ducts just behind the dental pad or incisor teeth, ones in the nasal passages, or both, to open and allows nonvolatile materials to enter.[59,60] Flehmen is shown most often in association with anestrous urine and postcopulation—evidence that the behavior is for priming and maintenance of sexual interest.[61,62] This conclusion is also supported by the fact that castration reduces the frequency of flehmen.[63]

Horses lack ductal openings in the mouth, having only the nasal ones.[64–67] Because of this, horses seldom show the licking behaviors common in other livestock species and are more likely to show urine marking instead.[31,61,67]

The oral/nasal ducts lead to a pair of cigar-shaped sacs, called the vomeronasal organs, which are located on the dorsal aspect of the hard palate. Each of these structures is lined with olfactory epithelium that responds to specific types of odor molecules associated with reproductive behaviors. Nerves leaving the vomeronasal organs go to the accessory olfactory areas of the brain to trigger innate responses not associated with the primary olfactory bulbs.[64]

The flehmen behavior can appear as early as the first day of life. Young colts show the behavior as often as once every 1.2 h, but fillies do so much less frequently, about once every 5–10 h.[68] The frequency of flehmen by colts peaks during the first 4 weeks of life, followed by a linear decrease up to 20 weeks.[69] In adult horses the behavior is often in response to exposure to urine, particularly mare urine. Prepubertal foals are the exception because they show no difference in responses to estrus or anestrus urine.[70] For them, 26% of the incidents occur in association with urination by another horse and 7% in association with the foal's own urination.[69] Why the high frequency of this behavior in foals compared to older horses remains unknown.

Taste

The sense of taste in horses is not well studied other than relative to dietary likes and dislikes. Comparing papillae on the tongue and their known function in other species is a logical place to start. The horse has filiform, fungiform, vallate, and foliate papillae.[71] Of those, filiform papillae are primarily mechanical in function and the other three are associated with taste buds. Filiform papillae are the small conical projections on the dorsum of the tongue that give it the rough feel. Fungiform papillae are interspersed among the filiform papillae on the rostral one-third of the tongue and along the lingual sides. They are considerably fewer in number than filiform ones. Two large, circular vallate papillae are located at the back of the tongue and associated with a number of taste buds. Horses also have a pair of foliate papillae located on the border of the tongue near the palatoglossal arch. Laryngeal taste buds line the wall of the larynx and are likely activated by various stimuli such as water, carbon dioxide, and perhaps chemical stimuli coming from the nasal cavity.[72] Horses lack conical papillae.

It is well known that horses can distinguish sweet and salt. Recent studies indicate they can also detect bitter. Sweet and bitter are experienced as the result of flavor molecules coupling to G protein–coupled receptors on taste buds. Salty sensations are associated with alkali metal ions triggering taste receptors.[73] Interestingly, horses respond to sucrose with a slight head bob, forward movement of the ears, and a slight tongue protrusion. Sucrose is a favored taste compared to water at concentrations between 1.25 and 10 g/100 mL, but horses are indifferent to it at higher or lower concentrations.[74] Horses respond to the bitter taste of quinine with a head extension, mouth gape, significant tongue protrusion, and backward movement of the ears.[75] Solutions that taste sour, bitter, or salty are treated indifferently until they reach a high concentration, when they are rejected.[74,76]

There are species differences in taste preferences, and even among herbivores there are differences in what they will eat and in how the taste receptors respond.[77] The digestive system of cattle and horses evolved differently, which probably explains why food vs. poison plant detection is also different.

As with humans, the panorama of tastes that could be experienced by a horse is also dependent on the sense of smell. The grinding of food releases volatiles that are sensed by smell and nonvolatile molecules that are detected by taste receptors.[78] Unlike humans, however, the taste and gustatory responses do not change with age, and consistency of behaviors associated with each type of food increases.[79]

Touch

What is known about the sense of touch in horses comes primarily from peripheral nerve mapping and histological nerve endings. These suggest that horses feel pain, pressure, warmth, and cold. The intricacies of touch are not well studied. Horses obviously react to a fly landing on its body hair, and they show a high level of sensitivity around their lips. Variations in stimuli responses are probably due to the location, size of area touched, timing, technique used, amount of warning, and angle and manner of the approach.[80] One unpublished study suggested that the sensitivity to touch on the horse's side is greater than on a human fingertip.[31,81]

Equipment that contacts a horse has obvious implications to touch, with the twitch being one. Twitching a horse's upper lip can be characterized by three phases of response (Figure 2-7).[82] Phase one is associated with a small amount of physical restraint of the head, and this is partially due to the mild distraction that results from some discomfort. It lasts 3–5 min. Quitting, or having the twitch slip off, during this first stage results in the horse learning to avoid future twitching. For the next 5–10 min, analgesia is associated with the release of endorphins. The horse typically drops its head, relaxes its eyes, and appears somewhat sedated. The response is an acupuncture-like reaction because the heart rate does not increase as it would for a painful procedure.[83] The third phase of twitching comes about 15 min after its application when the horse again starts fighting the twitch. For several minutes of application, the heart rate significantly decreases due to parasympathetic activity, but the last phase is associated with a dramatic, sympathetic-related increase.[84,85] Horses do not develop an aversion to subsequent nose twitching, if it is done properly each time.[86]

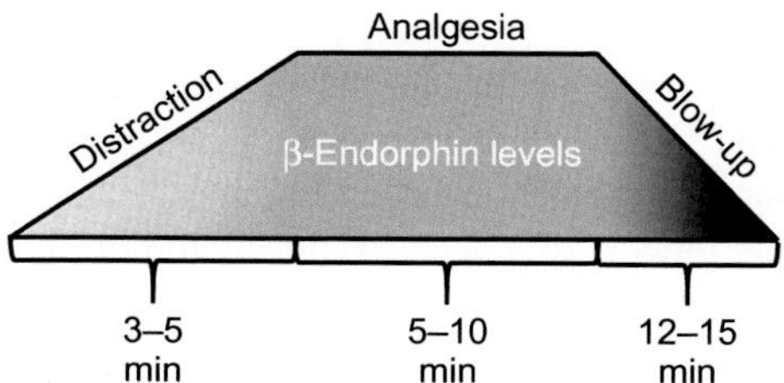

FIGURE 2-7 The three phases of response of behavior and β-endorphin levels to the use of a twitch.[82]

However, if twitching is only associated with aversive events such as ear clipping, over time the horse comes to expect negative events when the twitch appears.

In contrast, ear twitching is associated with an increased heart rate and other stress parameters, indicating increased sympathetic activity regardless of the length of the application.[85] Using an ear twitch will make a horse harder to handle and more difficult to touch afterward.

Nosebands and cavessons are commonly used to prevent a horse from opening its mouth. Tightness around the nose is a concern. While many associations have specific rules as to whether this equipment can be used or not and limitations as to tightness, abuse still happens. Eye temperature taken by infrared thermography has been correlated with cortisol concentrations, and thus with stress. Tight nosebands are associated with increased eye temperatures. Also, skin temperature distal to the noseband is lower than above it, suggesting compromised blood flow.[87]

Cinches are necessary to hold saddles in place. They can be problematic if improperly used. Tight cinches, especially those with multiple strands of mohair, can pinch skin unless applied carefully.

Pheromones

Pheromones are substances produced by one animal that carry a chemical signal to another animal. By definition, then, a pheromone could be something detected by taste or by smell, including by the vomeronasal organ.[88,89] Horses do not have superficial glands associated with pheromones. Studies relating to equine pheromones are done with a synthetic appeasing pheromone associated with the mammary region of a lactating mare. The commercially available equine appeasing pheromone may be mildly anxiolytic for timid or tense horses.

NEUROLOGICAL DEVELOPMENT OF FOALS

Neonates of prey species tend to be well developed at birth, and most ungulates are ready to follow their mothers within a few hours. This is true for foals. In modern times, however, the majority of foals are not raised on the American plains in free-ranging herds subjected to predator pressure, but rather on farms and ranches in more controlled environments. As a result, assessment of health and development of foals has become more sophisticated than simply being sure the foal can stand and follow. Prenatal and postnatal milestones can be used to evaluate the newborn's progress.[68]

Movement of the fetus is first detected during the third month of life, averaging two single movements every 10min. The number of single movements peaks around 16 within a 10-minute segment from the fourth to the ninth month of gestation and then decreases. During the tenth month, movements become more complex, and in the eleventh month, the frequency of complex

TABLE 2-1 Neurologic and Behavioral Milestones for Assessing Foal Development

Time of First Occurrence	Time Following First Occurrence	Expected Behavior
Newborn		Body temperature = 100–102°C
Newborn		Heart rate = 80–100 beats/min
Newborn		Respiratory rate = 60–80 breaths/min
0.5–3 min		Lifts and shakes head
3–13 min		Umbilical cord breaks
1–10 min		Sternal recumbency
10 min		Pupillary light reflex, light flash startle
2–10 min		Suckle reflex appears
5–10 min		Pupillary reflex appears
10–40 min		Head and ears can follow sound
0.25–2 h		Stands
	10 min	Walks well
1–3 h	(or) 30–90 min after standing	Nurses, meconium starts being passed
	30–90 min after nursing	Lies down
	80–100 min after nursing	Sleeps, usually on side
2–4 h		Trots, gallops, stretches
2–10 h		Defecates
	Within 4 days	Defecate yellowish feces (no meconium)
3–15 h		Urinates

movements increases to an average of 20 per hour. In the 3 days prior to birth, movements increase in frequency, becoming continuous for 10 or more minutes at a time. This may represent fetal positional adjustments.

Once delivery is complete, respiratory rhythm is rapidly established, and the pupillary light response can be detected within the first 10 min (Table 2-1).

Foals will typically fail in their early attempts to stand and are not successful for at least 15 min, with 30–60 min being common. Fillies tend to be successful sooner than breed matched colts. Failure to stand within 130 min is an indicator for concern.[68]

Nursing should follow standing, though not necessarily immediately afterward. The suckling reflex is usually present before the foal stands, so nursing can begin between 35 and 420 min after birth. Breed differences have been reported. Crossbred pony foals nurse in 32 min on average. The average is 90 min for Saddlebred foals and 111 min for Thoroughbred foals.[68]

Elimination behaviors are another monitor of health. Colts urinate about 6 h after birth and fillies do so around 10 h after being born. Defecation of meconium normally occurs within a few minutes after birth and is gradually replaced by the yellowish, milk-related feces within 4 days.[68]

Because of the relatively advanced state of development of newborn foals, many of the adult reflexes are already present.[90] The palpebral reflex causes the eyelids to blink when an area just above the eye is touched, and the lachrymal reflex results in tearing for foreign matter on the cornea. Bright light causes pupil constriction because of the pupillary light reflex, and stimulating ear hairs triggers the headshake reflex. Tapping the neck just behind the ear causes that ear to turn forward because of the cervico-auricular reflex. The extensor thrust reflex causes the limbs to extend when pressure is applied to the sole of the foot, helpful as the neonate first tries to stand. The sway reflexes cause near legs to bend and far legs to extend for balance when the foal is pushed from the side.

Imprinting is a sensitive period in development during which a youngster comes to identify its own species and the foal-mare bond is established. This lesson must occur during a very restricted period of time. One of the popularized handling techniques has been called "imprinting of foals" where the youngster is exposed to a variety of procedures it will encounter later—being sprayed, being rubbed with sacks or blankets, having feet handled, and tolerating clipper noise, to name a few. In reality, this is not imprinting at all, but rather desensitization to the techniques. As with other types of learning, desensitization lessons must be reinforced periodically in order to be retained. Imprinting, desensitization, and other foal behaviors will be discussed in more detail elsewhere.

TEMPERAMENT

"Temperament" is a human construct that is expressed as the foundation for an animal's personality, but it is hard to define.[91–93] It has been suggested that temperament is composed of the characteristics of energy, fearfulness, sensitivity, and adaptability.[94] "Personality" is a term used to describe the broader characteristics of what the animal may be feeling and thinking.[95,96] Personality has also been described as consisting of temperament and "character," with "character" being developed through learning to include submissiveness, aggression, human contact seeking, and self-reliance.[94] Unfortunately, the terms "temperament" and "personality" have been used interchangeably throughout the literature without clear definitions. "Temperament," as defined here, is innate and under neurological control. It is thought to play an important role in how an individual animal interprets and reacts to its environment, external stimuli, and handling because it deals with the inherent makeup of the horse and is not modified by learning. It may also play a role in the development of stereotypies.[97]

Ill-tempered horses are difficult to handle and potentially dangerous. If given a choice, riders avoid emotionally reactive mounts.[98] A great deal of effort has been invested in finding the genetic and heritability patterns of temperaments, but it is difficult.[99] Some characteristics are undoubtedly related to the breed of the horse, such as Arabians being flightier than Percherons. Other traits may concentrate within certain genetic lines within a breed.[100] Studies of the brain neurotransmitter serotonin have shown that a polymorphism in the serotonin receptor 1A gene (*HTR1A*) affects tractability and may have variations of expression based on the sex of the horse.[101] The frequency of the distribution of this variation is not yet known.

Some ways of interpreting temperament are showing validity, particularly identification of fearfulness ("flighty"), anxiousness, aggressiveness, and sociability.[95,102–105] Increases in the heart rate, neighing, frequency of defecation, and amount of walking are useful physiological and behavioral measures that indicate fearfulness and have been correlated with rider evaluations of temperament.[106–108] While fearfulness and anxiousness are concerning for riders, ancestral horses depended on them in moderation for survival. Being too flighty causes high stress and an excessive waste of calories escaping from things that should not trigger flight. Being not flighty enough could result in being eaten. Ultimately, horses with the right combination survived. The degree of flightiness for any one individual is consistent over time.[109,110] So too is the degree of fearfulness and tactile sensitivity; these independently predict success in show jumping competition.[109,111] Horses that are the least fearful, easiest to manipulate, and most sensitive to the rider's aids are the most successful. Fearfulness also affects performance.[105,112] A fearful horse shows its best performance under familiar, controlled conditions and its worst when stressed.

Most animals have specific types of sensory stimuli to which they are highly reactive, such as the sight of a fluttering plastic caught on a fence or odor of a rotting carcass. This reactivity is consistent over time and does not predict reactivity to other types of sensory stimuli.[81] Vision plays a part in reactivity within breeds where a third of adult horses are either near- or farsighted.[113] Warmbloods and Shires tend to be farsighted, and Thoroughbred crosses are more likely to be nearsighted. Reactivity in horses to novel objects is correlated to nervousness when being ridden, and half siblings tend to behave the same way.[114,115]

Speculation exists about possible interconnections between coat color, whorls, and temperament in several species. As an example, gray horses are considered to be docile, and yet if one thinks about survival of light-colored horses in the wild, it should result in questions about that deduction. A predator would be able to see a light-colored horse more easily, making a white or gray horse stand out significantly more and result in survivors having flightier personalities. People remember animals that fit the perception and forget those that do not. While speculation is common, science is slower to prove what relationships, if any, exist. Embryological origins of the skin and nervous system are the same. This suggests that a connection between color and behavior is not unreasonable. The base coat color is determined by polymorphisms of the melanocortin receptor 1 (*MC1R*) and Agouti signaling protein (*ASIP*) loci.[116] In several wild animal populations, darker-colored animals are more aggressive, sexually active, and resistant to stress than are lighter ones.[117] In other animals, *ASIP* polymorphism is associated with black coats, docility, and less overall activity. This relationship exists because melanin pigments and stress response hormones have similar developmental pathways.[116]

Temperament differences might have an association with color. Chestnuts are more likely to approach objects and other animals, compared to bays, regardless of familiarity.[118] Fear-like reactions in Icelandic horses are significantly greater in those carrying the Silver gene mutation than in horses of other colors. A follow-up to the Icelandic horse study addressed the genetic component. It showed that sires with the trait for fearful reactions passed the reactivity on to their progeny regardless of the foal's color.[119] A large study of Tennessee Walking Horses that were genotyped found no significant difference in temperament traits by genetic loci related to base coat color.[116]

Whorls generate similar speculations and have been historically identified as significant relative to temperaments by Native Americans of many generations. These twirling patterns occur in several locations on the body, but the ones on the head have had the most discussion (Figure 2-8). Interviews with experienced trainers show that each has a mental list of traits associated with whorl locations. Scientific reports of behavior as related to whorls on the face tend to refer back to work done in cattle, but the results do not always correlate with studies results from horses.[120–122] Horses with multiple facial whorls were more likely not to be friendly to other horses, not cooperate with people, and not be patient during training compared to horses with a single facial whorl.[120,122] They are also more prevalent in racing and jumping horses. Konik horses with whorls above the level of the eyes are less manageable, and those with an elongated whorl take longer to approach unfamiliar objects.[120,123]

Another factor to consider relative to whorls is heritability. It can be speculated that they are coincidentally related to a highly heritable temperament or laterality trait, because hair whorl positions are highly heritability (0.753 ± 0.056).[124] Two of three studies looking for a connection between whorls and side preferences (laterality) have shown that horses with whorls going from base to tip in a clockwise direction have a stronger right bias than do those with counterclockwise whorls.[125–127]

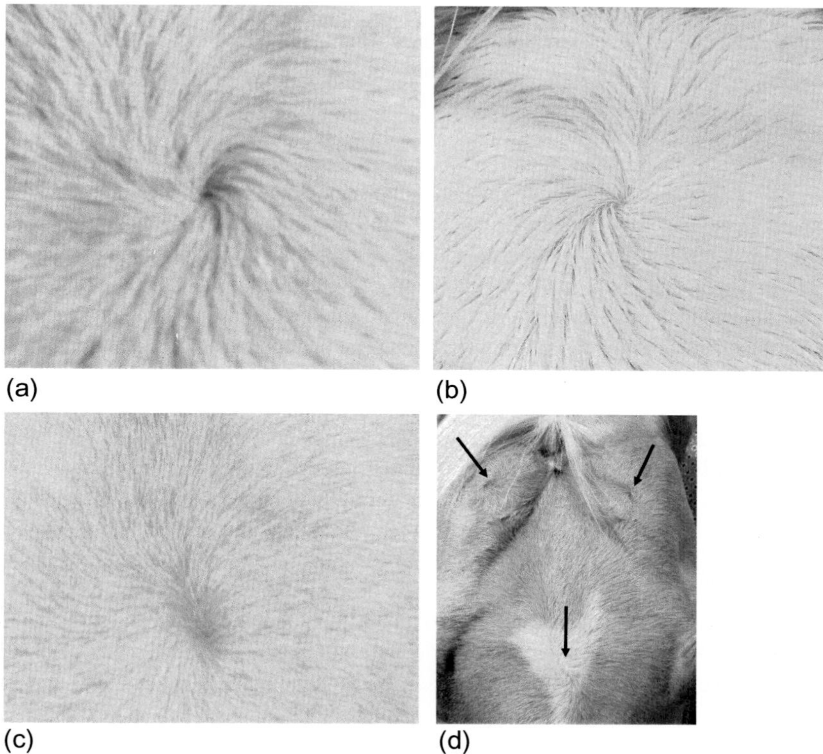

FIGURE 2-8 Facial whorls can direct hairs in a (a) clockwise, (b) counterclockwise, or (c) straight direction from their origin. They can also be located above, even with, or below eye level and be singular or multiple. The mare in (d) has two counterclockwise whorls below each ear and an elongated clockwise one at eye level. Horses with clockwise whorls are more likely to show a right-side bias than those with counterclockwise whorls.

NEUROLOGIC ORIGINS OF BEHAVIOR

The brain is the obvious origin of behavior and the intent of this section is not to describe associations between brain nuclei and their related behaviors, but rather to investigate related information that will help with assessment or treatments. The two areas that relate most closely to the expression of behavior and the pharmacological initiation and treatment of behavior problems are stress/distress and neurotransmitters.

Stress and Distress

The term "stress" can be defined as the emotional and physiological response to adverse circumstances. What is harder to define is when an adverse circumstance triggers a specific response and to what degree it affects the individual animal. At one extreme, some individual horses seem to do better when there is stress.[128] "Good stress" is called "eustress." As an example, show jumpers gradually become more stressed during competition, but those horses with the highest cortisol stress responses actually do better. The opposite was true for their riders, however. At the other extreme, stress becomes "distress," resulting in either a total shutdown to the outside world or extreme panic. Distress responses are inappropriate for the situation in which they occur and indicate that the stressor has gone to the point of affecting the horse's long-term welfare.

Small stresses are a normal part of life, whether they be the constant presence of flies in a pasture, the separation from a stablemate taken for a ride, or the strong afternoon sun of a summer day in Arizona. Over time, though, small stresses can become distressful, as could happen when the minor distress of the introduction of a new horse into a herd continues because the new horse has a very domineering personality. As indicated by cortisol levels and heart rate variability, show horses undergo stress starting with loading and transport. It continues for them when getting ready for the competition, and during the horse show as well.[129] By day three of competition, horses have slightly lower cortisol levels and heart rate variability, but all days appear less stressful than in transport.[129]

Individual variation in responses to any given situation are well documented in several species. Some individuals do not cope well with any change. Others show more curiosity than stress to the same trigger. Individuality aside, there are two broad types of stress responses—active and passive coping.[130] Active coping is characterized by reactivity. Passive copers show behavioral inhibition, including movement. Which response occurs relates to both genetic and situational components. The genetic relationship of flighty personalities was previously discussed. Past learning plays a role too. Bad experiences bring back memories that are likely to trigger avoidance. Experienced horsemen and women can feel the change in their mount as stress levels increase, but science looks at what is happening inside too, particularly as a way to evaluate the animal's welfare.

The list of things that can stress an individual horse is long. Most commonly, it is a startle response for a suddenly appearing object. This response is short-lived, lasting until the horse has time to figure out whether the occurrence is potentially serious. In a species that flees predators, the shying response should be expected. Longer term stress can be physiological, psychological, or both. Many times it is caused by humans. Examples would be competition, unpredictable feeding schedules, individual housing, temperature extremes, and inconsistent or unpredictable rewards and punishments. Stalled horses have higher cortisol levels than do pastured ones, and even exercise is associated with increased cortisol levels.[131] Social stresses from an incompatible herdmate, temperature extremes, overwork, illness, dehydration, and pain can also trigger stress responses.

Medical sources of stress must also be considered. Prey species tend to be stoic to not become an obvious target to a predator. Experienced horsemen and women notice the subtle behavioral changes, such as a mild reluctance to do a certain maneuver or the onset of a "grumpy" mood. While the presenting sign is "he just is not acting right," finding the cause can be a significant challenge to a veterinarian. The equine gastric ulcer syndrome is common in horses. Ulcers can be expressed as either equine squamous gastric disease (ESGD) or equine glandular gastric disease (EGGD), corresponding to the regions of the stomach in which they occur.[132] While the pathophysiology of ESGD is better understood, evidence is showing that both are probably related to chronic stress. The highest incidence of ESGD occurs in Thoroughbreds, in which 37% of untrained horses have these ulcers, and that number goes up to 80%–100% within 3 months of race training. Similarly, in endurance horses and racing Standardbreds, the prevalence of ESGD increases from approximately 45% when they have been idle to as high as 93% during active stages of their careers. Horses used for show or pleasure have an incidence between 40% and 60%. Besides the increased incidence of ESGD associated with use, there is a parallel increase in severity.[133] EGGD also increases in relationship to activity, although it has a higher incidence in leisure and sport horses than in racing ones. This suggests that factors besides stress may be contributing.[132]

Because of the horse's stoic nature, determining whether a horse is under stress is not always easy. For years, physiological assessment of stress has been heavily dependent on cortisol levels. Comparative studies showing increased cortisol levels are interpreted to suggest one handling method is better than another. Most conclusions, however, fail to account for circadian rhythms relating to cortisol levels, which peak in the late morning.[134,135] Peak times also vary depending on whether sampling is measuring salivary or serum cortisol.[134] There are also exercise-induced changes.[136] Because cortisol levels are not always elevated in potentially stressful situations, other measures of stress need to be included in assessments. The sum of all measures should be used to determine whether there is or is not stress (Table 2-2). As an example, a startle response will cause an increased heart rate and a behavioral reaction, but cortisol does not fluctuate.[144]

Behavioral assessments of stress include symptoms of physical discomfort such as temperament changes, pinning the ears back, general crabbiness, aggression, and decreased or increased responsiveness. Some horses show atypical motor movements like turning or flipping the head when no visual stimulus is present, teeth grinding, trembling, abnormal postures, and frequent lying and rising.[145] In general, the signs of increasing restlessness and fidgeting suggest increasing stress levels. A numerical scale of indicators has been developed that parallels the fluctuation in cortisol.[146] A low score of 1 represents no stress, where the horse is alert and interested in its surroundings. The numbers increase to 10 as stress increases. Evaluations include behaviors like increasing

TABLE 2-2 Comparative Activities That Have Been Evaluated for the Presence of Stress

Activity Comparisons Made	Stress Indicators Measured	Most Stressful
Stabled horses vs. pastured horses[131]	Cortisol	Stabling horses increases cortisol
Exercised vs. stalled horses[131]	Cortisol	Exercise
First saddling in round pen vs. not in round pen[137]	Cortisol, heart rate	No difference in cortisol, heart rate higher in round pen
Hyperflexion vs. competition frame vs. looser frame head carriage[138]	Behavior, cortisol, heart rate, rein tension	Hyperflexion changed cortisol and behavior
Hyperflexion vs. competition frame[139]	Behavior, cortisol, heart rate	Hyperflexion changed cortisol and behavior
Transport stress on naïve and previously transported horses[140]	ACTH, cortisol	Naïve to transport and poorly handled horses
Pre and post show jumping competitions[128]	Cortisol	Lower cortisol levels in horses and higher levels in riders
Pre, during, and post air transport[141]	Behavior, heart rate	Ascending and descending increased heart rate and aggression
Long term transport vs. no transport[142]	Appearance, body weight, dehydration, body temperature	24 h transport without access to water
Three day eventing[143]	Acute phase proteins, lysozymes, 4 protein adducts, CBC, lymphococytes	Lymphocytes CD4+ and CD8+ increased, and CD21+ decreased
Pre and post reining training[136]	Cortisol	Linear relationship with hematological parameters

restlessness to the point of agitation, acting uncomfortable, and even aggression. Using such a scale has the advantage of being noninvasive, easy to use, and fairly objective, and more accurate than physiological measures.[131] The other advantage is that it is not situation specific.

Determining the best way to relieve the stress or distress is situationally dependent; however, there are some general recommendations. Because horses are social, the presence of a calm companion can be helpful.[147] The first trail ride for a two-year-old is best done with an older, more experienced horse. Then, as the youngster learns what the wider world is about and how to trust the rider, it can eventually gain enough confidence to be ridden alone. Identifying stimuli that trigger undesired behavior is important. Gradual introductions should be part of any new situations that might be startling. Desensitization to fear-inducing stimuli, if done properly, can be extremely beneficial: something equine practitioners could find useful for many situations. Unpredictable fear or panic is difficult to treat, especially if eliciting stimuli are unknown. Careful evaluation of related events may identify the cause, and desensitization, perhaps with an appropriate medication, can then be helpful.

Anxiety and Fear

People have no way of actually knowing if a horse actually experiences anxiety or fear. There are changes in body language and physiological parameters that suggest the horse experienced something negative, and if it is fear or anxiety, we do not know whether they experience it in the same way humans do. For purposes of discussion, the reactions will be broadly defined as "fearful." Experiences vary in severity and by event, and the relative significance varies by individuals. Why a horse becomes fearful is dependent on several factors that relate to how it copes with stress in general. Genetic influences will influence certain bloodlines and are related to the origin of "hot-" and "cold-blooded" horse breeds. Genetic reactions can override environmental factors and training. The relationship of the hypothalamus-pituitary–adrenal (HPA) axis to cortisol production indicates that pathology within this axis could affect mood disorders. Added to the role of the HPA axis is the epigenetic association with subsequent generations when either the sire or dam is stressed preconception or during the pregnancy. The relationship of the sire's sperm to the offspring's brain regions associated with stress has been established.[148] Additionally, chronic stress of the dam results in cortisol crossing the placenta and negatively affecting the fetus's adrenal glands.[149]

Anxiety

Managing the fearful horse requires an understanding of what anxiety and fear actually are. Anxiety is defined as a feeling of apprehension that is the result of anticipation of some unidentified threat or danger. For a teenager, worry of what the father will say when he finds the damage done to his car is a perfect example of anxiety. It is not known if horses "think" the same type of thoughts, but there are situations when anxieties seem to occur.

Horses are most likely to experience social anxiety in two situations. A submissive animal forced into the presence of a very dominant horse is the first. In this case, the anxiety is associated with the uncertainty of whether the high-ranking horse will come near to threaten or attack the low-ranking one. The second situation is when a stablemate "friend" is removed. The resulting anxiety comes from "not knowing" if, or when, the stablemate will return. Behaviors associated with anxiety expression include increased locomotion, restlessness, vocalization, and contact-seeking.[150] This is not unlike the behavior profile for anticipatory behavior, so context is important.[151]

Repeated stressful, anxiety-producing events can lead to the development of stereotypies or self-directed psychogenic behavior, such as self-mutilation by stallions. The environment must be considered as a possible stressor, particularly stalls that restrict viewing or touching another horse. Long-term anxiety might be associated with paddocks or pastures too, especially those with an electric fence. Behaviorally, horses roll and move less in both small paddocks and electrically fenced ones. While the animals do not touch the fence as they would if it were metal or board, cortisol levels and heart rate measurements do not support associated stress responses.[152]

Treatment for the anxiety related to social stresses depends on the specific situation when the problem occurs. Punishment does not stop the behavior and will actually increase the amount of stress for the horse. First-line treatment should be a medical and environmental evaluation to identify contributing issues. Corneal clouding impairs vision; windy days confuse scent inputs.

Social stressors should be identified and removed. If the horse has difficulty around certain other horses, it should be removed from that environment and either put with nonthreatening horses or turned out at times when the difficult animal is not present. If the offending horse is a problem for several horses, it should be removed instead. For the horse that becomes anxious when a stablemate leaves (*separation anxiety*), there are several options. The horse might be fine around other friendly horses or if it can see other nearby horses. Another option would be to give the anxious horse something else to think about. This could be done by riding or groundwork. If the two tightly bonded horses are traveling together to a show, it may be possible to stable them out of sight of each other. Another option is to try the equine appeasing pheromone since it may be helpful for specific situational anxieties. Ultimately, the best way to handle separation anxiety is desensitization. While the specifics will be discussed elsewhere, the process involves gradually separating the two horses. This is a many-day and many-session training that begins with separation for very short distances and very short times. Then distances and times are gradually increased over several sessions at a rate slow enough so that the distress behaviors do not appear. The horse learns to tolerate separation gradually.

If the anxiety is primarily associated with being ridden, it is important to consider the rider. The problem is common when amateurs show horses. The horse does really well at home or for a professional, but the anxiety of the nonprofessional is transmitted to the horse. Professional help is needed so that the rider can gain confidence in their ability to ride and become consistent between trail riding, pen riding, and arena riding. Similarly, when a horse is asked to perform at a level higher than its training or a new rider uses aids that the horse has not learned, anxiety can result. Practice and repetition with gradual progression of learning is key to reducing the anxiety.

Certain horses seem to live under the "life is stressful" motto and are hyperreactive to any little thing. They are chronically anxious. This horse needs a structured environment with a specific routine. They do better with experienced riders with whom they can develop a trust. Anxiety increases when changes are made. If the overarching anxiety is chronic, excessive, and problematic, the horse will benefit from prolonged use of medications that increase serotonin levels, like the selected serotonin reuptake inhibitors. Shorter acting drugs can be used for situational anxieties, such as the first trailer ride or a new, fearful environment.

Fear

Fear is the feeling of apprehension resulting from the nearness of a threatening object or situation such as a natural danger or confusing sensory perception. Horses evolved to react to real or potential predators and to other dangers within their environment. They shy from things that move fast, buzzing clippers, or rotting carcasses. Plastic bags flapping on a fence are hard to get into focus, so instinctively the horse tries to get far enough away to be able to evaluate the seriousness of the threat. Social fears occur in response to the threat or act of aggression by a high-ranking horse, particularly when that horse is close. Single traumatic events can make horses afraid of everyday equipment to the point that they cannot safely be tied, led, or even touched.[153]

Ocular input goes to the superior colliculus of the brain and through associated pathways to the amygdalae. This paired area of the brain plays a major role in memory, decision-making, and emotional reactions. Events that trigger fear responses, if bad enough, can result in development of generalized fear memory in a two-step process. Something bad happens and the sensory input triggers the survival centers in the amygdala to react. At the same time, there is sensory input to memory areas. That input may be the vision of a horse trailer, a certain gate, a specific location, an odor, or other feature directly or indirectly related to the experience. The next time a similar object is seen, subconscious memories are triggered, and the animal tries avoidance. That avoidance will then reinforce the fear. Each subsequent avoidance continues reinforcement, and over time, fear can escalate to phobia.

Treatment of fears is dependent on environmental management, desensitization, and medication. First, the specific trigger of the behavior must be identified. Then, management is important so that the stimulus of the fear is totally avoided. This might involve removing the horse from the herd or removing the aggressive horse instead. It could be avoiding trails that are likely to have flapping items and riding with a seasoned horse that does not react to potentially frightful objects. The latter is particularly important for young horses. Desensitization begins by introducing frightening objects in tiny, nontriggering ways so as not to provoke a fearful response. The stimulus is very gradually increased in intensity. If done correctly, fear is eliminated. It is the most rapid method for eliminating fright responses.[154] In severe cases, medication may improve the amount of success with the desensitization. Short-term antianxiety drugs may help for individual sessions, as might the equine appeasing pheromone.[155] For the horse with generalized anxiety issues, several months of a selected serotonin reuptake inhibitor may be indicated.

Phobia

Phobias take fear to a whole new level, because they represent a fear that is excessive and out of proportion to the actual threat. The most common phobia in horses is shown as scrambling in a horse trailer. A bad experience when traveling, such as a corner being taken too fast or bee stings from a disturbed nest in the trailer, causes the horse to try to regain its balance or escape. Subsequent trailer rides reinforce the fear, even though the driving may be better or the bee's nest removed. Eventually the horse starts fighting the trailer as soon as it begins to move.[156] Most phobic animals are not particularly afraid to load, but stress peaks as soon as movement is felt. Watching the facial expressions while these horses are scrambling leaves no doubt that there is significant stress. Phobias can also develop in a number of other situations like the noise of a plastic raincoat or walking over a plastic tarp.

Successful treatment of a phobia is almost impossible, so owners need to understand improvement, not cure, is the goal. While panicking, a horse will not learn, so the reaction must be reduced from phobia to fear. Then the fear is treated. The easiest treatment is to simply avoid the triggers. For the phobic horse that is scrambling in a side-by-side trailer with a center partition, it may haul just fine on the other side of the trailer or with the partition removed or swung to one side. Using a stock trailer is another option. Antianxiety medications can help reduce the phobia to a fear and then desensitization can be done in the revised trailer configuration. It begins as load, feed, unload and works up to movement of a few feet, gradually lengthening the road trips. The process is repeated as the horse is gradually weaned off the medication.

Neurotransmitters

Neurotransmitters are responsible for sending the messages from one neuron to the next. While they exist throughout the body, they are most prevalent in the brain. Understanding brain function and responses to various psychopharmacological agents depends on a basic understanding of these internal chemicals. Classifying neurotransmitters is complicated because there are over 100 different ones. Fortunately, the seven "small molecule" neurotransmitters (acetylcholine, dopamine, gamma-aminobutyric acid (GABA), glutamate, histamine, norepinephrine, and serotonin) do the majority of the work. Another complicating factor is that neurotransmitters may have a number of subtypes, serotonin having 15, as an example.[157] Endorphins and oxytocin are neuropeptides that are sometimes considered to be neurotransmitters, and β-endorphin is associated with the feeling of pleasure in humans. Exercise causes β-endorphin and serotonin levels to significantly increase in horses.[158] This is likely to happen in human runners too. It is also thought some neurotransmitters may play a role in stereotypies.[159] Neurotransmitters do have general functions that hold true even across species, as will be described later in the book.

The equine brain has not been well studied relative to which neurotransmitters are associated with which nuclei or specific functions. Each nucleus may have multiple neurotransmitters, and there can be considerable differences in their proportions between species. This explains why a drug that works in one species may not be as effective for a similar problem in another. It is important to understand that the choice of a psychopharmacological drug is based on empirical data, so depending on the patient's response, it might be necessary to modify doses or drugs used.

Other Influences

Brain plasticity is a topic of much current research and will provide a great deal of information about behavior in the future. Research is showing that everyday experiences can profoundly affect neurotransmitters and neurocircuitry. In rodent models, early life stress can change neurotransmitters and brain structure.[160–162] In humans, the hormonal variations of pregnancy result in changes in the brain structure that are visible on functional MRI (fMRI) images.[163] At this time, little is known about how such things might affect horses, but hormones and social relationships have long-term behavioral significance in multiple species, so their potential role in changing a horse's behavior should be remembered.

Displacement Behavior

Displacement behaviors are ones that are inappropriate for the stimulus presented and therefore seem out of context. They are usually shown when there is a conflict between different drives occurring at the same time. As an example, a horse that is being ridden tries to move forward but the rider insists that it remain still. The horse will start to fidget, perhaps moving its head first, then pawing or moving the rear end to one side (Figure 2-9). A hungry horse may start biting on the manger or pawing when the human takes too long, in the horse's opinion, to bring the food. In this case it wants the food but is prevented from getting it because of a physical barrier.

FIGURE 2-9 After running into position, a flag-bearing horse in this rodeo shows air-pawing as an indication of the lack of patience for having to stand still.

It is likely horses are easily and frequently put into conflicting situations. Over time this can result in the development of a displacement activity, such as pawing. If continued, the behavior becomes more rigid in its expression, even to the point of becoming a stereotypy. The connection between a frequently expressed displacement behavior and the development of a stereotypy exists because both are expressed when excitement or anticipatory levels increase and not during times of "boredom."

Vacuum Activity

A vacuum activity is one that appears without an apparent stimulus. Most natural behaviors have a threshold that a stimulus must reach before the behavior is triggered. When that behavior cannot happen for a long time, the threshold gradually lowers until a point is reached where the behavior spontaneously appears.

A certain amount of time must pass after a stallion breeds a mare before it shows an interest in breeding again (the refractory period). After the second breeding, the refractory period is longer than it was previously, and subsequent periods increase in length. The opposite occurs too. If the stallion does not have the opportunity to breed a mare, the amount of stimuli needed to trigger his behavior is reduced. As this nonbreeding period is extended, mounting behavior seems to "spontaneously" appear if any other horse is around, including geldings or other stallions.

Self-mutilation by stallions is an example of a behavior that appears as a vacuum activity. There is no apparent stimulus, and a specific trigger is unknown. Speculation suggests an internal stimulus might involve misfiring of neurons and/or a pathological abnormality at some of the neuromuscular junctions.

REFERENCES

1. Heffner HE, Heffner RS. Sound localization in large mammals: localization of complex sounds by horses. *Behav Neurosci* 1984;**98**(3):541–55.
2. Heffner RS, Heffner HE. Visual factors in sound localization in mammals. *J Comp Neurol* 1992;**317**(3):219–32.
3. Wolski TR, Houpt KA, Aronson R. The role of the senses in mare-foal recognition. *Appl Anim Ethol* 1980;**6**(2):121–38.
4. Prendergast A, Nansen C, Blache D. Responses of domestic horses and ponies to single, combined and conflicting visual and auditory cues. *J Equine Vet* 2016;**46**:40–6.
5. Hall C. The impact of visual perception on equine learning. *Behav Process* 2007;**76**(1):29–33.
6. Prince JH. *Comparative anatomy of the eye*. Springfield, IL: Charles C. Thomas; 1956. p. 418.
7. Gilger BC. *Equine ophthalmology*. 2nd ed. Maryland Heights, MO: Elsevier; 2010. p. 536.
8. Evans KE, McGreevy PD. The distribution of ganglion cells in the equine retina and its relationship to skull morphology. *Anat Histol Embryol* 2007;**36**(2):151–6.
9. Murphy J, Hall C, Arkins S. What horses and humans see: a comparative review. *Int J Zool* 2009;**2009**(Article ID 721798):1–14.
10. Timney B, Keil K. Visual acuity in the horse. *Vis Res* 1992;**32**(12):2289–93.
11. Miller RM. The senses of the horse. *J Equine Vet* 1995;**15**(3):102–3.
12. Harman AM, Moore S, Hoskins R, Keller P. Horse vision and an explanation for the visual behavior originally explained by the 'ramp retina'. *Equine Vet J* 1999;**31**(5):384–90.
13. Sivak JG, Allen DB. An evaluation of the "ramp" retina of the horse eye. *Vis Res* 1975;**15**:1353–6.
14. Hanggi EB, Ingersoll JF. Stimulus discrimination by horses under scotopic conditions. *Behav Process* 2009;**82**(1):45–50.
15. Macuda TJ. *Equine colour vision*. Dissertation; 2000. p. 136. http://proquest.umi.com/pqdweb?index=0&did=728862111&SrcMode=1&sid=1&Fmt=2&clientid=2945&RQT=309&VName=PQD[downloaded Feb. 8, 2007].
16. Carroll J, Murphy CJ, Neitz M, Ver Hoeve JN, Neitz J. Photopigment basis for dichromatic color vision in the horse. *J Vis* 2001;**1**(2):80–7.
17. Geisbauer G, Griebel U, Schmid A, Timney B. Brightness discrimination and neutral point testing in the horse. *Can J Zool* 2004;**82**:660–70.
18. Hanggi EB, Ingersoll JF, Waggoner TL. Color vision in horses (*Equus caballus*): deficiencies identified using a pseudoisochromatic plate test. *J Comp Psychol* 2007;**121**(1):65–72.
19. Phillips CJC, Lomas CA. The perception of color by cattle and its influence on behavior. *J Dairy Sci* 2001;**84**(4):807–13.
20. Roth LSV, Balkenius A, Kelber A. Colour perception in a dichromat. *J Exp Biol* 2007;**210**(16):2795–800.
21. Timney B, Macuda T. Vision and hearing in horses. *J Am Vet Med Assoc* 2001;**218**(10):1567–74.

22. Macuda T, Timney B. Luminance and chromatic discrimination in the horse (*Equus caballus*). *Behav Process* 1999;**44**(3):301–7.
23. Prince JH, Diesem CD, Eglitis I, Ruskell GL. *Anatomy and histology of the eye and orbit in domestic animals.* Springfield, IL: Charles C. Thomas; 1960. p. 307.
24. Timney B, Keil K. Local and global stereopsis in the horse. *Vis Res* 1999;**39**(10):1861–7.
25. Hanggi EB, Ingersoll JF. Lateral vision in horses: a behavioral investigation. *Behav Process* 2012;**91**(1):70–6.
26. Bartoš L, Bartošová J, Starostová L. Position of the head is not associated with changes in horse vision. *Equine Vet J* 2008;**40**(6):599–601.
27. von Borstel UU, Duncan IJH, Shoveller AK, Merkies K, Keeling LJ, Millman ST. Impact of riding in a coercively obtained Rollkur posture on welfare and fear of performance horses. *Appl Anim Behav Sci* 2009;**116**(2–4):228–36.
28. Dziezyc J, Taylor L, Boggess MM, Scott HM. The effect of ocular blinkers on the horses' reactions to four different visual and audible stimuli: results of a crossover trial. *Vet Ophthalmol* 2011;**14**(5):327–32.
29. Houpt KA, Houpt TR. Social and illumination preferences of mares. *J Anim Sci* 1988;**66** (9):2159–64.
30. Roth LSV, Balkenius A, Kelber A. The absolute threshold of colour vision in the horse. *PLoS ONE* 2008;**3**(11). https://doi.org/10.1271/journal.pone.0003711.
31. Saslow CA. Understanding the perceptual world of horses. *Appl Anim Behav Sci* 2002;**78** (2–4):209–24.
32. Ollivier FJ, Samuelson DA, Brooks DE, Lewis PA, Kallberg ME, Komáromy AM. Comparative morphology of the tapetum lucidum (among selected species). *Vet Ophthalmol* 2004;**7** (1):11–22.
33. Rehkämper G, Perrey A, Werner CW, Opfermann-Rüngeler C, Görlach A. Visual perception and stimulus orientation in cattle. *Vis Res* 2000;**40**(18):2489–97.
34. Hanggi EB. The thinking horse: cognition and perception reviewed, In: *Proceedings of the 51st annual convention of the American Association of Equine Practitioners, Seattle, Washington*; 2005. p. 246–55.
35. Farmer K, Krueger K, Byrne RW. Visual laterality in the domestic horse (*Equus caballus*) interacting with humans. *Anim Cogn* 2010;**13**(2):229–38.
36. Larose C, Richard-Yris M-A, Hausberger M. Laterality of horses associated with emotionality in novel situations. *Laterality* 2006;**11**(4):355–67.
37. Sankey C, Henry S, Clouare C, Richard-Yris M-A, Hausberger M. Asymmetry of behavioral responses to a human approach in young naïve vs. trained horses. *Physiol Behav* 2011;**104** (3):464–8.
38. Dien J. Looking both ways through time: the Janus model of lateralized cognition. *Brain Cogn* 2008;**67**(3):292–323.
39. Robins A, Phillips C. Lateralised visual processing in domestic cattle herds responding to novel and familiar stimuli. *Laterality* 2010;**15**(5):514–34.
40. De Boyer Des Roches ADB, Richard-Yris M-A, Henry S, Ezzaouïa M, Hausberger M. Laterality and emotions: visual laterality in the domestic horse (*Equus caballus*) differs with objects' emotional value. *Physiol Behav* 2008;**94**(3):487–90.
41. Parker R, Watson R, Wells E, Brown SN, Nicol CJ, Knowles TG. The effect of blindfolding horses on heart rate and behavior during handling and loading onto transport vehicles. *Anim Welf* 2004;**13**(4):433–7.
42. Heffner HE, Heffner RS. The hearing ability of horses. *Equine Pract* 1983;**5**(3):27–32.

43. Heffner RS, Heffner HE. Hearing in large mammals: horses (*Equus caballus*) and cattle (*Bos Taurus*). *Behav Neurosci* 1983;**97**(2):299–309.
44. Yeon SC. Acoustic communication in the domestic horse (*Equus caballus*). *J Vet Behav* 2012;**7**(3):179–85.
45. Wilson WJ, Mills PC, Dzulkarnain AA. Use of BAER to identify loss of auditory function in older horses. *Aust Vet J* 2011;**89**(3):73–6.
46. Basile M, Boivin S, Boutin A, Blois-Heulin C, Hausberger M, Lemasson A. Socially dependent auditory laterality in domestic horses (*Equus caballus*). *Anim Cogn* 2009;**12**(4):611–9.
47. Heffner RS, Heffner HE. Localization of tones by horses: use of binaural cues and the role of the superior olivary complex. *Behav Neurosci* 1986;**100**(1):93–103.
48. Houpt K, Marrow M, Seeliger M. A preliminary study of the effect of music on equine behavior. *J Equine Vet* 2000;**20**(11):691–693 and 737.
49. Magdesian KG, Williams DC, Aleman M, LeCouteur RA, Madigan JE. Evaluation of deafness in American paint horses by phenotype, brainstem auditory-evoked responses, and endothelin receptor B genotype. *J Am Vet Med Assoc* 2009;**235**(10):1204–11.
50. Hauswirth R, Haase B, Blatter M, Brooks SA, Burger D, Drögemüller C, Gerber V, Henke D, Janda J, Jude R, Magdesian KG, Matthews JM, Poncet P-A, Svansson V, Tozaki T, Wilkinson-White L, Penedo MCT, Rieder S, Leeb T. Mutations in MITF and PAX3 cause "splashed white" and other white spotting phenotypes in horses. *PLoS Genet* 2012;**8**(4). https://doi.org/10.1371/journal.pgen.1002653.
51. Christensen JW, Zharkikh T, Ladewig J, Yasinetskaya N. Social behavior in stallion groups (*Equus przewalskii* and *Equus caballus*) kept under natural and domestic conditions. *Appl Anim Behav Sci* 2002;**76**(1):11–20.
52. Hothersall B, Harris P, Sörtoft L, Nicol CJ. Discrimination between conspecific odour samples in the horse (*Equus caballus*). *Appl Anim Behav Sci* 2010;**126**(1–2):37–44.
53. Kimura R. Volatile substances in feces, urine and urine-marked feces of feral horses. *Can J Anim Sci* 2001;**1**(3):411–20.
54. Ma W, Klemm WR. Variations of equine urinary volatile compounds during the oestrous cycle. *Vet Res Commun* 1997;**21**(6):437–46.
55. Krueger K, Flauger B. Olfactory recognition of individual competitors by means of faeces in horses (*Equus caballus*). *Anim Cogn* 2011;**14**(2):245–57.
56. Péron F, Ward R, Burman O. Horses (*Equus caballus*) discriminate body odour cues from conspecifics. *Anim Cogn* 2014;**17**(4):1007–11.
57. Christensen JW, Rundgren M. Predator odour *per se* does not frighten domestic horses. *Appl Anim Behav Sci* 2008;**112**(1–2):136–45.
58. Lindsay FEF, Burton FL. Observational study of "urine testing" in the horse and donkey stallion. *Equine Vet J* 1983;**15**(4):330–6.
59. Crump D, Swigar AA, West JR, Silverstein RM, Müller-Schwarze D, Altieri R. Urine fractions that release flehmen in black-tailed deer, *Odocoileus hemionus columbianus*. *J Chem Ecol* 1984;**10**(2):203–15.
60. Ladewig J, Hart BL. Flehmen and vomeronasal organ function in male goats. *Physiol Behav* 1980;**24**(6):1067–71.
61. Johns MA. The role of the vomeronasal system in mammalian reproductive physiology. In: Müller-Schwarze D, Silverstein RM, editors. *Chemical signals: vertebrates and aquatic invertebrates*. New York: Plenum Press; 1980. p. 341–64.
62. Ladewig J, Price EO, Hart BL. Flehmen in male goats: role in sexual behavior. *Behav Neural Biol* 1980;**30**(3):312–22.

63. Hart BL, Jones TOAC. Effects of castration on sexual behavior of tropical male goats. *Horm Behav* 1975;**6**:247–58.
64. Crowell-Davis SL. Flehmen. *Compend Equine* 2008;**3**(2):91–4.
65. Estes RD. The role of the vomeronasal organ in mammalian reproduction. *Mammalia* 1972;**36** (3):315–41.
66. Salazar I, Quinteiro PS, Cifuentes JM. The soft-tissue components of the vomeronasal organ in pigs, cows and horses. *Anat Histol Embryol* 1997;**26**(3):179–86.
67. Stahlbaum CC, Houpt KA. The role of the flehmen response in the behavioral repertoire of the stallion. *Physiol Behav* 1989;**45**(6):1207–14.
68. Crowell-Davis SL. Developmental behavior. *Vet Clin N Am Equine Pract* 1986;**2**(3):573–90.
69. Crowell-Davis S, Houpt KA. The ontogeny of flehmen in horses. *Anim Behav* 1985;**33** (3):739–45.
70. Weeks JW, Crowell-Davis SL, Heusner G. Preliminary study of the development of the flehmen response in *Equus caballus. Appl Anim Behav Sci* 2002;**78**(2–4):329–35.
71. Nickel R, Schummer A, Seiferle E, Sack WO. *The viscera of the domestic mammals.* Berlin: Springer-Verlag; 1973. p. 29–31.
72. Yamamoto Y, Atoji Y, Hobo S, Yoshihara T, Suzuki Y. Morphology of the nerve endings in laryngeal mucosa of the horse. *Equine Vet J* 2001;**33**(2):150–8.
73. PubMed Health. *How does our sense of taste work?* http://www.ncbi.nlm.nih.gov/pubmedhealth/PMH0072592/; 2016[downloaded Sept. 22, 2016].
74. Randall RP, Schurg WA, Church DC. Response of horses to sweet, salty, sour and bitter solutions. *J Anim Sci* 1978;**47**(1):51–5.
75. Jankunis ES, Whishaw IQ. Sucrose bobs and quinine gapes: horse (*Equus caballus*) responses to taste support phylogenetic similarity in taste reactivity. *Behav Brain Res* 2013;**256**:284–90.
76. Houpt KA. *Domestic animal behavior for veterinarians and animal scientists.* 3rd ed. Ames: Iowa State University Press; 1998. p. 495.
77. Taniguchi K, Koida A, Mutoh K. Comparative lectin histochemical studies on taste buds in five orders of mammals. *J Vet Med Sci* 2008;**70**(1):65–70.
78. Roura E, Humphrey B, Tedó G, Ipharraguerre I. Unfolding the codes of short-term feed appetence in farm and companion animals. A comparative oronasal nutrient sensing biology review. *Can J Anim Sci* 2008;**88**(4):535–58.
79. Bottom SH. *Age-related changes in taste and gustatory response and feeding behavior in the stabled horse.* http://ethos.bl.ukOrderDetails.do?uin=uk.bl.ethos.510181; 2008 [downloaded Sept. 22, 2016].
80. McGreevy P. *Equine behavior: a guide for veterinarians and equine scientists.* New York: Saunders; 2004369.
81. Lansade L, Pichard G, Leconte M. Sensory sensitivities: components of a horse's temperament dimension. *Appl Anim Behav Sci* 2008;**114**(3–4):534–53.
82. McDonnell S. Using the twitch properly. *The Horse* 2004;**XXI**(11):105–6.
83. Lagerweij E, Nelis PC, Wiegant VM, van Ree JM. The twitch in horses: a variant of acupuncture. *Science* 1984;**225**(4667):1172–4.
84. Flakoll B. Twitching in veterinary procedures: how does this technique subdue horses? In: *Poster #24, International Society of Exposure Science Meeting, Utrecht, The Netherlands*; 2016.
85. Flakoll B, Ali AB, Saab CY. Twitching in veterinary procedures: how does this technique subdue horses? *J Vet Behav* 2017;**18**:23–8.
86. Ali ABA, Gutwein KL, Heleski CR. Assessing the influence of upper lip twitching in naïve horses during an aversive husbandry procedure (ear clipping). *J Vet Behav* 2017;**21**:20–5.

87. McGreevy P, Warren-Smith A, Guisard Y. The effect of double bridles and jaw-clamping crank nosebands on temperature of eyes and facial skin of horses. *J Vet Behav* 2012;**7**(3):142–8.
88. Broom DM, Fraser AF. *Domestic animal behaviour and welfare*. 4th ed. Cambridge, MA: CAB International; 2007. p. 438.
89. Rekwot PI, Ogwu D, Oyedipe EO, Sekoni VO. The role of pheromones and biostimulation in animal reproduction. *Anim Reprod Sci* 2001;**65**(3–4):157–70.
90. Waring GH. *Horse behavior*. 2nd ed. Norwich, NY: Noyes Publications; 2003. p. 442.
91. Buss AH. *Personality: temperament, social behavior, and the self*. Needham Heights, MA: Allyn & Bacon; 1995. p. 420.
92. Goldsmith HH, Buss AH, Plomin R, Rothbart MK, Thomas A, Chess S, Hinde RA, McCall RB. Roundtable: what is temperament? Four approaches. *Child Dev* 1987;**58**(2):505–29.
93. McCrae RR, Costa Jr PT, Ostendorf F, Angleitner A, Hřebíčková M, Avia MD, Sanz J, Sáánchez-Bernardos ML, Kusdil ME, Woodfield R, Saunders PR, Smith PB. Nature over nurture: temperament, personality, and life span development. *J Pers Soc Psychol* 2000;**78**(1):173–86.
94. Suwata M, Górecka-Bruzda A, Walczak M, Ensminger J, Jezierski T. A desired profile of horse personality—a survey study of Polish equestrians based on a new approach to equine temperament and character. *Appl Anim Behav Sci* 2016;**180**:6–77.
95. Gosling SD. From mice to men: what we can learn about personality from animal research. *Psychol Bull* 2001;**126**(1):45–86.
96. Pervin LA, John OP. *Personality: theory research*. 7th ed. New York: John Wiley & Sons; 1997. p. 609.
97. Ijichi CL, Collins LM, Elwood RW. Evidence for the role of personality in stereotypy predisposition. *Anim Behav* 2013;**85**(6):1145–51.
98. Visser EK, van Reenen CG, Zetterqvist Blockhuis M, Morgan EKM, Hassmén P, Rundgren TMA, Blochuis HJ. Does horse temperament influence horse-rider cooperation? *J Appl Anim Welf Sci* 2008;**11**(3):267–84.
99. Gauly M, Mathiak H, Hoffmann K, Kraus M, Erhardt G. Estimating genetic variability in temperamental traits in German Angus and Simmental cattle. *Appl Anim Behav Sci* 2001;**74**(2):109–19.
100. Sellnow L. Heritability of behavior. *The Horse* 2003;**XX**(5). 49, 50, and 52.
101. Hori Y, Tozaki T, Nambo Y, Sato F, Ishimaru M, Inoue-Murayama M, Fujita K. Evidence for the effect of serotonin receptor 1A gene (*HTR1A*) polymorphism on tractability in thoroughbred horses. *Anim Genet* 2016;**47**(1):62–7.
102. Bulens A, Sterken H, Van Beirendonck S, Van Thielen J, Driessen B. The use of different objects during a novel object test in stabled horses. *J Vet Behav* 2015;**10**(1):54–8.
103. Graf P, von Borstel UK, Gauly M. Practical considerations regarding the implementation of a temperament test into horse performance tests: results of a large-scale test run. *J Vet Behav* 2014;**9**(6):329–40.
104. Mills DS. Personality and individual differences in the horse, their significance, use and measurement. *Equine Vet J* 1998;**30**(S27):10–3.
105. Peeters M, Verwilghen D, Serteyn D, Vandenheede M. Relationships between young stallions' temperament and their behavioral reactions during standardized veterinary examinations. *J Vet Behav* 2012;**7**(5):311–21.
106. Lansade L, Bouissou M-F, Erhard HW. Reactivity to isolation and association with conspecifics: a temperament trait stable across time and situation. *Appl Anim Behav Sci* 2008;**109**(2–4):355–73.

107. McCall CA, Hall S, McElhenney WH, Cummins KA. Evaluation and comparison of four methods of ranking horses based on reactivity. *Appl Anim Behav Sci* 2006;**96**(1–2):115–27.
108. Momozawa Y, Ono T, Sato F, Kikusui T, Takeuchi Y, Mori Y, Kusunose R. Assessment of equine temperament by a questionnaire survey to caretakers and evaluation of its reliability by simultaneous behavior test. *Appl Anim Behav Sci* 2003;**84**(2):12i–38.
109. Lansade L, Bouissou M-F, Erhard HW. Fearfulness in horses: a temperament trait stable across time and situation. *Appl Anim Behav Sci* 2008;**115**(3–4):182–200.
110. Visser EK, van Reenen CG, Hopster H, Schilder MBH, Knaap JH, Barneveld A, Blokhuis HJ. Quantifying aspects of young horses' temperament: consistency of behavioural variables. *Appl Anim Behav Sci* 2001;**74**(4):241–58.
111. Lansade L, Philippon P, Hervé L. Development of personality tests to use in the field, stable over time and across situations, and linked to horses' show jumping performance. *Appl Anim Behav Sci* 2016;**176**:43–51.
112. Valenchon M, Lévy F, Fortin M, Leterrier C, Lansade L. Stress and temperament affect working memory performance for disappearing food in horses, *Equus caballus*. *Anim Behav* 2013;**86**(6):1233–40.
113. Bracun A, Ellis AD, Hall C. A retinoscopic survey of 333 horses and ponies in the UK. *Vet Ophthalmol* 2014;**17**(S1):90–6.
114. Le Scolan N, Hausberger M, Wolff A. Stability over situations in temperamental traits of horses as revealed by experimental and scoring approaches. *Behav Process* 1997;**41**(3):257–66.
115. Wolff A, Hausberger M, Le Scolan N. Experimental tests to assess emotionality in horses. *Behav Process* 1997;**40**(3):209–21.
116. Jacobs LN, Staiger EA, Albright JD, Brooks SA. "A sorrel is hot…": a genetic investigation of the horseman's myth. *J Equine Vet* 2015;**35**(5):383.
117. Ducrest A-L, Keller L, Roulin A. Pleiotropy in the melanocortin system, coloration and behavioural syndromes. *Trends Ecol Evol* 2008;**23**(9):502–10.
118. Finn JL, Haase B, Willet CE, van Rooy D, Chew T, Wade CM, Hamilton NA, Velie BD. The relationship between coat colour phenotype and equine behavior: a pilot study. *Appl Anim Behav Sci* 2016;**174**(1):66–9.
119. Brunberg E, Gille S, Mikko S, Lindgren G, Keeling LJ. Icelandic horses with the silver coat colour show altered behavior in a fear reaction test. *Appl Anim Behav Sci* 2013;**146**(1–4):72–8.
120. Deesing MJ, Grandin T. Behavior genetics of the horse (*Equus caballus*). In: Grandin T, Deesing MJ, editors. *Genetics and the behavior of domestic animals*. 2nd ed. Waltham, MA: Academic Press; 2014. p. 237–90.
121. Grandin T, Deesing NJ, Struthers JJ, Swinker AM. Cattle with hair whorl patterns above the eyes are more behaviorally agitated during restraint. *Appl Anim Behav Sci* 1995;**46**(1–2):117–23.
122. Makiguchi K, Momozawa Y, Erb HN, Perry P, Kusunose R, Houpt HA. Can whorls on forehead in horses tell their temperament? In: *Proceedings of the American veterinary Society of Animal Behavior Annual Meeting, Denver, Colorado*; 2003. p. 46–50.
123. Górecka A, Golonka M, Chruszczewski M, Jezierski T. A note on behavior and heart rate in horses differing in facial hair whorl. *Appl Anim Behav Sci* 2007;**105**(1–3):244–8.
124. Górecka A, Słoniewski K, Golonka M, Jaworski Z, Jerierski T. Heritability of hair whorl position on the forehead in Konik horses. *J Anim Breed Genet* 2006;**123**(6):396–8.
125. Murphy J, Arkins S. Facial hair whorls (trichoglyphs) and the incidence of motor laterality in the horse. *Behav Process* 2008;**79**(1):7–12.

126. Savin H, Randle H. The relationship between facial whorl characteristics and laterality exhibited in horses. *J Vet Behav* 2011;**6**(5):295–6.
127. Shivley C, Grandin T, Deesing M. Behavioral laterality and facial hair whorls in horses. *Equine Vet Sci* 2016;**44**:62–6.
128. Peeters M, Closson C, Beckers J-F, Vandenheede M. Rider and horse salivary cortisol levels during competition and impact on performance. *J Equine Vet* 2013;**33**(3):155–60.
129. Becker-Birck M, Schmidt A, Lasarzik J, Aurich J, Möstl E, Aurich C. Cortisol release and heart rate variability in sport horses participating in equestrian competitions. *J Vet Behav* 2013;**8**(2):87–94.
130. Budzyńska M. Stress reactivity and coping in horse adaptation to environment. *J Equine Vet* 2014;**34**(8):935–41.
131. Young T. *Physiological and behavioural measures of stress in domestic horses*. University of Liverpool Thesis; 2011. *http://hdl.handle.net/10034/606598*[downloaded Sept. 26, 2016].
132. Sykes BW, Hewetson M, Hepburn RJ, Luthersson N, Tamzali Y. European College of Equine Internal Medicine consensus statement—equine gastric ulcer syndrome in adult horses. *J Vet Intern Med* 2015;**29**:1288–99.
133. Hartmann AM, Frankeny RL. A preliminary investigation into the association between competition and gastric ulcer formation in non-racing performance horses. *J Equine Vet* 2003;**23**(12):560–1.
134. Bohak Z, Szabo F, Beckers JF, de Sousa NM, Kutasi O, Nagy K, Szenci O. Monitoring the circadian rhythm of serum and salivary cortisol concentrations in the horse. *Domest Anim Endocrinol* 2013;**45**(1):38–42.
135. Bruschetta G, Di Pietro P, Miano M, Zanghì G, Fazio E, Ferlazzo AM. Daily variations of plasma serotonin levels in 2-year-old horses. *J Vet Behav* 2013;**8**(2):95–9.
136. Casella S, Vazzana I, Giudice E, Fazio F, Piccione G. Relationship between serum cortisol levels and some physiological parameters following reining training session in horse. *Anim Sci J* 2016;**87**:729–35.
137. Kędzierski W, Wilk I, Janczarek I. Physiological response to the first saddling and first mounting of horses: comparison of two sympathetic training methods. *Anim Sci Paper Rep* 2014;**32**(3):219–28.
138. Christensen JW, Beekmans M, van Dalum M, VanDierendonck M. Effects of hyperflexion on acute stress responses in ridden dressage horses. *Physiol Behav* 2014;**128**:39–45.
139. Zebisch A, May A, Reese S, Gehlen H. Effect of different head-neck positions on physical and psychological stress parameters in the ridden horse. *J Anim Physiol Anim Nutr* 2014;**998**(5):901–7.
140. Fazio E, Medica P, Cravana C, Feriazzo A. Pituitary-adrenocortical adjustments to transport stress in horses with previous different handling and transport conditions. *Vet World* 2016;**9**(8):856–61.
141. Stewart M, Foster TM, Waas JR. The effects of air transport on the behaviour and heart rate of horses. *Appl Anim Behav Sci* 2003;**80**(2):143–60.
142. Friend TH, Martin MT, Householder DD, Bushing DM. Stress responses of horses during a long period of transport in a commercial truck. *J Am Vet Med Assoc* 1998;**212**(6):838–44.
143. Valle E, Zanatta R, Odetti P, Traverso N, Furfaro A, Bergero D, Badino P, Girardi C, Miniscalco B, Bergagna S, Tarantola M, Intorre L, Odore R. Effects of competition on acute phase proteins and lymphocyte subpopulations—oxidative stress markers in eventing horses. *J Anim Physiol Anim Nutr* 2015;**99**(5):856–63.

144. Villas-Boas JD, Dias DPM, Trigo PI, dos Santos Almeida NA, de Almedia FQ, de Medeiros MA. Behavioural, endocrine and cardiac autonomic responses to a model of startle in horses. *Appl Anim Behav Sci* 2016;**174**:76–82.
145. McDonnell SM. Detecting discomfort. *The Horse* 2011;**XXVIII**(12). 58 and 60.
146. Young T, Creighton E, Smith T, Hosie C. A novel scale of behavioural indicators of stress for use with domestic horses. *Appl Anim Behav Sci* 2012;**140**(1–2):33–43.
147. Christensen JW, Malmkvist J, Neilsen BL, Keeling LJ. Effects of a calm companion on fear reactions in naïve test horses. *Equine Vet J* 2008;**40**(1):46–50.
148. Rodgers AB, Morgan CP, Bronson SL, Revello S, Bale TL. Paternal stress exposure alters sperm microRNA content and reprograms offspring HPA stress axis regulation. *J Neurosci* 2013;**33**(21):9003–12.
149. Weinstock M. The potential influence of maternal stress hormones on development and mental health of the offspring. *Brain Behav Immun* 2005;**19**(4):296–308.
150. Reid K, Rogers CW, Gronqvist G, Gee EK, Bolwell CF. Anxiety and pain in horses measured by heart rate variability and behavior. *J Vet Behav* 2017;**22**:1–6.
151. Peters SM, Bleijenberg EH, van Dierendonck MC, van der Harst JE, Spruijt BM. Characterization of anticipatory behavior in domesticated horses (*Equus caballus*). *Appl Anim Behav Sci* 2012;**138**(1–2):60–9.
152. Glauser A, Burger D, van Dorland HA, Gygax L, Bachmann I, Howald M, Bruckmaier RM. No increased stress response in horses on small and electrically fenced paddocks. *Appl Anim Behav Sci* 2015;**167**:27–34.
153. Duxbury MM. Animal behavior case of the month. *J Am Vet Med Assoc* 2006;**229**(5):678–9.
154. Christensen JW, Rundgren M, Olsson K. Training methods for horses: habituation to a frightening stimulus. *Equine Vet J* 2006;**38**(5):439–43.
155. Falewee C, Gaultier E, Lafont C, Bougrat L, Pageat P. Effect of a synthetic equine maternal pheromone during a controlled fear-eliciting situation. *Appl Anim Behav Sci* 2006;**101**(1–2):144–53.
156. McDonnell SM. Unbalanced behavior. *The Horse* 2008;**XXV**(12):57–8.
157. Upadhyay SN. Serotonin receptors, agonists and antagonists. *Indian J Nucl Med* 2003;**18**(1 & 2):1–11.
158. Bruschetta G, Medica P, Fazio E, Cravana C, Ferlazzo AM. The effect of training sessions and feeding regimes on neuromodulator role of serotonin, tryptophan, and β-endorphin of horses. *J Vet Behav* 2018;**23**:82–6.
159. Broom DM, Kennedy MJ. Stereotypies in horses: their relevance to welfare and causation. *Equine Vet Educ* 1993;**5**(3):151–4.
160. Chocyk A, Bobula B, Dudys D, Przyborowska A, Majcher-Maślanka I, Hess G, Wędzony K. Early-life stress affects the structural and functional plasticity of the medial prefrontal cortex in adolescent rats. *Eur J Neurosci* 2013;**38**(1):2089–107.
161. Makinodan M, Rosen KM, Ito S, Corfas G. A critical period for social experience—dependent oligodendrocyte maturation and myelination. *Science* 2012;**337**:1357–60.
162. Veenema AH, Blume A, Niederie D, Buwalda B, Neumann ID. Effects of early life stress on adult male aggression and hypothalamic vasopressin and serotonin. *Eur J Neurosci* 2006;**24**(6):1711–20.
163. Hoekzema E, Barba-Müller E, Pozzobon C, Picado M, Lucco L, García-García D, Soliva JC, Tobeña A, Desco M, Crone EA, Ballesteros A, Carmona S, Vilarroya O. *Pregnancy leads to long-lasting changes in human brain structure. Nat Neurosci* 2006;https://doi.org/10.1038/nn.4458http://www.nature.com/neuro/journal/vaop/ncurrent/full/nn.4458.html [downloaded December 21, 2016].

Chapter 3

Learning

Learning has been important for survival of equids during their evolutionary history—changing their behaviors through experience and adapting to new environments and threats. They had to habituate to some stimuli but yet remain alert for others, suppress instincts, make connections between events, learn tasks that were not natural behaviors, and respond fast to reinforcers.[1–3] Learning is what the horse industry is all about. Trainers, owners, riders, farriers, and veterinarians are all dependent on the concepts of learning, regardless of whether they understand the specifics. Studies reported at the 2015 International Society of Equitation Science (ISES) conference noted that the most dangerous job for occupational injuries in the United Kingdom was that of an equine veterinarian.[4] The report also noted that techniques using learning theory were used the least even though they were the very ones that would actually make the horse safer to handle. This deficit in understanding the correct use of learning is especially true relative to negative reinforcement.[5] Dangers to equine practitioners continue to increase because the equine background of those entering the profession is changing. More and more veterinary students come from urban environments having limited horse experience. This makes the understanding of how horses learn even more important for human safety, proficiency of the horse for tasks it is asked to do, and horse welfare.[6]

Many lessons a horse needs to know to be a good veterinary patient and safe to handle are the responsibility of the owner to teach, yet that does not always happen. The horse owner is changing. Women over 50 years of age, many of whom have never had a horse before, are the fastest growing segment of the horse industry. In addition, a high percentage of horse owners consider the animal to be a family member, pet, or companion.[7,8] Owners and riders can interpret behavior problems incorrectly and make them worse because of misinterpreted motivations and misapplied learning techniques.[9] They do not understand learning theory or how it should be applied.[10,11] Proper handling, including application and timing of reinforcement and punishment, can eliminate undesirable behavior without the need for twitching or chemical restraints.[12] To promote better welfare for horses, the ISES developed the eight Principles of Learning Theory in Equitation in 2006 and revised them in 2016 to include two additional ones (Table 3-1).[13]

Equine Behavioral Medicine. https://doi.org/10.1016/B978-0-12-812106-1.00003-6

TABLE 3-1 The Principles of Learning Theory in Equitation[13]

1. Train according to the horse's ethology and cognition.
2. Use learning theory appropriately.
3. Train easy-to-discriminate signals.
4. Shape responses and movements.
5. Elicit responses one at a time.
6. Train only one response per signal.
7. Form consistent habits.
8. Train persistence of responses (self-carriage).
9. Avoid and dissociate flight responses (because they resist extinction and trigger fear problems).
10. Demonstrate minimum levels of arousal sufficient for training (to ensure absence of conflict).

Several concepts about animal behavior and learning are misunderstood. The first misconception is that a horse has a "hibernate" mode—that it only learns during training sessions. Animals learn every waking moment, so it is important that whenever a person is interacting with the horse, the person must be aware of what it might actually be learning. The lesson could be very different from what the person is intending to teach: outcomes, not intentions, define what is learned.[14] A second concept people forget is that behavior is dynamic. It is always changing toward or away from some goal. Although rewards and punishments are associated with learning, they constitute a small part of the total picture. The third concept is that if the behavior is rewarded in some way, internally or externally, it will happen again. Lastly, there is a misconception that what a horse learns with its right eye, it must learn separately with the left one. Interhemispheric transfer of visual information has been shown in horses and other domestic animals too.[1,15–17] Anatomically, there is a decussation of about 17% of optic nerve fibers to the contralateral side of the brain, so information from each eye goes to both sides. Fibers also cross through the corpus callosum to transfer visual information between the optic areas and reacting nuclei on both sides of the brain.[18]

Learning is complex and influenced by many things.[19] It is related to several identifiable factors. The first is timing. Daily sessions facilitate successful learning better than sessions interrupted by several days without any lessons.[20] That is not to say horses cannot or should not have breaks. As has been shown for school children, a series of short breaks is better for retention than one long summer break. The same applies for horses.

Genetics influence a horse's ability to learn, but studying the specifics is not easy. The sire, the dam, and the sire-dam combination each can influence learning efficiency, as can breed.[6,21–25] Offspring of certain individuals show greater efficiency at learning tasks than those of other horses in the same herd, and females are generally better learners than males.

Motivation is a third factor of learning and is important when considering the results of any test. If the horse is willing to work for an outcome, it will appear to be a better learner. When there is no interest, there is little learning. The hungry, grain-motivated horse is easier to train to load into a trailer than one that does not readily eat grain or has just eaten.

Mood has an intangible effect on learning. On windy days, numerous scents are swirling around. Evolutionarily, horses respond with increased alertness. This takes attention away from lessons to focus instead on the potential for danger.

Previous learning is a fifth factor affecting how easily horses learn. Yearlings with no previous handling and horses that already know quite a bit are slower learners than are horses that have been handled but not taught much.[26] When comparing naïve yearlings to ones with a lot of previous handling, the naïve yearlings do poorer. The phenomenon called *learning to learn* allows new learning to build on previous lessons. However, if the new material is somewhat contradictory or totally unique, previous learning can interfere with the process.

If plotted graphically, the horse's learning performance for a task or maneuver would be a straight line beginning at zero and rising at an angle as the task is learned. The shape of the graph can be significantly modified by a horse's mental arousal. The Yerkes-Dodson law says a certain level of mental arousal will enhance performance, but only to a point. When that arousal level is exceeded, the results are opposite.[6] Application of this law results in a bell curve, with lows when there is little or too much arousal and the peak coming in between. Arousal at either end of the bell curve can be reflective of fear, stress, anxiety, or lack of motivation, making it important to understand how each might affect learning.

TYPES OF LEARNING

While there is general agreement about the types of learning, classification systems may vary. Knowing the exact label attached to a technique may not be critical, but understanding the techniques is.

Learning can be divided into three major categories: associative learning, nonassociative learning, and complex learning (Table 3-2). Each of these has an underlying concept that links the various subtypes listed under it.

TABLE 3-2 Classification of Learning
Associative learning
Classical conditioning
Operant conditioning
Trial and error learning
Imprint learning
Latent learning
Observational learning
Chaining (shaping)
Generalization
Nonassociative learning
Habituation
Extinction
Sensitization
Desensitization
Counterconditioning
Complex learning
Social learning
Concept learning

Associative Learning

The first of three major categories of learning is associative learning, and it is divided into three subtypes. The commonality within this category is that the response is dependent on the horse receiving a specific cue prior to the beginning of the behavior.

Classical Conditioning

Classical conditioning, also called Pavlovian conditioning, was the first type of learning described scientifically. In Ivan Pavlov's initial study, he was able to get a dog to drool when he rang a bell by pairing the bell with food presentation. The concept behind classical conditioning is that an unconditioned stimulus results in an involuntary response, usually controlled by the nervous system. When food was presented (conditioned stimulus), the Pavlov's dogs would drool (response). Then he started ringing a bell (unconditioned stimulus which has no normal relationship to the response) at the same time the food was presented. Initially when he did this, the dogs would drool because of the association of food. Eventually, the dogs made the connection between the bell and food, and they drooled even when the bell was the only stimulus. In some cases, as with the bell and food, learning is the result of an intentional pairing, but many times the horse makes an unintended connection based on proximity of conditioned and unconditioned stimuli. While Pavlov consciously taught the dogs to connect the two stimuli, animals we work with tend to learn the same thing on their own. Cats come running when the can opener signals it is meal time, and the dog dutifully sits for a word that has no real meaning except that the dog was taught the connection of sitting when a treat and the "sit" happen at the same time. The horse begins salivating and nickers when noise begins in the feed room. A horse connects the painful zap from an electric fence with the click of the fence charger so eventually either will cause an increased heart rate and avoidance behavior.

While the strict definition of classical conditioning means the response is a physiological one, any behavioral response is now generally considered acceptable. Horses can also make the connection between words (commands) and responses initially taught in other ways. The word "back" becomes sufficient to get an appropriate response. The show horse learns to use the announcer's words or the keying of the microphone to change gaits. Another horse fights any attempt to give it an injection. Why? Sight of a syringe and needle (unconditioned stimulus) has been coupled to the tap, tap, tap (conditioned stimulus) on the skin with the resulting painful jabbing of the needle and injection sting.

Operant Conditioning

The second type of associative learning is operant conditioning. For this type of learning, two events are intimately connected. "If I do X, I get Y." Operant conditioning is all about the consequences of an action. The horse manipulates its environment, and the outcome is either reinforced or punished. The former results in an increased likelihood the behavior will occur again, while the latter reduces the likelihood.

Reinforcement and Punishment

The consequences—it's all about the consequences. In operant conditioning, external events can drive a behavior, like a spur in the horse's side or a food treat. So too can internal events like the perception of good or bad associations made in the horse's mind. The internal consequence is much harder to identify and to overcome. Fear of a flapping plastic bag results in an attempt to escape, and if successful, the horse experiences the internal reward of relief. It is difficult to convince the animal that successfully escaping a fearful object is an undesired response.

Understanding what is meant by *consequences* comes before a discussion of the various types. People think in terms of rewards and punishment as useful to encourage or discourage a behavior, but it is more complicated. There is reinforcement and punishment, and both have positive (something is added) and negative (something is taken away) forms (Figure 3-1).

Positive reinforcement adds something of value to the receiver, and the higher the value, the stronger is the reward. This type of reinforcement is usually equated with a food treat, praise, or gentle stroking. Actually a "reward" can be internal too, such as in a feeling of accomplishment or success. Positive reinforcement is common in dog training, but it tends to be overlooked for horses. Techniques other than positive reinforcement may work, but they can also take the "want to" out of the horse. Many people feel that using food treats for reinforcement is inappropriate because it leads to horses biting hands or clothes. Food treats are associated with licking hands and searching clothing, but not nipping or biting.[27]

Negative reinforcement removes something negative—a relief from discomfort or distress. This is a common technique used with horses. Its application may be very subtle and probably unrecognized as having a reward value.[28] Negative in this case does not imply bad, just a lessening of discomfort. As an example, the rider removes leg pressure when a horse begins to move forward. The negative—leg pressure—goes away when the animal begins to move. Forward movement is rewarded and discomfort relieved. Horses have been shown to have left and right side differences for the amount of stimulus pressure it

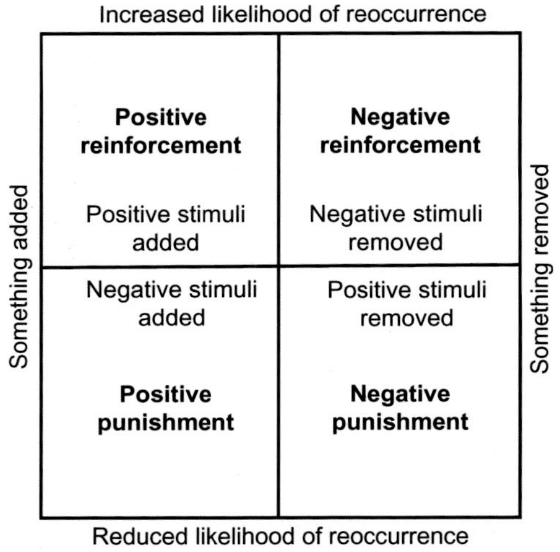

Increased likelihood of reoccurrence

	Something added	Something removed
Increased likelihood of reoccurrence	**Positive reinforcement** Positive stimuli added	**Negative reinforcement** Negative stimuli removed
Reduced likelihood of reoccurrence	Negative stimuli added **Positive punishment**	Positive stimuli removed **Negative punishment**

Reduced likelihood of reoccurrence

FIGURE 3-1 Operant conditioning is about the consequences and whether those consequences encourage or discourage a repeat of the behavior. Reinforcers reward the behavior, while punishers have the opposite effect.

takes to create a response.[29] Leg pressure on the right side to get the horse to move to the left may be significantly more than needed to have it go to the right. Negative reinforcement may not be particularly obvious to a person, as when a horse is tied to a fence and pulls back out of fear. If the lead rope or fence accidentally break, escape may be internal negative reinforcement. An alternative outcome occurs if part of the fence remains attached to the rope and keeps hitting the fearful horse. In this case, the horse might learn to fear being tied because of the extremely negative outcome.

With either positive or negative reinforcement, application techniques are important for learning and long-term retention. Initially, a constant reinforcement schedule is used, such that every time the horse responds correctly, it is reinforced. Once the behavior is well learned, reinforcement must continue to avoid extinction of the learning, but not all correct responses are rewarded. One of four intermittent schedules is applied. A *fixed ratio intermittent reinforcement schedule* (FR) applies the reward after a specific number of correct responses. For example, only every third (FR3) correct response is rewarded. This may have started as every other correct response being rewarded (FR2), and progress to FR3 and then FR4 as the horse demonstrates a good understanding and quick response to the cue. The *fixed interval schedule* (FI) rewards a behavior after a set time period. An FI would be useful for rewarding a horse for not pawing while being tied, as an example. It might start as an FI of 10 s and progress from there to 20 s, and then to 30 s as the horse gets better. *Variable ratio reinforcements* (VR) create more variation in when the reinforcement comes. This means various performances of the behavior are rewarded—sometimes the first, other times the third, sixth, second, or eighth. A VR4 indicates that on average the reward would be given every fourth occurrence. The fourth type of intermittent schedule is the *variable interval* (VI). A VI5 would suggest that on average, every 5 min the horse would receive a reward for a correct behavior.

Positive punishment adds a negative consequence for a behavior—typically just called "punishment." The horse touches as electric fence and gets zapped. The shock has added a negative result to the behavior of touching the wire. A horse rears up but the rider anticipates the behavior and spurs the horse as it starts rearing. The undesired rearing is positively punished with the application of the spur. The horse pulls its foot away from a farrier, and the farrier hits the animal on the belly with a rasp. Proper use of a lip chain is another example of positive punishment. Inappropriate application, timing, or intensity of any positive punishing technique can make it ineffective or even abusive. Naturally occurring, negative experiences can also be equated with positive punishments and result in a long-term memory of that experience.[30] Even a mild-mannered horse may suddenly become difficult to handle, trying to escape, to avoid something like vaccinations. After a few experiences with painful injections, the sight of a syringe and needle become associated with punishment.

Remote positive punishment has the advantage of letting the horse think the action or the environment is applying the punishment, not the nearby human. As an example, the nippy halter horse receives positive punishment from a thumb tack in the handler's glove. In this example, the horse might expect the person to hit them (a direct application of positive punishment). Instead, the person does not move, and the horse associates the act of biting with being stuck by a sharp point (remote positive punishment).

Negative punishment occurs with the removal of something good. This punishment technique can be particularly useful to correct certain behaviors such as the flattening of ears when a person approaches with feed. Negative punishment would happen if the person stops as soon as the ears are flattened and walks away before any feed leaves the bucket.

With four reinforcement/punishment options, the most effective for long-term results is positive reinforcement, especially when rehabilitating problem horses and during encounters with frightening objects.[31–33] It is important, though, that the reinforcement reward the desired calm behavior and not the signs of reactivity. If the horse is starting to pull away from scary plastic, using positive reinforcement is too late. It would reinforce the shying behavior, not the stay-and-check-it-out behaviors. The specific lesson may determine the best type of reinforcement or punishment to use, or the preferred learning style of the individual horse may be the determining factor.[1,34]

Inappropriately applied punishments have the worst outcomes.[35] There are many reasons both reinforcement and punishment can fail: application, intensity, timing, and technique. It is particularly important that punishment be of the proper intensity for the individual. Not severe enough means the horse learns to ignore it, even if the intensity is then gradually increased. Too severe means the animal learns to avoid the punisher instead. People are notoriously inconsistent in the application of punishments. In order to be effective, every occurrence of a behavior needs to be punished and with equal intensity. If not, the occurrence of the behavior without punishment is actually rewarding. Such intermittent rewards can be very strong motivators for continuance of a behavior as proven by intermittent payout of slot machines reinforcing the desire to play them. The type of "punishment" used can be viewed differently by the horse.[14] As an example, stopping punishment while the unwanted behavior occurs changes it from punishment to negative reinforcement. Similarly, if the bad behavior occurs to gain human attention, the interaction becomes the reward regardless of what else is happening.

The biggest error made in reinforcing or punishing a behavior is in the timing of its application. The direct connection between event and outcome occurs in less than 0.5 s. The longer the time from the event to the consequence, the less likely the horse is to connect the two. Even a 10-s delay will significantly increase the degree of difficulty of a task for horses.[36] This unpredictability of the stimulus-response relationship used in training will result in conflict behaviors and actually block learning.[37] Foals being taught to lead show faster learning when the pull on the lead is stopped (negative reinforcement) as soon as the foal takes its first step or two.[38]

Types of Operant Conditioning

While operant conditioning is about consequences, its subtypes do not always have obvious reinforcers or punishers. Internal positives and negatives must be considered.

Trial and Error Learning The first of the five subtypes of operant conditioning is trial and error learning. Here, the first occurrence of a behavior is a natural, spontaneous, random act that is reinforced. A horse fiddles with a gate latch and accidentally gets it open. Escape is positive reinforcement, which increases the likelihood the horse will try the behavior again. Each successful escape increases the frequency of reoccurrence. Trial and error learning is common in animals. A yearling's first experience with a large rubber ball in the pasture is to jump away when he first touches it, but the ball moves away too. The second time he touches it, he gives a little push and the ball rolls. A few more cautious encounters and what started as a slightly fearful object ends with the ball becoming a toy for the colt.

Imprint Learning Imprint learning is associated with a specific time period (sensitive period) during which the animal learns with which species to identify. The foal is relatively well developed at birth, which was necessary for their ancestors to survive in the wild. A consequence of needing to avoid predators soon after birth is the newborn's need to quickly learn "mom" and others of its species. While imprinting for survival is less important today, the process remains critical to the horse's understanding of "self" for future mating and social wellbeing. It is so important that identification with one's own species takes priority even if the foal is around multiple species. When raised in isolation from other horses, a foal is negatively impacted throughout its life.

Imprinting involves species identification. It is different from socialization, which also involves a sensitive period but of a longer duration. During socialization, the foal learns to accept species other than its own, like humans and dogs. Imprinting teaches the foal it is a horse.

A great deal has been made about "imprint training" of foals, but the wrong label has been attached to the process made popular several years ago. At that time, specific guidelines were published that described how a person should expose the foal to a number of different stimuli, like handling its feet, rubbing it with a plastic bag or gunny sack, touching the inside of the mouth, ears, and nostrils, and spraying it with water from a sprayer bottle. This is not imprinting. It is habituation—teaching the foal to accept certain noxious stimuli.[39]

Latent Learning The third type of operant conditioning is latent, or innate, learning. The learning that occurs is not immediately obvious but becomes obvious at some later time. It also means the reward is not obvious. In many cases, the reward is actually an internal one—stress reduction. There are several excellent examples of latent learning in people who grew up around horses.

They watched what different animals did and now "instinctively" know what to expect. These people actually learned what to expect in a particular situation over time. A horse plays in the water trough, swishing its nose through the water and then raises its head and lifts its upper lip. People who have been around horses have seen the behavior—latently learned. They might not have known that "flehmen" is the name of the behavior, but it is easy for them to connect the word with the behavior.

Latent learning is more likely to occur in the horse's normal environment. The foal encounters objects and learns not to fear things like feed tubs, salt licks, wheelbarrows, or tractors. It sees these things frequently and sees that none of the other horses are reacting. This gained confidence carries over to new environments too. Horses pastured outdoors have been shown to complete trials and training programs more quickly than horses that are individually stalled.[40,41]

Latent learning is not just about positive experiences. If the experience is negative, such as exposure to loud, fast-moving people, the foal may come to shun all people because it did not like what it was first exposed to. There are no "do overs" in latent learning.

Observational Learning Observational learning is talked about in the context of a horse learning to crib by watching another crib. It is not quite that simple, however. Learning by watching is difficult to prove in animals because they rarely mimic exactly what they have seen. Studies to test for observational learning rely on how fast a horse can complete the behavior. Can the observer horse learn to perform a task by watching another horse and then successfully perform it faster than would happen by trial and error alone? The number of trials to successful completion should be less than by chance.

Owners often assume that when multiple horses in a barn show the same behavior, they learned it by watching others.[1] Much has been made about observational learning relative to the acquisition of stereotypies, especially cribbing. While stereotypies will be discussed elsewhere, the bottom line is that well controlled tests for observational learning in horses have failed to positively demonstrate that it occurs.[22,42,43] Horses are kept in the same environment, often have similar bloodlines, and are ridden in similar ways. These factors may be more important in the expression of similar behaviors and development of stereotypies.

If observational learning does occur, it is likely to relate to very specific behaviors under very specific circumstances.[44] Colts may learn some sexual behaviors by observing older breeding stallions.[37] Foals that are present when their mothers are brushed for 15 min a day will approach, allow stroking, and accept a saddle pad placed on their back more readily than foals with dams not brushed.[39] Horses that watch a familiar person open a feeder will do so faster than horses that do not get to watch.[45] Lastly, horses choose to eat from a feed bucket that the dominant horse did not use and avoid locations where a high-ranking horse typically eats.[46] Whether these actually represent observational learning or something else such as odors or latent learning is uncertain.

Chaining (Shaping) The fifth type of operant conditioning is *chaining* (also called *shaping*). As the name implies, the animal learns one behavior first, which then becomes a step to the next, and it to the next. Each lesson is linked together until the desired outcome is reached. B.F. Skinner popularized shaping when he taught chickens to peck a specific location. A light was turned on to signal a correct action for the smallest head move toward the pecking spot. The next steps required the head come closer and closer to the desired location for each subsequent reward. Eventually the body also had to face the goal. This procedure uses trial and error learning to progress in a stepwise fashion toward a desired goal.

Chaining is a good way to teach an animal to do various tricks, such as how to bow.[1] At first the horse must be accepting of having one of its front feet picked up and held. This is rewarded. For the second step, the raised foot and leg are pulled back slightly and the horse rewarded. The foot and leg are pulled back a little farther, and the horse begins to shift its weight backward. As the weight continues to be shifted backward, the chest is lowered and as it gets close to the ground, the raised foot is allowed to touch the ground. In addition to the high value rewards given to reinforce each small, sequential step, a specific cue is given. Classical conditioning pairs that cue with the bowing behavior.

The lessons of each step build on the one before it, and each subsequent step becomes easier, being accomplished in a shorter time (Figure 3-2).[1] This is "learning to learn." When lessons build on previous ones, older horses are more efficient learners.[21,47] However, if the previous learning is not consistent with the new lesson to be learned, old knowledge can actually interfere with new learning. This would explain why older animals might take longer to learn something compared to younger, naïve horses.[48,49]

Generalization

Generalization, the sixth type of associative learning, is the occurrence of a behavior in a new location or situation because of similarity of stimuli. A rider practices crossing a white wooden bridge for a trail class at home and the horse readily walks across it. At the horse show, the horse refuses a similar obstacle. The refusal occurs because the horse has not generalized that a raised wooden obstacle is similar regardless of its color or location. For consistency in performance, the horse should encounter flat wooden boards, raised flat and angled bridges, bridges of different colors and types of side rails, and bridges in lots of different locations. That takes the mystery out of apparently "new" objects just because they are in a different location.

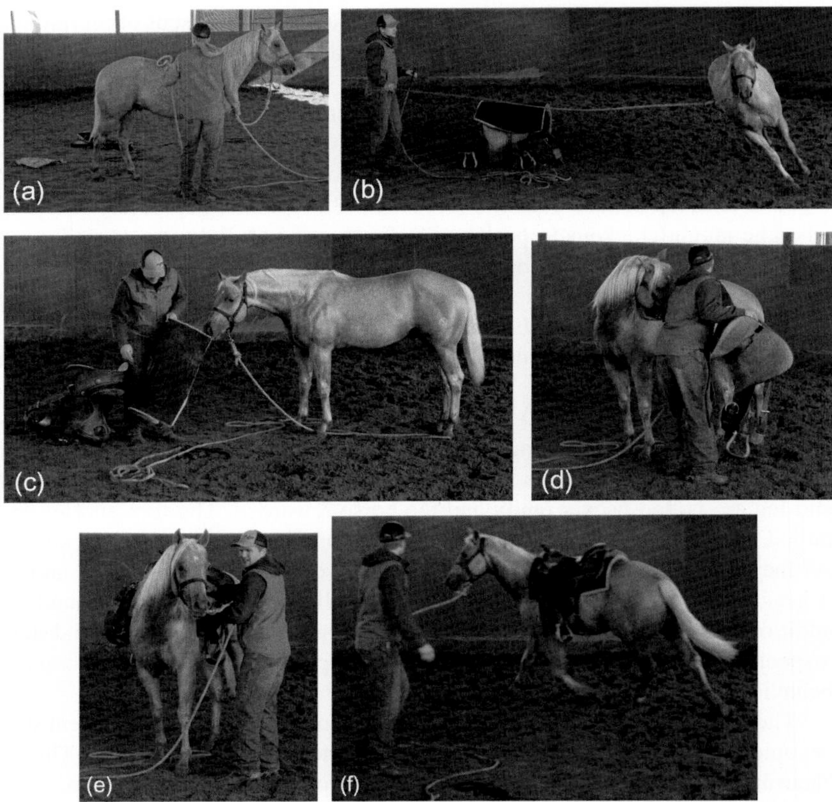

FIGURE 3-2 Chaining involves a series of small steps toward the desired goal. In this series: (a) the horse gets used to the rope touching parts of its body; (b) the rope is then used to lunge him in a circle in sight of the saddle and pad; (c) the colt is introduced to the saddle pad by smell and touch, and the pad will be placed on his back; (d) in a similar way, the horse gets to investigate the saddle before it is put on his back; (e) with the pad and saddle in place, the cinch is tightened enough to keep the tack in place; (f) movement while tacked teaches the feel of some weight, the flapping of saddle fenders, and sounds associated with the leather. Bucking can be discouraged by having the horse come in or change directions. Finally, a person will mount so the horse can learn to accept the added weight and gradually move on to learning higher skills.

A horse that readily loads into a trailer refuses to load into a friend's trailer even though both trailers are the same type. The generalization to "trailer" has not included "boxes" with different smells, step-up heights, ramps or no ramps, sounds when the foot first contacts the floor or ramp, or the amount of give when the weight is transferred to the trailer. Learning was limited to one "box." The enriched, diverse environment and range of experiences are valuable because they allow generalization through latent learning.

Target training is a technique that complements generalization. The horse is trained to touch a specific target, such as a hand or stick with a small ball on the end, by chaining. It is then classically conditioned to touch the target with a specific command, such as "touch it." This target can then be used to shape behaviors in new locations or situations like loading in a trailer or walking over bridges.[50]

Nonassociative Learning

The second major category of learning is called nonassociative learning. The similarity between the subtypes is that the response to a specific stimulus changes over time due to the repetition of that stimulus. The specific behavior could either increase or diminish in intensity. For the progressive increase or decrease to happen, it is extremely important not to stop the application of the stimulus too soon. Doing so will reinforce the very thing that is undesirable. The five subtypes discussed here are the most common, at least relative to horses.

Habituation

Habituation is a natural learning process that helps an animal subconsciously filter out normal, inconsequential events in its environment so it can focus on potentially dangerous ones. Repetition is the basis for habituation learning, gradually resulting in less response. Flapping plastic streamers on the refreshment stand by a horse show arena are eventually ignored through repeated exposure. If walked back and forth past the streamers, the horse habituates to the flutter and comes to ignore it. Sharp sounds and fluttering objects startle horses, causing them to jump and look toward the source. Assuming the horse gets a brief period to relax, the second time the noise or flutter happens the response is less dramatic. Subsequent events have even less effect until they are eventually ignored altogether. Trail horses can be habituated to walking over a plastic tarp through repeated exposure to it. Roping horses get used to the zinging noise and sudden appearance of a rope through repetition.

Yearlings that were raised in isolation from other horses react less and habituate faster to novel objects compared to group-raised yearlings. They are also better learners.[51] Whether this is due to being raised in a relatively enriched environment or because they generally show less emotional reactivity is not clear. While color is not of high significance to horses, there are indicators that the color, or perhaps reflectivity, of the novel object may be of more significance in habituation learning than the shape of the object.[52]

As described earlier, "imprint training" of foals is really habituation rather than imprinting. This misnomer has been frequently applied to early handling practices used on newborns and leads to confusion about terminology. Suggested procedures emphasize the importance of repetitions of things the foal will need to know at later ages relative to foot handling, blanketing, tolerating sprays, and other types of human handling.[53] While many horse farms like the results of this early handling, research has shown that without intermittent reinforcement, the effects of habituation are not lasting. Even by 3 months of age, there is no difference between "imprint trained" foals from those not handled at all.[39,54–56] Tractability to people is the other variable that is being taught during "imprint training." Evidence as to whether foals retain tractability from this early handling is split. Studies suggest there is some retention of manageability, or at least, less fear of people, for up to 18 months.[57–61] Other studies contradict this.[23,25,62] Most studies conclude that there are no obvious differences between "imprint trained" foals and controls after 6 months of no additional handling.[56,59,60,63] Continued handling is necessary for learning retention. There is evidence that postnatal handling for as little as 1 h may negatively impact foals. They show a stronger dependence on the mare, reduced play, and impaired social competences at all ages.[64]

Because "imprint training" can be associated with some negative outcomes, careful planning is necessary if it is to be done. The first concern is that the enthusiasm of doing the desensitization will interfere with mare-foal bonding. During the first few hours, it is best to leave the two alone.[65] There is a difference in the bond formed between the mares of foals that are handled during the first days compared to those not handled. The mares of early-handled foals are less active in keeping their foals nearby.[60] The significance of this is not understood, but it does show that handling does cause change. Another potential problem is that excited mares can stress their foal, causing them to struggle against handling more. Ultimately, this can teach the foal to fear human handling instead. Worse yet, the baby could be injured. If not properly done, the attempts to "imprint" the foal to certain procedures could end up sensitizing them, making the procedures harder to do in the future, not easier.

Extinction

The second type of nonassociative learning is extinction. This is the "unlearning" of a behavioral response. The horse has learned to respond a certain way to a certain cue, and the response has been reinforced by a high value reward. Once the response is well learned, all reinforcement is stopped. Without the reward, the animal eventually stops showing the behavior for the cue. The behavior has been extinguished. Extinction can work in a negative or positive way depending on what behavior is involved. The example most people are familiar with is a dog that is taught to sit for a special treat. Once it is doing a good job of sitting on cue, owners become lax about rewarding the correct response, eventually stopping all together. Gradually the dog lessens its response to the "sit" command, eventually ignoring it completely.

Undesirable behaviors sometimes have a reward that the owner does not realize. A horse has learned to open a stall door by wiggling the latch with its mouth through trial and error. The reward is the escape. The latch is lowered and a snap put on it so that the horse can no longer undo it. Initially the horse might try harder to open the latch, but eventually it quits trying. Extinction occurs because there is no longer an internal reward.

Sensitization

Sensitization is defined as an increased strength of a response as the result of repeated exposure to a specific stimulus. Phantom (stray) voltage will cause the horse to experience numerous shocks in an electrified stall. Gradually, the horse begins to refuse going back into that specific stall. Horses that are physically forced to load into a trailer can become harder and harder to load. The pain associated with ear twitching sensitizes the horse to people reaching toward the top of its head, and the degree of head-shyness increases each time an ear twitch is used.

While negative experiences are commonly cited, positive ones can sensitize a horse too. Stalled horses do not have the opportunity to be groomed in the withers area by herdmates, but people can rub along the horse's neck and withers. Most horses quickly learn which people will scratch them, and increasingly seek out the person or greet them in a significant way.

Desensitization

The fourth of the nonassociative learning subtypes is called desensitization, a technique commonly used to treat undesired behaviors. In a controlled setting, a specific stimulus is presented in such a low amount that the usual adverse reaction does not occur. If a noise is a problem, the initial volume is so low that there is no reaction. The intensity of the stimulus is very gradually increased over a long period of time in such a way that it is ignored and the behavior not triggered. In the shooting sports, it is important that a horse not startle with the gun goes off. The animal can be desensitized by first firing the gun far enough away from the horse that it does not react. The firing is repeated while the person very gradually moves closer, but only so slowly that the horse continues to not react. This last point—slowly—is important and is usually where failures occur. Humans are always in a hurry and move too close, too fast. The horse does not get enough time to allow it to learn to not fear the sound.

Desensitization can also happen to reinforcers or punishments. As an example, if positive punishment is used too lightly, the animal ignores it. The person then tries again but is a little harsher. The horse ignores it. Increasing levels of harshness are tried. Eventually a level will be reached where the desired response happens, but that level is significantly greater than would have been needed if the punishment was initially applied correctly.

Desensitization is similar to habituation in that the stimulus is repeated over and over again, but in desensitization, the strength of the stimulus increases gradually whereas in habituation, the strength does not change. To give a comparative example, the horse that is afraid to walk on plastic can be habituated if the stall floor is covered only with a plastic tarp. If, however, the plastic is covered with shavings and the amount of shavings used is gradually reduced over several days, the technique would be desensitization.

Counterconditioning

Counterconditioning (also called *response substitution*) occurs when the stimulus for an unwanted response becomes paired with a desired response instead. A special, high value food treat is paired with a potentially fearful event. The treat is generously offered while the fearful event is happening. Gradually the horse begins to look forward to the once frightening stimulus instead. As an example, the noise of a tractor moving down the aisle of the barn spooks a couple of yearlings stalled near the center of the barn. If the horses are fed immediately after the tractor is started, they quickly learn that the once fearful stimulus is now one that indicates food will follow shortly.

Desensitization and counterconditioning are often used together to increase the degree of success at a faster rate. This combination is particularly useful when working with fear-related problems. In the above example, desensitization can be applied by starting the tractor outside the barn for a few days at a distance where there is no reaction by the yearlings. They can barely hear the noise, and it is not distressful. Step two moves the tractor noise slightly closer, such as near the door for a few more days. Then the tractor moves to just inside the door for a few more. Gradually the tractor is moved forward, one stall at a time, over the next several days. This is desensitization. Counterconditioning is added, so that every time the tractor is started and regardless of where it is started, the yearlings are fed immediately.

Complex Learning

The third major type of learning is complex, or insight, learning, and it is usually equated with thinking. Because it is impossible to prove if or what a horse might be thinking, behaviors are used to infer possibilities. A classic anecdotal example that horses can reason was seen on a video recording of horses working in a "T" maze. The horse would enter at the base and move into one upper branch of the "T." At each tip of the "T's" crossbar would be either a black or white bucket, with only one containing food. The bucket positions were switched between trials so that the horse could learn that only the black (or white) bucket contained food. During one trial, a single horse was seen to stop immediately after it entered the stem of the "T." It looked left and then right before proceeding toward the appropriate goal. This does not prove the horse paused to assess the situation; it just implies it.

Then there is the mare that has been hauled thousands of miles during her show career, but is able to size up her new owner and refuses to load when they come to get her. An experienced horseman came over and in his firm voice said, "Dang it, Chipper, get in that trailer." She calmly walked in. How did she know she could get by with not loading? What was the mare thinking? Because we can never truly know what a horse is thinking, people tend to give a horse too much or not enough credit for their ability to think. The bottom line is that it is highly likely that horses think, but we do not know if it resembles human thought in any way. And it does not matter. When working with horses, it is best to set up situations so the horse can only respond as intended.

Social Learning

Social learning falls under the category of complex learning. Group membership requires some unique lessons such as the animal's place within the social hierarchy, individual recognition within and across species, and spatial features of the habitat.[41,66,67] Latent learning and internal reinforcement is involved as part of the process. Social interactions like mutual grooming and play release brain endorphins, the internal reward. The importance of social behaviors is confirmed by preventing horse-to-horse interactions and measuring the buildup of stress. When the behaviors are allowed again, the behaviors rebound excessively, supporting the concept of an internal reward.[68] Horses raised with other horses learn tasks faster than do those housed individually.[46] As would be expected, the young, low-ranking, and more inquisitive horses learn from older herdmates, but older horses are slower to learn this way because of the interference of previous learning.[69]

Special aspects of social life have been shown to influence learning, such as rank within the herd, familiarity, and visual contact. Horses that watch higher-ranking animals follow a human will copy the following behavior, while those that watch lower-ranking horses follow do not.[70] Visual contact is known to be important socially. A horse is more likely to go to a familiar or unknown person even if the person is not looking at them, than they are to the bucket that potentially has feed.[71] Contact has a higher reward value than does a possible meal. The horse that watches a familiar person open a feeder will open the feeder faster than a horse that does not get to watch, suggesting social learning across species.[72]

Concept Learning

Concept learning is another subtype of complex learning. A horse first is presented with items of different shapes or colors and is taught to pick out one specific item for its color. The items are then replaced with others, but only one of the new items has the same target color. The horse is to learn the concept of only picking items containing that specific feature. The horse is presented with a black, white, and blue ball, and the goal is to identify only blue items. The horse is rewarded for touching the blue ball until it consistently does so. Then the horse is presented with a green, yellow and blue cone and rewarded each time it touches the blue cone. Lastly, the horse is presented with a gray, orange, and blue cube, and is expected to pick the blue cube more consistently than it would do through trial and error. Instead of three cubes, the trial could also use an orange ball, red cone, and blue cube with the blue cube being the correct choice to demonstrate concept learning. The same concept can be used to pick a specific shape instead of color. Some horses apparently lack the ability to make these distinctions.[1,73–76]

Concept learning is helpful for generalization. Horses that learn to load into a specific horse trailer typically balk when asked to enter a strange trailer. Those that have been in several different trailers are comfortable going into other trailers as well as into other narrow areas. They have learned the concept of going into relatively tight spaces.

BRIDGING

Bridging (also called secondary or conditioned reinforcement) is a technique that quickly provides a message that the response was correct and positive reinforcement is coming soon. It bridges the time span between the response and the reward. Classical conditioning is used to give meaning to a bridging stimulus. The advantages of the bridging stimulus are better timing for reinforcement, more consistency in timing and tone, extra time to get the physical reinforcer to the animal, and facilitation of learning.[77] When people who are not proficient in their timing for treats or praise reinforcement use bridging, they increase the speed of learning because of the quick coupling of behavior to the reinforcer and the consistency of the sound.

The most common type of bridging is the clicker. This sound provides a bridge that can be activated quickly, has a distinct tone, and most significantly, is constant in its tone. The human voice is quite variable, and the tongue click is much like the giddy-up cluck. Consider riding a horse and asking it to do something new. It does. Now the rider must reach down to pat the horse or get a treat out of their pocket, stop the horse, and give it the treat (which actually rewards the stop, not the new behavior). Either way, it will take longer than the ideal 0.5 s. A bridge would reinforce the behavior and give the rider time to give the significant reinforcer to the animal. Useful bridges could be a "Good Boy," leg slap, tongue click, or clicker click.

To use bridging, the horse must first connect the click (or other bridge) with a treat. The food is presented, or the horse patted, as the clicking noise is made. This is repeated several times so that the animal associates a click with food or pat. Gradually the interval between the click and the reward is lengthened to several seconds. Once the connection is made, each time the horse shows the desired behavior, a click goes off immediately, followed shortly by the treat. If the treat following the use of the bridge is discontinued, the bridge will no longer have any value to the horse and its response will be extinguished.[78]

There are multiple times when bridging would be helpful, such as entering a wash rack. The horse approaches the wash rack and abruptly stops. Instead of pulling or pushing, the person patiently stands ahead of the horse. A forward step by any hoof is clicked and the horse treated. The animal can be coaxed with treats just out of reach to encourage forward movement, but the head is not pulled just steadied to look into the wash rack. Cautious steps are clicked and rewarded. Eventually the horse will enter. Let it stand there for a short time eating treats and then exit. After a little while, the process is repeated. Subsequent times entering will get shorter and without much, if any, hesitation. The same technique can be used to teach horses to load into trailers, walk over bridges, and walk over a tarp. It also speeds up target training.

MEMORY

As previously mentioned, the connection between a behavior and reinforcement or punishment must be made quickly for learning to occur, but memory is different. One type of memory is short-term retention, which in people lasts about how long one can remember a phone number without constantly repeating it or writing it down. Horses have good short-term memory. As an example, a horse can watch grain being placed in one of two buckets, be distracted, and still go to the correct bucket, even with release delayed 30 s.[79]

Long-term memory is responsible for reactions to similar negative events repeated sometime later. Negative memories intensify over time because of internal reinforcement. If the horse has a bad experience, the next time it remembers the bad and tries to avoid it. If avoidance is successful, the fearful memory is rewarded. Horses become needle shy or fight trailers, as examples. Long-term memory also occurs for neutral and highly rewarding things too. It is the basis for horse training.

Over time, long-term memories will fade and eventually be extinguished without periodic refreshing. And, the more frequently the memory is refreshed, the stronger it becomes. As an example, a 3-year-old mare in training as a reiner relaxed when the owner would rub the neck and withers areas. This would happen every 3–4 weeks. While the mare seemed to "enjoy" the moment, she did not show a recognition of the owner until the rubbing started. After several months with the first trainer, the mare was moved to a closer trainer to permit weekly visits. At this second facility, the horses were saddled and tied outdoors near the arena to make them ready for their turn. By the second week at this new location, the mare saw the owner's truck drive in and started pulling toward it. She repeated the behavior each week but only for the owner's truck and approaching owner. The association of a reward given by a person during any type of interactional context enhances learning, resulting in increased contact and interest toward the person immediately and even after 8 months of separation.[44]

MOTIVATION

Exposure to something new does not necessarily result in learning, even latent learning, unless there is some level of motivation. This level could vary from a passing curiosity to a high-level internal drive. If food is used as a reward but the animal is not hungry or does not like the food, the lesson fails. A box with small amounts of food was put in the stall of several horses. The horses learned that food was inside by a person lifting the lid and letting each horse eat a tiny amount.[25] Each horse was timed to see if it learned where the food was or if it used trial and error to lift the lid. The horse was retested the following day. Of 120 horses tested, 15 never opened the food box lid, showing that food was apparently not a strong motivator for those horses. For the 105 horses that succeeded on day 1, all were again successful on subsequent days at a rate 2–10 times faster. For them, food was a good motivator. This type of test is an example of good long-term memory.

Learned Helplessness

In psychological terms, learned helplessness occurs when an individual learns it has no control over an unpleasant or harmful situation and nothing they do will make a difference. As a result, they seem to "shut down" from any action and show no interaction with their environment. There is no longer any motivation to act. A trainer might be trying to apply habituation, but with improper timings and stimulus control, the horse might be giving up instead. This passive behavior might be viewed as desirable, but the horse may suddenly show an unpredictable action like a stereotypy or "explosive" behavior. There is a corresponding suppression of the immune system too. The condition has not been well studied in horses, but that does not mean it cannot occur.[80]

SPECIAL TECHNIQUES USING LEARNING

Taste Aversion

Taste aversion is a technique of positive punishment used to stop, or at least minimize, oral behaviors. A person who has Chinese food one evening and then spends the rest of the night vomiting it back up has a strong aversion to Chinese food for quite a while. This is the concept of taste aversion. There are two ways to use this concept for horses. The first uses a foul-tasting substance, perhaps one that also irritates the gums such as undiluted hot sauce and pepper sauce. Creosote has a taste that horses dislike and is used the most.[81] Capsaicin, the active ingredient in chili peppers, can also be used.[82] For the horse that chews the fence, owners will typically brush the stall or fence boards with the product and may or may not re-treat the area in the future. This technique rarely works. To be successful, taste aversion depends on coupling a "really bad" experience to the taste first. The horse smells the product and then 5 mL is squirted directly into the horse's mouth. The goal is for the animal to connect the smell of the product with the awful taste. Once this is done, the product is used to coat the chewed or licked surfaces. Subsequently, when the horse smells or tastes the product, it is repelled. Wind, rain, and air currents dilute the effectiveness of these products, so recoating surfaces on a fairly regular basis and occasional taste reminders for the horse are necessary.

The second taste aversion technique uses chemical administration, such as lithium chloride or apomorphine.[83,84] Lithium is administered via stomach tube and apomorphine is injected into a vein immediately prior to the animal being able to eat or chew the undesired item. The resulting nauseated feeling conditions the animal to avoid the item it was chewing on. There are obvious limitations to these chemicals because of routes of administration, timing of the administration, and difficulty in overcoming internal rewards if highly palatable feed is soon available.

Smell Aversion

Smell aversion is another positive punishment technique that depends on olfaction rather than taste. The horse smells the product odor and as it takes in a good whiff, there is a rapid, scary action and/or sound. The goal is to have the horse couple the frightful happening with a particular odor. As an example, a spray deodorant can be used to get a horse to stop licking a board. The horse smells the top of the can and as he does, the top is depressed to release a hissing noise. At the same time, the can is moved rapidly toward the horse. Shaking noisy cans can also be synchronized to the hiss. With the horse out of the stall, the board is sprayed with the aerosol deodorant. The horse couples the smell on the board with the hissing, rapid movement event and avoids the board. As another example, some horses can smell an electric fence and avoid touching it if there is a particular odor present. This is a learned aversion after getting zapped a few times while atmospheric conditions were such that a small amount of unique odor was produced by the hot wire. Because of the dilution of odor molecules by air and rain, smell aversion is not as successful as taste aversion, but is an alternative technique that can be used when taste aversion is not possible or appropriate.

NEGATIVE IMPACTS ON LEARNING

Many things influence learning in horses and measurement of it. Researchers commonly fail to consider that senses of prey species are quite different from those of humans, so results based on sensory perceptions can be flawed. Two things relating to visual discrimination are important to consider. The first is the effect of light intensity on whether the horse can see something well. The second is the position of the object or test relative to the horse's head.[85] Since the horse eye is rod-dominated, rods are more effective in low light, and the tapetum magnifies light entering the eye, it should be expected that the horse will not see as well in bright daylight as it will on cloudy days or nearer sunrise and sunset. As in other grazing animals, the visual streak of a horse, the retinal location of sharpest vision, is positioned so that closer objects are in sharpest focus. Those images enter the lower portion of the eye. Objects are seen better if they are in front of and at nose level instead of at eye level.

Stereotypic behaviors affect learning. While the main concern is relative to the horse's health and welfare, the development of any stereotypy lowers a horse's ability to learn.[86,87]

Excessive training can result in a horse passing peak performance and beginning to regress. While it is desirable to maintain appropriate physical fitness, it is common to overlook mental sharpness and continue training the same thing over and over again. There is individual variation as to what a particular horse will tolerate or need compared to another. It has been said that the Western pleasure horse has to be particularly dumb to tolerate going around and around in circles for years without any variation except for the three gaits. Symptoms that might be expected as the horse reaches the limit of "too much of one thing" would include irritability, malaise, fatigue, reduced interest in training, poor performance, reduced body weight, and even immune suppression.[88] Variations in the training routine and schedule are important to prevent and treat this problem.

Reinforcement, or lack of it, can present another problem for a horse's learning. While professional trainers are working with horses pretty much on a daily basis, horses that are used for recreation do not get the same consistency. Their riders typically ride a few days a week so any lesson has a prolonged interval that can be detrimental to learning. Recreational riders are often more concerned about staying on or looking good than they are about correctly reinforcing the horse's behavior. They may even reinforce or permit unacceptable behaviors, such as running back to the barn or pulling a foot away.

Fear, Stress, and Anxiety

In some countries, fear, anxiety, and other types of emotions are attributed to animals and the "emotional state" is taken into consideration in assessing welfare.[89] It is also part of the Five Domain technique of welfare assessment.[90] In order to discuss emotions, they must first be defined. Anxiety is the anticipation of something negative, while fear occurs when the threat is real and present. Stress is the physiological response. Complicating this is that neither anxiety nor fear have a universal response, and there is individual variation on what is causative. This creates a dilemma in understanding emotions in animals. The true meaning of "fear" as a feeling will vary between situations and individuals. In any unique situation, fear can only be known by introspection.[91] Everyone seems to know what an emotion is until they are asked to define it.[92] While physiological responses in animals may parallel those of humans in emotionally equivalent situations, it is not known if the mental experiences are the same. This is not a denial that animals experience equivalent emotional states. Rather, it is a reminder that assessing the circumstances that precipitate the reaction is more important in minimizing the avoidance behaviors.[93,94]

Horses can be quick to show stress when they are alone or at competitions. Behavior such as vocalizing, defecating, pawing, snorting, mouth gaping, difficultly being handled, and running are common. These behaviors also correlate with an increased heart rate.[95,96] There are also sex-related differences. Studies suggest that there are physiological reasons that males respond differently than females. In males but not females, production of corticotropin-releasing-hormone-binding protein (CRHBP) in the brain blocks the corticotropin-releasing hormone, and thus cortisol production.[97] Lower cortisol levels are associated with increased risk-taking, so stallions will tend to be bolder than mares in response to similar stress-producing stimuli.[98]

Expressions of "fear" are instinctive. Physiological responses are controlled by the sympathetic nervous system: increased heart and respiratory rates, dilation of the pupil, reduced salivary production, and increased cortisol production, for example. Behavioral indications include shying, halting, and snorting. How the animal responds to a "fearful" stimulus may or may not be fear related. It can be instinctive, developed for the survival of the species, and not truly fear inducing. Shying at a fluttering object is an example. Is there an emotional fear response, or does the fluttering simply trigger a survival response?

Fear can also be learned. The reaction is not associated with the first experience, but builds as the result of learning and subsequent self-reward. Because sensory information is cumulative and several types of sensory input are of approximately equal importance, different types of negative experiences can result in an animal showing aversion to a specific stimulus. The first time a horse is trailered is not necessarily negative, until the driver takes a corner too fast and the horse has to scramble to keep its footing. The second time it is trailered, there might be a remembrance of the difficulty in maintaining footing and anxiety sets in. Again, a corner is taken a little too fast and there is an internal reinforcement that the anxiety was justified. The third time in the trailer, the horse might experience more anxiety and a higher degree of fear when motion begins. In another example, a horse having sharp points on its molars may mouth or fight a bit excessively and yet do well in a cavesson or bosal. If the horse comes to associate pain with insertion of a bit into its mouth, it may never willingly accept a bit.

Stressors change the focus of attention, influencing decision and learning speed.[19,99,100] Small amounts of glucocorticoids and catecholamine aid memory formation, but too much disrupts memory. Effects on recall are less clear. Single-stalled yearlings learn faster and are less reactive when tested in novel environments compared to yearlings kept in a group.[101] Long-term separation could, however, be detrimental.

Timing of stressors can affect learning. Horses stressed prior to training sessions learn a behavior faster than those not stressed. Highly active horses show this same tendency, so it is unclear whether the activity of reactive horses prepares them for learning or whether it is a result of their temperament. Conversely, horses experiencing stress after the initial learning session but before retrial are better learners only if they show the fewest signs of fearfulness and are the least gregarious.[102] Significant stress may cause the horse to revert to previous learning, rather than use "cognitive" memory to better evaluate the response.[103]

Not only are fearful horses less likely to learn desired goals, they can actually be dangerous. When a horse reacts to a fearful stimulus, people tend to respond with punishment. This escalates the fear even more.[100] Trying to calm a frightened horse with soothing voice tones is not effective. It might be helpful to the rider, but neither soothing nor harsh tones have a calming effect on horses.[104] How a fearful object is presented can make a difference. Objects held by a familiar person are more likely to be approached eventually than are ones on the ground.[71,105] Desensitization is the most helpful technique for reducing fearful reactions. Neither habituation nor counterconditioning reduces reactions and, in some cases, may actually sensitize the horses more.[106] While the behavior changes, the increased heart rate remains elevated in all three types of learning.

Temperament

Temperament is often described as a component of learning, but just how and when can vary. It turns out that the temperament relationship is task dependent.[107] Horses that are described as generally fearful or active will learn avoidance behaviors quickly. Those considered more fearless will learn food-rewarded tasks more readily.

RELATIONSHIP BETWEEN TRAINING AND LEARNING

Five characteristics that affect behavior besides genetics include disposition, effort, athletic ability, early experience, and intelligence.[108] Disposition is the "want to" that a horse displays during training. The effort a horse will put into a task and the ability to accomplish that behavior help determine ultimate success. Poor conformation will affect athletic ability and prevent a horse from excelling on the race track or cutting arena. Necessary behaviors can be learned, but the horse may be unable to express them at competition level. Early experience is important, particularly relative to not fearing humans. The longer a horse is allowed to be reactive to its environment, the longer it takes to undo that lesson. Finally, there is the individual's intelligence. Intelligence is hard to define, even in humans, and it certainly cannot be measured if it cannot be defined. Trying to measure intelligence across various species is a daunting challenge. We cannot define it, we cannot measure it, yet we know there are differences between horses in their ability to learn.[100] Those differences make the application of the principles of learning variable for each individual. Smarter horses learn faster but are harder to keep interested and keep from anticipating commands.

There is a great deal of anecdotal information about equine learning and training, especially compared to the amount of science done on the subject.[109] Several potential complicating factors exist for designing studies that reflect real-world conditions.[110] The task must be within the horse's behavioral repertoire, be physically possible, and take into account the horse's sensory capabilities. Another complicating factor is that training typically uses negative reinforcement strategies, while experimental designs rely almost exclusively on positive reinforcement.[37,48,110] This begs the question as to whether the long-term impact of one method is better than the other. The answer seems to depend on the type of lesson the horse is asked to learn and on the individual horse's personality. Positive reinforcements can speed learning of new lessons and the generalization of a specific lesson to new locations.[111,112] As positive reinforcers, food treats generally engage a horse's attention for longer periods compared to withers' grooming rewards, suggesting food is a stronger reinforcer.[113,114] Fear-based problems respond better to positive reinforcement. Horses with trailer loading problems showed faster learning and less stress when positive reinforcement was used, as an example.[115] Negative reinforcement certainly has proven over time to work well in teaching new things, and relief from an aversive stimulus is remembered for a very long time.[116] When working with wild or feral horses, mild forms of negative reinforcement, like a gentle touch, are better for learning than positive reinforcement, probably because these horses are less likely to recognize value in food treats.[117]

Horse trainers in the United States are starting to use catch phrases such as "natural horsemanship," "sympathetic horsemanship," "cooperative training," and "horse whisperers." The basic concept for each of these is the use of human body language to communicate an intention to the horse and chaining to build on successive lessons. It is just the applied name that varies by what might be the most popular or kindest sounding. Studies between the "natural horsemanship" technique and that of conventional training show that if the former is properly applied, it is associated with less stress for the horse. While both can accomplish the same thing, "natural horsemanship" training is associated with less body tension, lower head carriage, fewer lip movements, and less teeth grinding.[118] Horses trained this way also tend to approach people faster.[119,120] It must be acknowledged that trying to standardize real-world "conventional training" is difficult to impossible, because methods vary considerably. This makes it hard to compare different training strategies.

Round pen work is often the first training a young horse gets. The position of the person's body relative to the horse is important for a response. Initiating forward movement involves the person moving slightly into the horse's flight zone (see Figure 5-1 in Chapter 5). The horse moves away so that the person is no longer within that zone. The instinctive response is negatively reinforced. Unfortunately, trainers advocating "natural horsemanship" often incorrectly describe why a technique works by overemphasizing the role of social rank and predator-prey interactions instead of describing the type of learning that is occurring.[121]

Horses can respond to cues other than intrusion into their flight zone. They will respond to pointing at a goal and to marker placement by the goal. They do not respond to tapping the goal, human body orientation to face the goal, or the human gaze alternating between the horse and goal.[5,122] The relationship of these cues to "human predator-horse prey" or to human "dominance" in the horse's social order can never be known and does not matter. The correct application of the principles of learning will accomplish what needs to be done.

REFERENCES

1. Hanggi EB. The thinking horse: cognition and perception reviewed, In: *Proceedings of the 51st annual convention of the American Association of Equine Practitioners, Seattle, Washington*; 2005. p. 246–55.
2. Ladewig J. Clever Hans is still whinnying with us. *Behav Process* 2007;**76**(1):20–1.
3. Miller RM. The amazing memory of the horse. *J Equine Vet* 1995;**15**(8):340–1.
4. Beckstett A. The science behind horsemanship. *The Horse* 2016;**XXXIII**(1):16–25.
5. Proops L, Walton M, McComb K. The use of human-given cues by domestic horses, *Equus caballus*, during an object choice task. *Anim Behav* 2010;**79**(6):1205–9.
6. Olczak K, Nowicki J, Klocek C. Motivation, stress and learning—critical characteristics that influence the horses' value and training method—a review. *Ann Anim Sci* 2016;**16**(3):641–52.
7. American Veterinary Medical Association. *U.S. pet ownership & demographics sourcebook.* Schaumburg, IL: American Veterinary Medical Association; 2007. p. 159.
8. Ruse K, Bridle K, Davison A. Exploring human-horse relationships in Australian thoroughbred jumps racing. *J Vet Behav* 2016;**15**:95.
9. Hothersall B, Casey R. Undesired behavior in horses: a review of their development, prevention, management and association with welfare. *Equine Vet Educ* 2012;**24**(9):479–85.
10. Bornmann T. Riders' perceptions, understanding and theorietical application of learning theory. *J Vet Behav* 2016;**15**:79–80.
11. Telatin A, Baragli P, Green B, Gardner O, Bienas A. Testing theoretical and empirical knowledge of learning theory by surveying equestrian riders. *J Vet Behav* 2016;**15**:79.
12. Miller RM. Counter conditioning: a technique for quickly modifying equine behavior, In: *40th Annual AAEP Convention Proceedings: 131*; 1994.
13. International Society for Equitation Science. *Principles of learning theory in equitation.* http://www.equitationscience.com/learning-theory-in-equitation; 2016 [downloaded 11/22/2016].
14. Mills DS. Applying learning theory to the management of the horse: the difference between getting it right and getting it wrong. *Equine Vet J* 1998;**30**(S27):44–8.
15. Hanggi EB. Interocular transfer of learning in horses (*Equus caballus*). *J Equine Vet* 1999;**19**(8):518–24.
16. Hanggi EB, Ingersoll JF, Waggoner TL. Color vision in horses (*Equus caballus*): deficiencies identified using a pseudoisochromatic plate test. *J Comp Psychol* 2007;**121**(1):65–72.
17. Murphy J, Hall C, Arkins S. What horses and humans see: a comparative review. *Int J Zool* 2009;**2009**(Article ID 721798):1–14.

18. Timney B, Macuda T. Vision and hearing in horses. *J Am Vet Med Assoc* 2001;**218**(10):1567–74.
19. Mendl M. Performing under pressure: stress and cognitive function. *Appl Anim Behav Sci* 1999;**65**(3):221–44.
20. Kusunose R, Yamanobe A. The effect of training schedule on learned tasks in yearling horses. *Appl Anim Behav Sci* 2002;**78**(2–4):225–33.
21. Bonnell MK, McDonnell SM. Evidence for sire, dam, and family influence on operant learning in horses. *J Equine Vet* 2016;**36**:69–76.
22. Lindberg AC, Kelland A, Nicol CJ. Effects of observational learning on acquisition of an operant response in horses. *Appl Anim Behav Sci* 1999;**61**(3):187–99.
23. Mader DR, Price EO. Discrimination learning in horses: effects of breed, age and social dominance. *J Anim Sci* 1980;**50**(5):962–5.
24. Maros K, Boross B, Kubinyi E. Approach and follow behavior—possible indicators of the human-horse relationship. *Interact Stud* 2010;**11**(3):410–27.
25. Wilk I, Kędzierski W, Stachurska A, Janczarek I. Are results of crib opening test connected with efficacy of training horses in a round-pen? *Appl Anim Behav Sci* 2015;**166**:89–97.
26. Heird JC, Lennon AM, Bell RW. Effects of early experience on the learning ability of yearling horses. *J Anim Sci* 1981;**53**(5):1204–9.
27. Hockenhull J, Creighton E. Unwanted oral investigative behavior in horses: a note on the relationship between mugging behavior, hand-feeding titbits and clicker training. *Appl Anim Behav Sci* 2010;**127**(3–4):104–7.
28. McLean AN. The positive aspects of correct negative reinforcement. *Anthrozoös* 2005;**18**(3):245–54.
29. Ahrendt LP, Labouriau R, Malmkvist J, Nicol CJ, Christensen JW. Development of a standard test to assess negative reinforcement learning in horses. *Appl Anim Behav Sci* 2015;**169**:38–42.
30. LeDoux JE. Coming to terms with fear. *Proc Natl Acad Sci* 2014;**111**(8):2871–8.
31. Fox AE, Bailey SR, Hall EG, St. Peter CC. Reduction of biting and chewing of horses using differential reinforcement of other behavior. *Behav Process* 2012;**91**(1):125–1288.
32. Heleski CR, Bello NM. Evaluating memory of a learning theory experiment one year later in horses. *J Vet Behav* 2010;**5**(4):213.
33. Innes L, McBride S. Negative versus positive reinforcement: an evaluation of training strategies for rehabilitated horses. *Appl Anim Behav Sci* 2008;**112**(3–4):357–68.
34. Christensen JW, Ahrendt LP, Lintrup R, Gaillard C, Palme R, Malmkvist J. Does learning performance in horses relate to fearfulness, baseline stress hormone, and social rank? *Appl Anim Behav Sci* 2012;**140**(1–2):44–52.
35. Hockenhull J, Creighton E. Training horses: positive reinforcement, positive punishment, and ridden behavior problems. *J Vet Behav* 2013;**8**(4):245–52.
36. McLean AN. Short-term spatial memory in the domestic horse. *Appl Anim Behav Sci* 2004;**85**(1–2):93–105.
37. Murphy J, Arkins S. Equine learning behavior. *Behav Process* 2007;**76**(1):1–13.
38. Warren-Smith AK, McLean AN, Nicol HI, McGreevy PD. Variations in the timing of reinforcement as a training technique for foals (*Equus caballus*). *Anthrozoös* 2005;**18**(3):255–72.
39. Houpt KA. Imprinting training and conditioned taste aversion. *Behav Process* 2007;**76**(1):14–6.
40. Rivera E, Benjamin S, Nielsen B, Shelle J, Zanella AJ. Behavioral and physiological responses of horses to initial training: the comparison between pastured versus stalled horses. *Appl Anim Behav Sci* 2002;**78**(2–4):235–52.
41. Sigurjónsdóttir H. Equine learning behavior: the importance of evolutionary and ecological approach in research. *Behav Process* 2007;**76**(1):40–2.

42. Ahrendt LP, Christensen JW, Ladewig J. The ability of horses to learn an instrumental task through social observation. *Appl Anim Behav Sci* 2012;**139**(1–2):105–13.
43. Rørvang M, Ahrendt L, Christensen J. Horses fail to use social learning when solving spatial detour tasks. *Anim Cogn* 2015;**18**(4):847–54.
44. Sankey C, Richard-Yris M-A, Leroy H, Henry S. Positive interactions lead to lasting positive memories in horses *Equus caballus. Anim Behav* 2010;**79**(4):869–75.
45. Schuetz A, Farmer K, Krueger K. Social learning across species: horses (*Equus caballus*) learn from humans by observation. *Anim Cogn* 2017;**20**(3):567–73.
46. Krueger K, Flauger B. Social learning in horses from a novel perspective. *Behav Process* 2007;**76**(1):37–9.
47. Murphy J. Assessing equine prospective memory in a Y-maze apparatus. *Vet J* 2009;**181** (1):24–8.
48. McCall CA. A review of learning behavior in horses and its application in horse training. *J Anim Sci* 1990;**68**(1):75–81.
49. Sappington BKF, McCall CA, Coleman DA, Kuhlers DL, Lishak RS. A preliminary study of the relationship between discrimination reversal learning and performance tasks in yearling and 2-year-old horses. *Appl Anim Behav Sci* 1997;**53**(3):157–66.
50. Beckstett A. *Does target training help reduce horses' stress?* www.thehorse.com/articles/36757/does-target-training-help-reduce-horses-stress; 2015 [downloaded 1/13/16].
51. Lansade L, Neveux C, Levy F. A few days of social separation affects yearling horses' response to emotional reactivity tests and enhances learning performance. *Behav Process* 2012;**91**(1):94–102.
52. Christensen JW, Zharkikh T, Ladewig J. Do horses generalize between objects during habituation? *Appl Anim Behav Sci* 2008;**114**(3–4):509–20.
53. Miller RM. *Imprint training of the newborn foal*. Colorado Springs: Western Horseman; 1991. p. 137.
54. Goodwin D. Equine learning behavior: what we know, what we don't and future research priorities. *Behav Process* 2007;**76**(1):17–9.
55. Mal ME, McCall CA, Cummins KA, Newland MC. Influence of preweaning handling methods on post-weaning learning ability and manageability of foals. *Appl Anim Behav Sci* 1994;**40**(3–4):187–95.
56. Williams JL, Friend TH, Toscano MJ, Collins MN, Sisto-Burt A, Nevill CH. The effects of early training sessions on the reactions of foals at 1, 2, and 3 months of age. *Appl Anim Behav Sci* 2002;**77**(2):105–14.
57. Lansade L, Bertrand M, Boivin X, Bouissou M-F. Effects of handling at weaning on manageability and reactivity of foals. *Appl Anim Behav Sci* 2004;**87**(1–2):131–49.
58. Lansade L, Bertrand M, Bouissou M-F. Effects of neonatal handling on subsequent manageability, reactivity and learning ability of foals. *Appl Anim Behav Sci* 2005;**92** (1–2):143–58.
59. Simpson BS. Neonatal foal handling. *Appl Anim Behav Sci* 2002;**78**(2–4):303–17.
60. Søndergaard E, Jago J. The effect of early handling of foals on their reaction to handling, humans and novelty, and the foal-mare relationship. *Appl Anim Behav Sci* 2010;**123**(3–4):93–100.
61. Spier SJ, Pusterla JB, Villarroel A, Pusterla N. Outcome of tactile conditioning of neonates, or "imprint training" on selected handling measures in foals. *Vet J* 2004;**168**(3):252–8.
62. Williams JL, Friend TH, Nevill CH, Archer G. The efficacy of a secondary reinforce (clicker) during acquisition and extinction of an operant task in horses. *Appl Anim Behav Sci* 2004;**8** (3–4):331–41.

63. Williams JL, Friend TH, Collins MN, Toscano MJ, Sisto-Burt A, Nevill CH. Effects of imprint training procedure at birth on the reactions of foals at age six months. *Equine Vet J* 2003;**35**(2):127–32.
64. Henry S, Richard-Yris M-A, Tordjam S, Hausberger M. Neonatal handling affects durably bonding and social development. *PLoS ONE* 2009;**4**(4) https://doi.org/10.1371/journal.pone.0005216.
65. Houpt KA. Formation and dissolution of the mare-foal bond. *Appl Anim Behav Sci* 2002;**8**(2–4):319–28.
66. Leblanc M-A, Duncan P. Can studies of cognitive abilities and of life in the wild really help us to understand equine learning? *Behav Process* 2007;**76**(1):49–52.
67. Nicol CJ. The social transmission of information and behavior. *Appl Anim Behav Sci* 1995;**44**(2–4):79–98.
68. Van Dierendonck MC, Spruijt BM. Coping in groups of domestic horses—review from a social and neurobiological perspective. *Appl Anim Behav Sci* 2012;**138**(3–4):194–202.
69. Krueger K, Farmer K, Heinze J. The effects of age, rank and neophobia on social learning in horses. *Anim Cogn* 2014;**17**(3):645–55.
70. Krueger K, Heinze J. Horse sense: social status of horses (*Equus caballus*) affects their likelihood of copying other horses' behavior. *Anim Cogn* 2008;**11**(3):431–9.
71. Krueger K, Flauger B, Farmer K, Maros K. Horses (*Equus caballus*) use human local enhancement cues and adjust to human attention. *Anim Cogn* 2011;**14**(2):187–201.
72. Schacht S, Houpt KA. Horsey see, horsey do.... *Equus* 2001;**283**:128–9.
73. Gardner LP. The responses of horses in a discrimination problem. *J Comp Psychol* 1937;**23**(1):13–34.
74. Hanggi EB. Categorization learning in horses (*Equus caballus*). *J Comp Psychol* 1999;**113**(3):243–52.
75. Hanggi EB. Discrimination learning based on relative size concepts in horses (*Equus caballus*). *Appl Anim Behav Sci* 2003;**83**(3):201–13.
76. Nicol CJ. Equine learning: progress and suggestions for future research. *Appl Anim Behav Sci* 2002;**78**(2–4):193–208.
77. McCall CA, Burgin SE. Equine utilization of secondary reinforcement during response extinction and acquisition. *Appl Anim Behav Sci* 2002;**78**(2–4):253–62.
78. Lansade L, Calandreau L. Secondary reinforcement did not slow down extinction in an unrelated learned task in horses. *J Vet Behav* 2016;**15**:85.
79. Hanggi EB. Short-term memory testing in domestic horses: experimental design plays a role. *J Equine Vet* 2010;**30**(11):617–23.
80. Hall C, Goodwin D, Heleski C, Randle H, Waran N. Is there evidence of learned helplessness in horses? *J Appl Anim Welf Sci* 2008;**11**(3):249–66.
81. Edwards L, Houpt KA. A taste for plastic? *Equus* 2016;**471**:66.
82. Aley JP, Adams NJ, Ladyman RJ, Fraser DL. The efficacy of capsaicin as an equine repellent for chewing wood. *J Vet Behav* 2015;**10**(3):243–7.
83. Houpt KA, Zahorik DM, Swartzman-Andert JA. Taste aversion learning in horses. *J Anim Sci* 1990;**68**:2340–4.
84. Pfister JA, Stegelmeier BL, Cheney CD, Ralphs MH, Gardner DR. Conditioning taste aversions to locoweed (*Oxytropis sericea*) in horses. *J Anim Sci* 2002;**80**(1):79–83.
85. Hall C. The impact of visual perception on equine learning. *Behav Process* 2007;**76**(1):29–33.
86. Hausberger M, Gautier E, Müller C, Jego P. Lower learning abilities in stereotypic horses. *Appl Anim Behav Sci* 2007;**107**(3–4):299–306.

87. Malamed R, Berger J, Bain MJ, Kass P, Spier SJ. Retrospective evaluation of crib-biting and windsucking behaviours and owner-perceived behavioural traits as risk factors for colic in horses. *Equine Vet J* 2010;**42**(8):686–92.
88. Geor R. Overtraining—can your horse get too much of a good thing? *The Horse* 2000;**XVII**(2):72.
89. Stratton R, Cogger N, Beausoleil N, Waran N, Stafford K, Stewart M. *Indicators of good welfare in horses*. Wellington, New Zealand: Ministry for Primary Industries; 2014. p. 48.
90. Mellor DJ, Beausoleil NJ. Extending the 'five domains' model for animal welfare assessment to incorporate positive welfare states. *Anim Welf* 2015;**24**(3):241–53.
91. Tinbergen N. *The study of instinct*. New York: Oxford University Press; 1951. p. 228.
92. LeDoux JE. Emotion: cues from the brain. *Annu Rev Psychol* 1995;**46**:209–35.
93. LeDoux JE. The slippery slope of fear. *Trends Cogn Sci* 2013;**17**(4):155–6.
94. LeDoux JE. Feelings: what are they & how does the brain make them? *J Am Acad Arts Sci* 2015;**144**(1):96–111.
95. Ali AB, Gutwein K, Heleski CR. Exploring the relationship between heart rate variability and behavior—social isolation in horses. *J Vet Behav* 2016;**15**:82–3.
96. Collyer PB, Harris BC. Comparison of exercise and stress parameters in barrel racing horses in the home and competitive environment. *J Vet Behav* 2016;**15**:93–4.
97. Li K, Nakajima M, Ibañez-Tallon I, Heintz N. A cortical circuit for sexually dimorphic oxytocin-dependent anxiety behaviors. *Cell* 2016;**167**(3):60–72.
98. van den Bos R, Harteveld M, Stoop H. Stress and decision-making in humans: performance is related to cortisol reactivity, albeit differently in men and women. *Psychoneuroendocrinology* 2009;**34**(11):1449–58.
99. Fenner K, Webb H, Starling M, Frieire R, Buckley P, McGreevy P. The effects of pre-conditioning on behavior and physiology of horses during a standardized learning task. *J Vet Behav* 2016;**15**:79.
100. Randle H. Welfare friendly equitation—Understanding horses to improve training and performance. *J Vet Behav* 2016;**15**:vii–viii.
101. Lesté-Lasserre C. Isolated yearlings learn better, study says. *The Horse* 2010;**XXVII**(6):15.
102. Valenchon M, Lévy F, Prunier A, Moussu C, Calandreau L, Lansade L. Stress modulates instrumental learning performances in horses (*Equus caballus*) in interaction with temperament. *PLoS ONE* 2013;**8**(4) https://doi.org/10.1371/journal.pone.0062324.
103. Schwabe L, Wolf OT, Oitzl MS. Memory formation under stress: quantity and quality. *Neurosci Biobehav Rev* 2010;**34**(4):4–591.
104. Heleski C, Wickens C, Minero M, DallaCosta E, Wu C, Czeszak E, von Borstel UK. Do soothing vocal cues enhance horses' ability to learn a frightening task? *J Vet Behav* 2015;**10**(1):41–7.
105. Cliffe S, Scofield RM. Safely introducing horses to novel objects: a pilot investigation into presentation techniques. *J Vet Behav* 2016;**15**:86.
106. Christensen JW, Rundgren M, Olsson K. Training methods for horses: habituation to a frightening stimulus. *Equine Vet J* 2006;**38**(5):439–43.
107. Lansade L, Simon F. Horses' learning performances are under the influence of several temperamental dimensions. *Appl Anim Behav Sci* 2010;**125**(1–2):30–7.
108. Johnson JA, Heird JC. Approaching the equine mind. *Q Horse J* 1988;**40**(7):66–9.
109. Cooper JJ. Equine learning behavior: common knowledge and systematic research. *Behav Process* 2007;**76**(1):24–6.
110. McCall CA. Making equine learning research applicable to training procedures. *Behav Process* 2007;**76**(1):27–8.

111. McDonnell S. Positive reinforcement. *The Horse* 2007;**XXIV**(7):63–4.
112. Slater C, Dymond S. Using differential reinforcement to improve equine welfare: shaping appropriate truck loading and feet handling. *Behav Process* 2011;**86**(3):329–39.
113. Ellis S, Greening L. Positively reinforcing an operant task using tactile stimulation and food—a comparison in horses using clicker training. *J Vet Behav* 2016;**15**:78.
114. Rochais C, Henry S, Sankey C, Nassur F, Goracka-Bruzda A, Hausberger M. Visual attention, an indicator of human-animal relationships? A study of domestic horses (*Equus caballus*). *Front Psychol* 2014;**5**.
115. Hendriksen P, Elmgreen K, Ladewig J. Trailer-loading of horses: is there a difference between positive and negative reinforcement concerning effectiveness and stress-related signs? *J Vet Behav* 2011;**6**(5):261–6.
116. Visser EK, van Reenen CG, Schilder MBH, Barneveld A, Blokhuis HJ. Learning performances in young horses using two different learning tests. *Appl Anim Behav Sci* 2003;**80**(4):311–26.
117. Lesté-Lasserre C. Using learning theory techniques in equine welfare cases. *The Horse* 2013;**XXX**(1):15.
118. Visser EK, Van Dierendonck M, Ellis AD, Rijksen C, Van Reenen CG. A comparison of sympathetic and conventional training methods on responses to initial horse training. *Vet J* 2009;**181**(1):48–52.
119. Baragli P, Mariti C, Petri L, De Giorgio F, Sighieri C. Does attention make the difference? Horses' response to human stimulus after 2 different training strategies. *J Vet Behav* 2011;**6**(1):31–8.
120. Fureix C, Pagès M, Bon R, Lassalle J-M, Kuntz P, Gonzalez G. A preliminary study of the effects of handling type on horses' emotional reactivity and the human-horse relationship. *Behav Process* 2009;**82**(2):202–10.
121. Henshall C, McGreevy PD. The role of ethology in round pen horse training—a review. *Appl Anim Behav Sci* 2014;**155**:1–11.
122. Maros K, Gácsi M, Miklósi Á. Comprehension of human pointing gestures in horses (*Equus caballus*). *Anim Cogn* 2008;**11**(3):457–66.

Chapter 4

Equine Communicative Behavior

Humans are a verbal species, so we have become heavily dependent on listening for messages and are relatively poor observers of body language. Animals, on the other hand, are the reverse. For them, it is all about physical expressions, supplemented with auditory inputs. There is a strong desire to be able to communicate with horses, at least in a manner that lets the person know what the horse is thinking. While horses do not have a human's vocabulary, that does not stop researchers from trying to make a connection.[1]

VOCAL

As *Equus ferus* separated from their *Equus asinus* relatives, unique sound patterns developed. A spectrographic analysis of separation calls made by Przewalski's horses and nonhorse equids shows the differences, at least in some of the vocalization patterns.[2–4] With the exception of the squeal, sounds made by domestic horses have a noise pattern rather than a pure tonal one. "Noise" indicates that they contain a variety of harsh and rough combinations of tone frequencies that are generally unpleasant to listen to for any length of time.

Voiced Emissions

Voiced communications are those produced in the larynx. This implies a degree of neurologic control over the muscles of the larynx in order to change the size of the glottis and tension on the vocal folds. Each of the various sounds produced by a horse also has different volumes and durations as part of the message being sent. Spectrographic studies of each vocalization have helped define their auditory characteristics.[5,6]

Whinny

The whinny (neigh) is the sound most associated with the horse. It is the one mimicked by children and used on fake soundtracks in old Western movies. Of horse sounds, the whinny is the loudest and longest, capable of being heard

Equine Behavioral Medicine. https://doi.org/10.1016/B978-0-12-812106-1.00004-8

about 0.6 miles (1 km).[5,6] The initiation of a whinny starts with the horse blinking the eyes and dilating its nostrils. The head is elevated with the ears and gaze directed forward. As the sound begins, the corners of the mouth are drawn back and the mouth open. The teeth remain covered.[5,7]

Graphically, a spectrogram shows the whinny consist of two fractions—a long first part that suggests an emotional state, and a short overlapping end fraction that is lower in pitch and indicates intensity.[7,8] Whinnies that have a positive association, such as a herdmate approaching, are shorter and lower pitch. Those with a negative association are longer and higher pitch. The total length of each neigh is typically 1–3 s.[7]

Two main functions are attributed to the whinny. The first is individual recognition. Maintaining or regaining social contact is the second. Long-distance social contact with herdmates is facilitated too, because the sound is audible for a great distance.[6] Spectrographs of stressed horses, such as a mare separated from her foal, differ from those where a horse is anticipating something good, such as food.[9]

There are sexual implications in the pitch of the whinny, much like feather color has for birds. Stallions with the lower pitch whinnies are given more attention by a mare, regardless of whether she is in heat or not. Her head and body orient toward those stallions significantly more than toward stallions having a higher pitch.[10] In addition to the caller's sex, its body size and identity are also apparent in the whinny.[6,10]

Squeal

Squeals are mostly associated with agonistic or aggressive situations and serve as a warning or threat that the aggression will escalate if provocation continues. It is mostly voiced during olfactory investigations or with mock or serious fighting.[7] The head position is variable and the mouth is usually closed with the corners retracting as the sound begins.

Sound characteristics show the squeal to be unique among vocalizations because of its harmonic, tonal quality. The fundamental frequency is approximately 1 kHz.[5] This is close to the sound of high C on the piano, one musical octave higher than middle C. The sound is single and variable in duration, typically lasting between 0.1 and 1.7 s.[5]

Loudness varies. Some squeals are so soft as to only be heard for about 6 ft (2 m), as when a mare protests her foal's attempts to nurse.[5] At the other extreme, some are intense and occasionally called a "scream." These are very loud and typically associated with episodes of serious aggression.[7]

Nicker

The nicker is a low-pitch, guttural sound that occurs in three different patterns.[5,6] The first is what can be termed an "anticipatory" or "begging" nicker. It is commonly expressed just before a horse is fed or to indicate the animal's presence or anticipation. The primary sound for this nicker is approximately 2 kHz (two octaves up from middle C).[5] It is relatively continuous and quite loud, audible for about 100 ft (30 m).[5]

The "sexual interest" nicker is used by stallions in association with approaching a mare. Graphically it shows as alternating bands of loud and soft noise. This too is audible for at least 100 ft (30 m).[5]

The third nicker type, the "mother-foal" nicker, is used by a mare showing concern for her foal, including with the appearance of potential danger. It alternates in volume, with loudness peaking every tenth of a second.[5] It is only audible within a few feet of the mare.

Groan/Grunt/Sigh

Groans have a negative context, typically being associated with physical effort, prolonged discomfort, or situations where there might be mental conflict. They are also associated with events similar to those where the horse might squeal. Groans are low-pitched vocalizations made while the mouth is closed.[7] The primary frequency is less than 300 Hz, putting them one note above middle C on the musical scale. Although pulsations show graphically, the sound is monotone to humans because the pulses are so rapid.[5] The groan can be heard for approximately 5 ft (1.5 m), and is generally very short in duration, lasting only 0.1–1.7 s.[5]

The sigh is a variation of the groan that some classify separately. It is usually expressed as a prolonged exhalation after a quick, deep inhalation.[7] Sighs occur most often when the horse lies down.

Grunts are either classified as a separate vocal pattern or considered to be the shortest duration groans.[5] There are horses that naturally grunt when they lie down. While grunting is a normal sound, it can be worrisome if done frequently. Some horses grunt a lot when they become tired, almost as an involuntary sound. Others do it as they settle into a routine or while eating. Owners will occasionally associate grunts with times when their horse seems "bored." It seems the grunt bothers the rider more than the horse.

The grunt-like noise is also associated with windsucking, so a good history is important to differentiate normal from problematic grunting. Windsucking is much more significant and worrisome than if the horse is just making noise.

Nonvoiced Sounds

Nonvoiced sounds are not associated with the larynx, but rather with other body parts. Just as with laryngeal sounds, these too are used in a variety of contexts.

Snort

The forceful exhalation of air through the nostrils causes them to flutter in a pulsating manner, resulting in a snort.[5] The sound is short in duration, typically just under 1 s, and it is loud, being audible at 40–50 ft (12–15 m).[5] Snorting is most common during vigorous exercise, especially as the horse becomes tired, when it is restless but constrained, and when the nasal passages are irritated.

Blow

The blow is an alarm, serving to alert herdmates and warn intruders. This broadband sound results from the forceful expulsion of air through the nostrils, which are completely dilated.[5] Duration of the blow is typically less than a half second, but it can go longer.[6] The majority of the sound's energy is below 3 kHz (7 keys down from the highest note on the piano keyboard), but parts are much higher.[5] The suddenness and loudness distract from its high frequency level and catch the attention of others. Longer blows, lasting up to 1.3 s, can be emitted after a horse has investigated something by sniffing.[5]

Snore

The snore is a raspy sound lasting 1–2 s.[5,6] It can be associated with labored breathing while lying and also be heard immediately prior to the alarm blow due to the rapid intake of air.[5] This latter type of snore is short lived, lasting less than a half second.

Mouth Smacking

Mouth smacking is made by a mare while nursing her foal. She will occasionally turn her head toward the foal and suddenly open her mouth, resulting in a smacking noise. It appears that this is a mild threat, perhaps a variation of the bite threat.[5,6]

Hoof-Substrate Sounds

Hoof sounds vary by what the horse is doing. They are associated with locomotion, as with the galloping horse, and the surface the horse is on, as with the clip-clop of the carriage horse on asphalt. They can accompany visual threats such as the stomp and expressions of "frustration" such as pawing.

Incidental Sounds

Incidental sounds are those that happen associated with other events. They would include coughing, sneezing, sighing, flatulence, intestinal sounds, and "gelding gurgle" as examples.

POSTURAL

Body language has significant meaning for most species of animals. Friendly messages are shown with the general meaning of "it is all right to come close" or to reduce the distance between the approacher and the one being approached. Distance-increasing signals suggest the horse does not want interactions, perhaps even to the point of aggression. There are times, however, when there is internal conflict about what message the horse should send, resulting in ambivalent body language. Part of the message is "come closer," but part of it says "go away." This often happens with fearful situations.

Large animal species are somewhat handicapped in body language signals because their bodies are not particularly flexible, especially in the thoracic and abdominal regions. For dogs and cats, threats are usually associated with the appearance of increased size by arching the back or piloerection. The size of the horse is intimidating enough, so head and rear threats are sufficient.

Distance-Reducing Body Language

Distance-reducing body language indicates the horse is comfortable with the situation at hand, or perhaps, anticipating a positive interaction. The head and neck can show a number of indicators of friendly behaviors, beginning with their relative position. The relaxed horse will carry its head in a neutral position. That means that the top of the neck is roughly parallel to the ground. There are variations based on conformation and natural head carriage among different breeds, but the general position is relatively flat. Horses can also be taught to lower their head and neck so that the nose is almost touching the ground as a "go to" posture when stressed or conflicted. A horse interested in what is approaching will raise the head and neck above the neutral position.

Structures on the head are strong indicators of the general message a horse is transmitting, because horses "talk with their face." Facial features can be used as a method for coding observations for consistency in interpretation.[11]

The ears are strong indicators of the general mood of the horse. In friendly encounters, the ears point forward when the horse has an interest in what is approaching. They can also be held in a neutral side-facing or pointed slightly backward direction. This is associated with the animal watching its surroundings but paying minimal attention to it (Figure 4-1).

The eyes are frequently described as being "soft" when the lids are rounded in their neutral oval shape (Figure 4-2). The pupils are appropriate for light levels, and the sclera is barely visible, if at all. Eyelids that are closed or half closed are common in resting horses.[7,12]

Several features around the nose and mouth area suggest distance-reducing messages. The nostrils are relaxed, retaining their original shape (Figure 4-2). They are not flared as would happen with excitement or fear. The lips can show a lot of expression. The upper lip can extend forward with a slightly squared off appearance. If there is no tension in it, the posture signals that the horse is apparently enjoying a grooming session. The lower lip droops in the tired or very relaxed horse (Figure 4-3). Another behavior, lip licking, is a sign of submission or "giving up" and is often accompanied by a sigh or deep inhalation and exhalation. This is a signal given by a tense horse when it stops fighting a situation, making it useful to assess the reduction of stress. Licking of people or inanimate

FIGURE 4-1 Ear positions indicate where attention is directed. From left to right, the foal has its ears forward so it can direct attention toward the camera. The ears slightly back suggests this mare is paying attention to her foal and the other horses she does not want to interact with the foal. The ears pointed sideways indicate a neutral position with no particular interests. The ears slightly forward are also in a relatively neutral position but with some interest in the horse to its right.

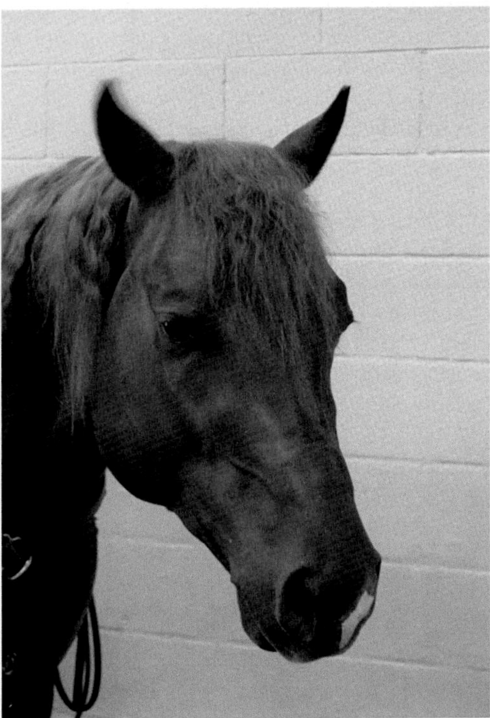

FIGURE 4-2 The roundish shape of the eyelids, elongated oval of the nostril, and slightly angled position of the ears indicate the lack of tension in the face of this horse.

FIGURE 4-3 The mare is very relaxed as indicated by the drooping lower lip and resting hind foot. Horses commonly use this position during slow wave sleep.

objects can be a sign of insecurity as might happen at a horse show, or it could indicate there is a tasty leftover on the person's hand. Horses that crave salt can also lick objects. The last oral behavior associated with distance-reducing body language is *jaw chomping* (also called *snapping* or *mouth clapping*). The mouth is opened but the teeth remain covered. Jaw movements partially open and close the mouth several times in rapid succession while the head and neck are extended forward (Figure 4-4). The teeth do not usually meet. The behavior is commonly shown by foals and horses less than 2 years old and directed toward older, high-ranking horses. This suggests it is an appeasement or submissive behavior to avoid aggression from the higher ranking horse.[7, 13–16] Jaw chomping is particularly common in orphan foals that are raised on a bottle.[16] On rare occasions, the behavior is shown by introverted older horses toward a high-ranking herdmate.

Resting horses and those in light sleep often remain standing with three feet flat on the ground and the fourth resting on the point of the hoof (Figure 4-3). Horses may actively approach a person or other familiar horse as a distance-reducing indicator, or they may also remain still when being approached. Both indicate the horse is comfortable with the person or horse coming toward them.

Tails may be relaxed and hang straight down, with occasional swishing to remove insects. Tail positions elevate in association with estrus, mounting, greetings, suckling, and play. Positive excitement results in an elevation of the tail until it is about 45 degrees from vertical (Figure 4-5). This high tail position is synchronous with head carriage and is an indicator of preparation for running, increasing speed, and confident approach.[17])

FIGURE 4-4 The facial expression of jaw chomping (snapping), shown by the yearling stallion on the left, is directed at a higher-ranking yearling in an adjacent pen. The display involves the corners of the mouth pulled back while the mouth opens and then partially closes multiple times.

FIGURE 4-5 The elevated tail of an excited yearling that is playing with a pasture mate.

Anticipatory behaviors for potentially favorable events such as an evening's feed are characterized by a significant increase in activity and heart rate.[18] Most horse owners have seen the increased vigilance and movement of pastured horses around feeding time, particularly if the person is late. Maintenance behaviors like fly swatting are also reduced during anticipation.

Juvenile play behaviors are similar to adult behaviors but occur out of context and out of the normal adult sequence. Play allows the young to practice behaviors they might need for survival in the wild, tone muscles, and develop coordination. In young animals of several species, a special gesture indicates behaviors that follow are play. For dogs it is the play bow. For young colts, it is dropping to its knees and biting. This "play invitation" or "sparring initiation gesture" will become part of the aggressive behaviors in fighting adults.[19] Touching or brushing the forelimbs of a colt will occasionally trigger play aggression, just as would happen with another colt nipping at the legs.

Distance-Increasing Body Language

"Go away" is the message associated with distance-increasing body signs. In general, the muscles begin to tighten and tension is seen throughout the body. The head and neck are raised initially to assess what might be approaching. The neck can be arched if the other horse is nearby, as when a new horse is introduced into the pasture (Figure 4-6). An arched neck threat is associated with stallions posturing before a fight.[20] If aggression and a chase become more likely, the neck is lowered and the head extended so biting can be attempted. At the extreme, the head and neck are lowered toward the ground, but the nose is extended forward. This posture is used by stallions to drive their harem group and occasionally when going after dogs or other potential predators.

The horse's face is very expressive relative to distance-increasing messages. The most obvious sign will be ears pinned flat back onto the neck (Figure 4-7).

FIGURE 4-6 An arched neck is shown by the horse on the left when newly introduced into a pen of yearlings. They are investigating each other by smelling breath.

FIGURE 4-7 The facial expressions on both mares indicate a "get away from my foal" message, including the pinned-back ears, tension in the lip and nose, and elevated upper eyelids.

As the intensity of the "go away" message increases, the upper eyelid elevates near its midpoint to take on an approximately 90-degree angle, and the lighter colored sclera is visible. The upper lip becomes tense and the roundness becomes angulated instead. The head and neck can be extended with teeth showing, especially if the horse is attempting to bite. Biting the opponent's leg is common among bachelor stallions and is the primary fighting tactic of horses.[20] The horse can also extend its head and neck in an elevated manner such that the sclera shows, making the horse look like it has rolled its eyes. These signs would be accompanied by tension in the nose and ears flat against the neck, unless the elevated head is associated with play.[7]

The horse will use its feet and legs in distance-increasing messages. Front legs are used to strike, or the horse may rear to strike with both front feet. The hind limbs are also dangerous. One may be raised as a threat, or the horse can kick back with one or both feet.

The tail is carried down and tense in fearful situations such as when startled or chased. It can also be clamped tightly to the body and either held there or if aggression is likely to follow, lashed back and forth. Lashing of the tail is common when the horse is spurred often or hard or when receiving conflicting signals from the rider.[12] Some horses naturally flip their tails when changing leads, especially when it is done multiple times in rapid succession.

Fighting is fairly ritualized in horses and will be discussed under social behavior. Certain postures are used to send the "go away" message before actual fights begin. Striking out with a front foot is one of the first postural changes seen, and if a greater threat is needed, the horse will rear and strike out with both front feet. The elevated rearing position would strategically position a stallion to strike an opponent or predator on the head or back to inflict serious injuries. It might also be a threat by demonstrating superior height.[20] Strikes and rearing are more often used to threaten rather than injure, though a follow-up behavior has the harem stallion chasing away other stallions. If the fight escalates instead, the "goal" seems to be to pin the opponent to the ground. Stallions will bite at the forelimb or flank of the opponent. Then each will drop to its knees and lean toward the adversary to try to prevent a leg bite. They will rise again, reposition, and continue all forms of aggression toward each other until one runs away or stops fighting.

Ambivalent Body Language

In fearful situations, mixed messages are common. If the discomfort continues, the distance-increasing component is likely to prevail over the come-closer part. Working with a horse showing ambivalence requires special handling for the human to remain in control.

When each ear points in a different direction, the horse is watching different things on each side of its body, suggesting its attention is divided. The direction each points is a guide to what has the attention. Body tension and a head held high are postures suggesting a fear component.

The tail can provide another clue that the horse might be conflicted. When the top is held tight to the body and the lower portion directed away, it suggests the environment is fear-provoking at the same time the horse is trying to trust the rider.

Yawning also has an ambivalent message. While humans associate it with "I'm sleepy," it is more likely to have a social meaning in animals. In dogs, for example, the behavior is common in stressful situations. Although the specific meaning has not been determined in horses, differences between wild and domestic horses is puzzling. Przewalski's horses frequently yawn after a threat to bite or chase, perhaps to defuse tension. On the other hand, pastured domestic horses are more likely to yawn following a positive or neutral interaction, like sniffing.[21] Stallions and stalled horses are also more likely to show the behavior.

SCENTS AND PHEROMONES

Because horses rely on a combination of senses for survival, no particular one of them seems to have a higher priority than the others, but all except vision are keener than for humans. The introduction of the commercial synthetic equine appeasing pheromone has increased interest in equine olfaction.

Scent communication discussions typically center on odors associated with reproduction, for which there is quite a bit of information. During estrus, mares have significant chemical differences in their urine and feces.[22,23] There are also changes in the concentration of certain substances in stallion urine during the breeding season. It has been suggested this occurs so when stallions overmark the mare's feces, it will mask her odor from competing stallions.[22,24] In other species, pheromones may suppress the hormonal status of low-ranking males and synchronize the reproductive cycles of females.[25] We do not know if this is true in horses, but there are suggestions it might be.

Scents produced by horses for communicating with other horses are more extensive than those related to sexual behavior. They are important is social behavior too. There is a significant correlation between a horse's social rank and the time and frequency it spends smelling urine and feces.[24] In addition to social rank information, urine and feces marks will communicate trails, minimize aggression, and perhaps help with orientation.[26] Horses are capable of detecting the body odors of other horses. Studies of the chemical compositions of sweat and saliva have identified unique proteins, one of which is a lipocalin.[27] This class of protein is recognized for its role in olfaction and pheromone transport,[28] suggesting it plays a major role in bringing the individual's unique scent to the skin surface. Horses pick rolling areas based on odors left where others previously rolled.[25] Because of social implications of equine body odors, humans need to be mindful that they may carry specific odor messages between horses on clothing and unwashed hands.[25]

Pheromones are even more difficult to understand than traditional odors. Appeasing pheromones are associated with the mammary region of a lactating female and are associated with a calming effect. A logical use would be as an aid in weaning; however, they do not change behavioral responses or cortisol measures.[29] Instead, synthetic versions are finding a use for horses in mildly stressful situations. After being treated with the commercial product, horses show fewer signs of behavioral and cardiac stress, although their ability to learn during those situations does not change.[30,31]

PAIN ASSOCIATED BODY LANGUAGE

Animals that evolved as prey species use a survival technique of being able to hide illness and pain. This makes it hard to evaluate the subtle changes, although extremes like colic are generally obvious. Ideally, assessment is easiest if comparisons can be made to the individual's normal behavior as the control, although this is generally not possible. Instead, we are dependent on an observant owner or trainer knowing the horse is "just not acting right." Increases in heart and respiratory rates may be associated with pain, but alone are not always good indicators.[32] Digestive sounds and rectal temperature as physiological pain parameters can vary from normal, but not always. The ideal is to have a scale that pulls together the total of possible observations into a meaningful indicator of pain and that correlates with cortisol levels.[32–35]

Acute pain is characterized by a series of changes, not all of which may be present depending on where and when the pain is occurring, severity of the pain, and sensitivity of the individual animal. Horses in pain tend to be hypersensitive to sound, sights, and/or touch. There are changes in body posturing including weight bearing, sweating, looking or kicking at the abdomen, teeth grinding, restlessness, rolling, and posturing.[32,33] The ears are offset or back. There can be reluctance to put pressure on an affected foot or to move the affected part. Pawing and kicking out are also common. Horses may head press or press an affected part against an object. They may stand in an unusual way such as with the legs stretched in a way similar to the stance of a foundered or urinating animal. Changes in disposition can also occur. These might include general grumpiness, constant or intermittent reduction in appetite, changes in responses to sound or touch, and threats or aggressiveness toward herdmates not previously seen. Horses being ridden may suddenly rear or kick out at the change of gaits or start of an intricate maneuver such as a lead change or rollback. This is because of pain or anticipation of pain, not "stubbornness."

Specific facial expressions are other indicators of pain. Ears can be stiff and offset from each other, or pointed caudally. Tension of the upper eyelid results in a squarer appearance to the eye. Pain of the eye itself produces squinting. There can be tension of the lips and nostrils such that the nostrils transition from the normal elongated opening to a square one.[34,36]

Chronic pain can be more difficult to identify, and the horse may be presented as a behavior problem instead. This makes it important to rule out physical pain first. The most common sign of chronic pain is reduced performance. This may be a reduced ability to perform certain tasks, asymmetry in maneuvers, reluctance to change leads, or a "bunny hopping" canter (lope). The horse may also show strong resistance at certain times by bucking or kicking out.

Horses that experience chronic pain are more likely to show signs of depression: the flat neckline parallel to the ground; the absence of ear, head, and tail movement; and general inattention to surrounding activities. The latter has been correlated with the level of back pain.[37] Intermittent episodes suggesting acute pain may occur with the waxing and waning of pain perception.

Subtle pain behaviors are difficult to recognize, as demonstrated in videotaped studies of horses before and after elective surgery. The videos also show how significantly signs can vary between individuals. To aid in the assessment, researchers have developed scoring systems that take into account behavioral differences. Observations can be scored on a numerical scale such as 0–3 or 1–5, and the sum of all the parameters provides an indication of the severity of the pain being experienced.[32, 33, 38]

Two of the scoring systems are based on changes noted in facial expressions.[33, 39, 40] Analysis scores facial features from 0 (absent) to 2 or 3 (obviously present), and sums them into a total score. The first facial feature relates to ear position, in which the strongest indicator of pain is when the ears are stiff and turned slightly backward. The eyelids figure into two of the features. If the eye is closed by more than half, pain is strongly suggested. Tension above the eyes is indicated as increased prominence of underlying bone structures there and visibility of the sclera. Muscle tension proximal to the commissure of the lips represents straining of the chewing muscles and indicates pain. Another indicator is evidenced by the upper lip drawn back, making the lower lip evident as a pronounced "chin." The nostrils are the sixth feature evaluated for strain, slight dilatation, or wrinkles between them. The result is a flatter nose profile and lips that seem to be elongated. The head can also be twisted and the jaw crossed.

Other scoring systems are broader in scope, evaluating whole body behavior and physiological parameters.[35, 38, 41] For the "composite pain scale" in one study, assessment ranges from 0 (no pain) to 3 (very painful).[35] Nine behavior factors are evaluated, the first of which is appearance. It ranges from bright and alert with no reluctance to move, and goes to continuous movement and abnormal facial expressions. Sweating ranges from none to water actually running off the horse. Kicking at the abdomen was the third factor and, if present, is evaluated by frequency during a 5-min period. Similarly, pawing as an indicator of pain is evaluated for severity by the frequency of occurrence within 5 min. Posture is evaluated by the presence or absence of weight bearing and muscle tremors. Head movements become increasingly frequent with pain. Appetite, interactive behavior with the observer, and response to palpation of the potentially painful area round out the behavior factors evaluated. Physiological measures—heart rate, respiratory rate, rectal temperature, and digestive motility—are added to behavior scores for a total composite pain score. The second study evaluates eight criteria.[41] Head position and movement, ear position, eye expression, movement of the mouth, and whether the bit was pulled through the mouth were similar to the previous facial evaluations. The position or movement of the tail, obedience during movement, and gait abnormalities were added.

Each of the scoring systems has merit and provides veterinarians and owners with an objective indicator of the quantity of pain a horse might be experiencing. Because signs are often subtle, having several criteria to evaluate in a simple format can prove useful. At this time, these evaluation criteria need continued refinement and validation.[42]

HUMAN-HORSE COMMUNICATION

The possibility of interspecies communication has fascinated people for years. Horses obviously can learn what certain cues mean and how to respond. Over time, cues can become very subtle as the horse's experience with and trust in the rider increases. Science continues to study how well horses can actually read the human body language, not just leg and voice cues. Early studies paralleled those in dogs, where researchers tried to determine whether a horse would use a person's pointing gesture to find hidden food.[43,44] If the gesture continues as the horse approaches the treat, the horse was able to find the treat at a level significantly greater than trial and error. However, if the person points and then stops pointing, the horse is not successful unless the person was quite close to the treat. And then, "success" might really be accidental if the familiar person is near the correct choice.[45] Familiarity with and positioning of the person has to be taken into account because horses are more likely to approach people they know. Horses are also capable of using the human's focus of attention as a cue, doing so significantly better if the person is familiar.[45] The familiarity factor is supported in a comparison of horses under 3 years of age to older horses. The older ones were much better at using human body orientation and pointing or tapping cues to find hidden treats.[46]

Horses are also capable of using the human gaze, facial expressions, and nods/shakes in their decisions.[46–48] Clever Hans could successfully "solve" simple math problems if his owner was near by reading the tension in the person's neck. When the pawing reached the correct answer, the owner relaxed and the pawing stopped. This was an unconscious and very subtle cue. Horses are also able to differentiate between happy and angry human facial expressions presented in photographs. Angry expressions cause the horse to switch to the left-gaze bias associated with potentially negative inputs. At the same time, there is an increase in heart rate.[48]

Since communication is a two-way process, the question remains as to whether the horse will actively try to communicate with humans. Recent findings suggest the answer is "yes," at least in certain situations. This represents more than a whinny that brings food. It is a subtle intent to convey a message. When a horse knows where a food treat is located but needs human help to get it, it communicates with the human using looks and touches. These efforts increase when the human is unaware of the location of the treat.[49] The continued gaze at the person, instead of glancing back and forth between person and goal as a dog would do, suggests the horse's behavior is more of a request than an attempt to show a location.[49] The horse acts differently when the person knows the treat location compared to when they do not. This hints that horses are capable of understanding whether the human does or does not know the location of the hidden object.[49] These results are from one study, and while interesting, they need to be confirmed or refuted by advanced research. The ability to make this distinction has been shown in primates, but is rare in other species.

REFERENCES

1. Giving animals a voice—computer software that could tell us what they are thinking. http://phys.org/news/2016-07-animals-voicecomputer-software.html [downloaded July 12, 2016].
2. Alberghina D, Caudullo E, Bandi N, Panzera M. A comparative analysis of the acoustic structure of separation calls of Mongolian wild horses (*Equus ferus przewalskii*) and domestic horses (*Equus cabalus*). *J Vet Behav* 2014;**9**(5):254–7.
3. Alberghina D, Caudullo E, Chan WY, Bandi N, Panzera M. Acoustic characteristics of courtship and agonistic vocalizations in Przwewalskii's wild horse and in domestic horse. *Appl Anim Behav Sci* 2016;**174**(1):70–5.
4. Policht R, Karadžos A, Frynta D. Comparative analysis of long-range calls in equid stallions (Equidae): are acoustic parameters related to social organization? *Afr Zool* 2011;**46**(1):18–26.

5. Waring GH. *Horse behavior*. 2nd ed Norwich, NY: Noyes Publications; 2003. p. 442.
6. Yeon SC. Acoustic communication in the domestic horse (*Equus caballus*). *J Vet Behav* 2012;**7**(3):179–85.
7. McDonnell S. *A practical field guide to horse behavior: the equid ethogram*. Lexington: The Blood-Horse; 2003. p. 375.
8. Briefer EF, Maigrot A-L, Mandel R, Freymond SB, Bachmann I, Hillmann E. Segregation of information about emotional arousal and valence in horse whinnies. *Sci Rep* 2015;**4**:9989.
9. Pond RL, Darre MJ, Scheifele PM, Browning DG. Characterization of equine vocalization. *J Vet Behav* 2010;**5**(1):7–12.
10. Lemasson A, Boutin A, Boivin S, Blosi-Heulin C, Hausberger M. Horse (*Equus caballus*) whinnies: a source of social information. *Anim Cogn* 2009;**12**(5):693–704.
11. Wathan J, Borrows AM, Waller M, McComb K. The equine facial action coding system. *PLoS ONE* 2015;**10**(8).
12. Kiley-Worthington M. *The behaviour of horses: in relation to management and training*. London: J.A. Allen; 1987. p. 265.
13. Christensen JW, Zharkikh T, Ladewig J, Yasinetskaya N. Social behavior in stallion groups (*Equus przewalskii* and *Equus caballus*) kept under natural and domestic conditions. *Appl Anim Behav Sci* 2002;**76**(1):11–20.
14. Crowell-Davis SL, Houpt KA, Burnham JS. Snapping by foals of *Equus caballus*. *Z Tierpsychol* 1985;**69**:42–54.
15. Henshall C, McGreevy PD. The role of ethology in round pen horse training—a review. *Appl Anim Behav Sci* 2014;**155**:1–11.
16. Williams M. The effect of artificial rearing on the social behavior of foals. *Equine Vet J* 1974;**6**(1):17–8.
17. Kiley-Worthington M. The tail movements of ungulates, canids and felids with particular reference to their causation and function as displays. *Behaviour* 1976;**56**(1–2):69–116.
18. Peters SM, Bleijenberg EH, van Dierendonck MC, van der Harst JE, Spruijt BM. Characterization of anticipatory behavior in domesticated horses (*Equus caballus*). *Appl Anim Behav Sci* 2012;**138**(1–2):60–9.
19. McDonnell S. Play invitation. *The Horse* 2004;**XXI**(5):107–8.
20. McDonnell SM, Haviland JCS. Agonistic ethogram of the equid bachelor band. *Appl Anim Behav Sci* 1995;**43**(3):147–88.
21. Górecka-Bruzda A, Fureix C, Ouvrard A, Bourjade M, Hausberger M. *Investigating determinants of yawning in the domestic (*Equus caballus*) and Przewalski (*Equus ferus przewalskii*) horses. Naturwissenschaften* 2016;**103**:72. https://www.ncbi.nlm.hih.gov/pmc/articles/PMC4992016 [downloaded October 18, 2016].
22. Kimura R. Volatile substances in feces, urine and urine-marked feces of feral horses. *Can J Anim Sci* 2001;**1**(3):411–20.
23. Ma W, Klemm WR. Variations of equine urinary volatile compounds during the oestrous cycle. *Vet Res Commun* 1997;**21**(6):437–46.
24. Jezierski T, Jaworski Z, Sobczyńska M, Kamińska B, Górecka-Bruzda A, Wallczak M. Excreta-mediated olfactory communication in Konik stallions: a preliminary study. *J Vet Behav* 2015;**10**(4):353–64.
25. Saslow CA. Understanding the perceptual world of horses. *Appl Anim Behav Sci* 2002;**78**(2–4):209–24.
26. Turner Jr JW, Perkins A, Kirkpatrick JF. Elimination marking behavior in feral horses. *Can J Zool* 1981;**59**(8):1561–6.

27. D'Innocenzo B, Slazano AM, D'Ambrosio C, Gazzano A, Niccolini A, Sorce C, Dani FR, Scaloni A, Pelosi P. Secretory proteins as potential semiochemical carriers in the horse. *Biochemistry* 2006;**45**:13418–28.
28. Flower DR. The lipocalin protein family: structure and function. *Biochem J* 1996;**318**(1):1–14.
29. Berger JM, Spier SJ, Davies R, Gardner IA, Leutenegger CM, Bain M. Behavioral and physiological responses of weaned foals treated with equine appeasing pheromone: a double-blinded, placebo-controlled, randomized trial. *J Vet Behav* 2013;**8**(4):265–77.
30. Cozzi A, LaFont Lecuelle C, Monneret PL, Articlaux F, Bougrat L, Bienboire Frosini C, Paget P. The impact of maternal equine appeasing pheromone on cardiac parameters during a cognitive test in saddle horses after transport. *J Vet Behav* 2013;**8**(2).
31. Falewee C, Gaultier E, Lafont C, Bougrat L, Pageat P. Effect of a synthetic equine maternal pheromone during a controlled fear-eliciting situation. *Appl Anim Behav Sci* 2006;**101** (1–2):144–53.
32. Bussières G, Jacques C, Lainay O, Beauchamp G, Leblond A, Cadoré J-L, Desmaizières L-M, Cuvelliez SG, Troncy E. Development of a composite orthopaedic pain scale in horses. *Res Vet Sci* 2008;**85**(2):294–306.
33. Dalla Costa E, Minero M, Lebelt D, Stucke D, Canali E, Leach MC. Development of the horse grimace scale (HGS) as a pain assessment tool in horses undergoing routine castration. *PLoS ONE* 2014;**9**(3) https://doi.org/10.1371/journal.pone.0092281.
34. Gleerup KB, Forkman B, Lindegaard C, Andersen PH. An equine pain face. *Vet Anaesth Analg* 2015;**42**(1):103–14.
35. Van Loon JPAM, Back W, Hellebrekers LJ, van Weeren PR. Application of a composite pain scale to objectively monitor horses with somatic and visceral pain under hospital conditions. *J Equine Vet* 2010;**30**(11):641–9.
36. Chambers CT, Mogil JS. Ontogeny and phylogeny of facial expression of pain. *Pain* 2015;**156** (5):798–9.
37. Rochais C, Fureix C, Lesimple C, Hausberger M. Lower attention to daily environment: a novel cue for detecting chronic horses' back pain? *Sci Rep* 2016;**6**.
38. Gleerup KB, Lindegaard C. Recognition and quantification of pain in horses: a tutorial review. *Equine Vet Educ* 2016;**28**(1):47–57.
39. Dyson S, Berger J, Ellis AD, Mullard J. Can the presence of musculoskeletal pain be determined from the facial expressions of ridden horses (FEReq)? *J Vet Behav* 2017;**19**:78–89.
40. Mullard J, Berger JM, Ellis AD, Dyson S. Development of an ethogram to describe facial expressions in ridden horses (FEReq). *J Vet Behav* 2017;**18**:7–12.
41. Dyson S, Berger J, Ellis AD, Mullard J. Development of an ethogram for a pain scoring system in ridden horses and its application to determine the presence of musculoskeletal pain. *J Vet Behav* 2018;**23**:47–57.
42. Gleerup KB, Andersen PH, Wathan J. What information might be in the facial expressions of ridden horses? Adaptation of behavioral research methodologies in a new field. *J Vet Behav* 2018;**23**:101–3.
43. Maros K, Gácsi M, Miklósi Á. Comprehension of human pointing gestures in horses (*Equus caballus*). *Anim Cogn* 2008;**11**(3):457–66.
44. McKinley J, Sambrook TD. Use of human-given cues by domestic dogs (*Canis familiaris*) and horses (*Equus caballus*). *Anim Cogn* 2000;**3**(1):13–22.
45. Krueger K, Flauger B, Farmer K, Maros K. Horses (*Equus caballus*) use human local enhancement cues and adjust to human attention. *Anim Cogn* 2011;**14**(2):187–201.
46. Proops L, Rayner J, Taylor AM, McComb K. The responses of young domestic horses to human-given cues. *PLoS ONE* 2013;**8**(6) https://doi.org/10.1371/journal.pone.0067000.

47. Malavasi R, Huber L. Evidence of heterospecific referential communication from domestic horses (*Equus caballus*) to humans. *Anim Cogn* 2016;**19**(5):899–909.
48. Smith AV, Proops L, Grounds K, Wathan J, McComb K. Functionally relevant responses to human facial expressions of emotion in the domestic horse (*Equus caballus*). *Biol Lett* 2016;**12**(2) https://doi.org/10.1098/rsbl.2015.0907.
49. Ringhofer M, Yamamoto S. Domestic horses send signals to humans when they face with an unsolvable task. *Anim Cogn* 2017;**20**(3):397–405.

Chapter 5

Equine Social Behavior

The horse evolved as a social species, living in herds of several animals. Early domestication did not change that relationship much, but the more people controlled the husbandry of their animals, the smaller and more artificial the groups became. Now, the common practice is to singly house horses, particularly stallions, and artificially control their diets, exercise, and breeding. While these changes can be good for health and long-term survival, they can also be problematic. Understanding the normal social behavior of the horse is beneficial for developing a management program that incorporates practices that benefit the mental health of each horse.

CRITICAL PERIODS

In the development of a foal, there are three critical lessons that must be learned for survival: who is "mom," "I am a horse," and "other species to be accepted." When during development each of these occurs can vary depending on circumstances, but for each, there is a critical period when the foal is most sensitive to learning the important concept. In some cases, when the sensitive period ends, so does the opportunity for that lesson to be learned.

During the first critical period, a newborn foal must bond to its mother, and she to it. This is important for access to milk and protection. Because of the importance of this two-way attachment, it begins immediately after the foal's birth. The mare will spend considerable time contacting the foal and licking fetal membranes, especially within the first 10 min.[1,2] There are approximately 80 foal-directed activities carried out by the mare within the first 30 min. Licking and touching the foal continue for the first 3 days, gradually diminishing in frequency.[1] By studying mare-foal reactions when briefly separated, researchers have been able to estimate when the bonds form. Mares show distress with separation very quickly, suggesting a rapid bonding. By the third day, she is able to recognize her own foal.[1] Foals, however, show an increasing amount of distress for the first 2 weeks, suggesting their attachment to the mare takes a couple of weeks to form completely.[2] The foal's attachment to its dam is important so that it can follow her, but it also allows the foal to learn about the mare's home range environment. The foal is a "follower" that remains close to

Equine Behavioral Medicine. https://doi.org/10.1016/B978-0-12-812106-1.00005-X

its mother, as opposed to a calf or fawn that is a "hider," staying in one place while the mother grazes elsewhere.[3] Prolonged separation during these first few days or delayed nursing can have significantly negative effects for the foal's later social attachments. Deprived foals tend to remain exceptionally close to the mare, not socializing to other horses. Later, they do not interact well with peers and can even be aggressive to them.[4]

Imprinting is species identification, and it occurs in a second critical period. The foal must learn what other horses are, and that it is a horse. This is necessary so the foal will identify as a herd member as it matures. For foals that remain close to their dams or play with other foals, imprinting is not a problem. On the other hand, it can be a significant problem for orphan foals lacking an equine role model. Animals raised with their own species and another species will preferentially identify with their own kind. They only imprint on a different species if there are no others like them.[5]

Much has been written about foal "imprinting" in popular books, which then go on to describe the importance of repetitions of actions the foal will need to know later, such as tapping on feet, rubbing with towels, and spraying water.[6] For foals that are worked with before they first stand and suckle, the procedures can have a long-lasting effect. As yearlings and two-year-olds, those that were handled show less active locomotion when separated from other horses, but they do whinny as much as nonhandled horses.[7] Improperly applied, these lessons can be problematic. Unfortunately, the wrong term has been applied to the repetitive lessons. These things have nothing to do with imprinting and everything to do with habituation.[8–10]

Introduction to Humans

Socialization is the third critical period a foal goes through and involves the process of an individual learning to recognize and accept other species, like humans or dogs. The term is sometimes applied to imprinting too, but technically imprinting refers to learning its own species. Socialization has been well studied in dogs and cats, but poorly investigated in other species. Dogs socialize to humans, horses, cats, sheep, and other species best between 3 and 12 weeks of age.[11] For cats, the socialization period is between 3 and 9 weeks of age.[12] Both of these species are immature when born and must wait for their eyes and ears to open before they can interact with their environment. At the end of this sensitive period, the life lessons shift from an interest in living beings to an interest in the environment.

Socialization is not well studied in horses or other livestock species because it is difficult to do and expensive. Since foals are relatively well developed at birth, it should be expected they are capable of learning about other nearby animal species within a few weeks of birth. However, foals living in the wild would not need to be comfortable around other species, so the complexity of the process may be quite different from that in dogs and cats. Domestication has changed this need, because horses are often around humans, dogs, cats, cattle, and small ruminants. The exact socialization period is not known for horses, but a sensitive period that results in ease of handling of foals has been found—sometime in the first 42 days of life.[13,14] Recent research in rodents has shown that alterations in brain structure and function occur after the socialization period ends that make later introductions to new species very difficult.[15] This would explain the significance of sensitive periods. Similar studies have not been done in other species, but the possibility must be considered. Once the socialization period ends, the animal can gradually learn to accept single individuals of other species, but it will remain uncomfortable around large numbers and unfamiliar individuals from those species.

Another factor that influences a foal's acceptance of humans is the behavior of its dam around people. When the mare is used to being handled and readily accepts brushing and hand feeding, her foal will be more comfortable close to and touching people. It has a shorter flight distance and more quickly accepts a saddle pad put on its back compared to foals with mothers not used to being touched.[16,17] Over time, these foals generalize their ease of handling from familiar people to unfamiliar ones too. Having calm, easily handled mares has enduring effects on the foals.[17]

Studies that looked at "imprint training" have provided information about foal socialization to people. Most studies show that these foals are somewhat more manageable, generally less reactive, and willing to approach people sooner than foals that have had limited human contact during their early life.[18]

By the time of weaning, foals that ran with their mothers in pasture without being socialized to people provide an interesting example of what it takes to tolerate people. Weanlings that are touched but not physically handled will retain fear reactions toward humans. Those that are forced to have contact by being haltered, restrained, and touched are less defensive, are more manageable, and can be touched easier. However, the results only last a few months.[19]

The human-horse relationship can enhance a horse's ability to cope with novel environments. Young horses handled fairly intensely or housed individually have lower heart rates and will approach people more readily than those that are not handled or are kept in a group.[18,20]

SOCIAL (REACTIVE) DISTANCES

Surrounding each animal is a series of invisible boundaries that affect where they are and what they do. These are social distances. Another set of distances, reactive distances, relate to how an animal reacts when approached. Both are important to horses and to the human understanding of why animals respond as they do. In theory, many of these distances form a circle around the horse, but the limits of vision to the rear of the animal make it more likely that the shape is closer to an oval and somewhat unevenly distributed around the animal (Figure 5-1).

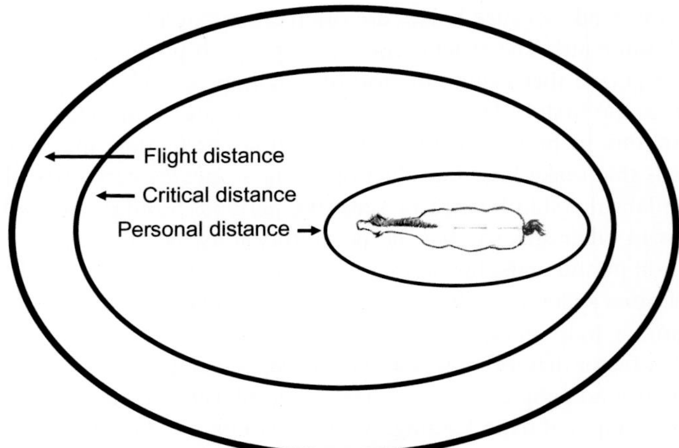

FIGURE 5-1 Three of the social (reactive) distances around a horse.

Home Range

The home range is the largest social distance and refers to all the area a horse would normally travel over a several month period. For most horses, this area is artificially limited, being a stall, paddock, or fenced pasture. When we think of free-ranging horses, this area would be much larger. The exact size will vary considerably based on several environmental factors: the location of water and grass; the presence of shelter, ravines, or rocky outcrops; yearly weather patterns; fences; and other harem groups in the area. Horses in the lush North Carolina Shackleford Banks have a home range of about 2.3 mi^2 or 6 km^2 and have an average population density of 28.5 horses/mi^2 (11 horses/km^2). In contrast, the Wassuk Ridge horses in the mountainous area of Nevada use 12 mi^2 (31 km^2) and have an average population density of 0.26 horses/mi^2 (0.1/km^2).[21] The size of the home range will also change year to year and season to season.[22] Where there are multiple groups of horses, home ranges overlap, especially around watering holes. These overlaps occur for harem groups and for harem-bachelor groups.[23] Even though the animals seem to work out a schedule so that the groups avoid each other, encounters are frequent, and the stallions work to maintain the integrity of their band.[24] The home range appears to be significant to horses. If removed and then turned loose, the horse will travel several miles to return to its home area. Whether it is the draw of the location or of the companions is uncertain, however.

Territory

Territories are the second largest invisible space. By definition, each is an area actively defended against strangers. In some species, like donkeys, the territory is a geographic space that is important to defend. If they chase an intruder away, the chase ends at the invisible boundary of the territory. Horses are not territorial, and so do not show this type of defense.[25] Although it has been said that a harem stallion might protect the area around others in his harem, it is more accurate to say he is protecting individuals, not space.[21]

Reactive Distances

When a person or animal approaches a horse, its reaction will vary by the amount of threat the horse perceives. The first *perceptive distance* (alert point) is the distance at which the horse becomes aware that something is coming. While this can vary with the type of activity the horse is doing and the position of its head relative to the approacher, the usual distance at which a horse will notice it is being approached is approximately 650 ft (200 m).[26] Once the approacher is recognized, the horse may move a step or two toward it—a short distance called the *approach distance*.

Free-ranging horses are good models to describe a few other reactions. If the strange animal continues its approach, the horse eventually will flee. The distance between the intruder and horse when it starts to move away is called the *flight distance* (Figure 5-1). The actual distance between the two animals will vary somewhat based on previous experience, the individual horse's temperament, and weather conditions, but movement away would start when the stranger is approximately 450 ft (146 m) away. The flight distance to a person approaching is considerably less, about 55 ft (16.5 m) if the horse has previously been near people.[26]

The flight distance can be used in a controlled manner to move a band of wild horses in a desired direction. A slow approach causes the animals to slowly begin to move away, keeping the flight distance as a minimum space of separation. A step backward will cause the horses to stop because that step puts the person just outside the horses' flight distance. As the person moves forward again, passing just into the flight distance, the animals will begin moving away. Controlled movement continues by taking a small step directly toward the animal just into the flight distance and then stepping back again to release stimulus pressure on the animals. Shifting forward and back will result in the desired movement.

Once the horse begins to flee, it is potentially dangerous to run a great distance without evaluating whether the new area is safe. To prevent that from happening, there is an innate stopping point. The total distance that the horse has moved is called the *withdrawal distance*. While the flight distance remains relatively constant for each individual, the withdrawal distance does not. In most situations horses withdraw 160–325 ft (50–100 m) if there is no pursuit.[21,25] This distance tends to lessen if the stranger approaches slowly, but the horse goes faster and farther when the stranger uses an active approach, such as when a person swings a rope, waves their arms, or walks fast.[21,27]

If for some reason, the horse is not immediately aware that it is being approached or if it is cornered and unable to flee, the stranger may pass the flight distance and reach another invisible zone around the horse—its *critical distance* (Figure 5-1). This is the point where fleeing is no longer an option and the horse attacks instead. In most cases, the attack is used to gain enough space so that escape is possible, but the severity of the attack increases if the horse perceives a threat of harm.

If the animal or person approaching is not threatening to the horse, the *personal,* or *individual, distance* becomes important. It is the 10–15 ft (3–5 m) of space immediately around the horse into which it allows "close friends."[27] Just as people are not comfortable having a stranger walk up and stand nose to nose, animals avoid that too. Taming individual horses allows humans to enter that personal space. The perimeter of the personal distance is also the distance a horse will maintain to avoid being caught by a person carrying a halter or bridle.

SOCIAL ORGANIZATION OF FREE-RANGING AND WILD HORSES

The social nature of the horse is important for developing of motor skills, gathering environmental information, and conforming socially.[28] For this, free-ranging horses have developed a complex social system quite different from that of horses that interact closely with humans. Studying free-ranging horses provides the best understanding of natural equine behavior. A herd of horses is a loosely affiliated group of individuals that live in a large geographical area with overlapping home ranges. The overall population is approximately half male and half female, and the makeup of a herd is approximately 17% foals, 12% yearlings, 12% 2-year-olds, 10% 3-year-olds, and 52% 4 years and older, or about three adults to every one juvenile.[21,24,29,30]

Within a herd, most horses live in small bands or harem groups. These groups consist of a stallion, about three mares, and their foals, yearlings, and 2-year-old offspring. The total size of a band will vary but usually consists of 4–17 individuals.[21,23,30,31] In areas where young stallions are artificially removed, the size of each band can increase, even exceeding 30 horses.[21] The number of mares varies between one and eight, depending on the number of stallions present and population density.[30] Social activities within a band tend to center around the mares, while the stallion defends the harem members and retains a patriarchal position. Adult membership within harem bands is relatively stable. Although a stallion may occasionally add an adult mare or a mare might leave or die, mares that stray are herded back by the stallion, strangers are not commonly admitted, and approaching stallions are chased away. An occasional bachelor stallion will accompany a harem group and even be integrated into it by forming an alliance with the dominant stallion.[32] The bachelor might breed a few mares, but his primary role is to help protect the group.[33,34] The presence of a subordinate alliance stallion also reduces overall foal mortality, so more live past the difficult first week.[34] Multiple-male harem bands tend to be larger than single stallion ones, but alliances typically do not last more than several months. When an alliance lasts longer, the subordinate male is much older than the harem stallion—a stallion well past his prime.[33]

In an effort to control population size in free-ranging horses, immunocontraception is being tried. One of the unintended consequences is the disruption of harem groups.[35] During the nonbreeding season, contracepted mares are more likely to move to other social groups than are noninjected mares. This can disrupt social ties and normal social functioning.

Stallions use specific behaviors with their mares such as moving around the periphery as members graze.[36] In casual movement, the stallion is more likely to be at the back of the group. *Herding* is used to control the direction and movement of the mares by driving them using a lowered head posture with ears back.[37] This posture also involves the side-to-side movement of the head, called *snaking*.[21,36,37] *Harem tending* refers to behaviors used to maintain the harem, including both recruitment of mares and defense by a stallion positioning himself between a perceived threat and the mares. Also included within this definition is the behavior where the stallion covers mare urine and feces with his own and defecates on specific cumulative spots used by stallions, called *stud piles*.[37]

Free-ranging stallions engage in more social interactions and remain closer together than do domestic stallions kept in pastures.[38] On the range, stallions stay in bachelor groups consisting of two to eight individuals until they are 5 or 6 years old.[21,23] This results in a slowing of their maturation and prolonging of the development period before they establish their own band.[39]

Life expectancy is influenced by a number of things, particularly the environment. Mortality occurs most often in foals and older horses. Estimates as high as 25% mortality of foals can happen in the first year. The age of death of adult horses is quite variable and exact figures are unknown. Some horses, particularly lone males, can live 18–25 years.[24,30,40]

Social Orders

Within a group of horses, members establish a social order, commonly called a "peck" or "dominance" order. This provides a stable but subtle ranking between individuals, minimizing the need for aggressive actions in competitive situations.[41] Threats replace physical aggression if a lower ranking horse does not respond at first.[21] Monitoring threats is a more accurate predictor of social ranks between horses than is fighting.

The social order within a band of free-ranging horses is linear-tending. That means there is a trend toward a numerical progression from the highest-ranking individual to the lowest member. In larger groups, there can be shared positions or even a triangular relationship. These typically occur in the middle of the social order.[21,42,43] The harem stallion may be, but is not necessarily, the highest-ranking member.[24,43,44] In part, this is the result of time spent guarding the edges of his group. It also seems to depend on the stallion's length of time within that band.[31] Rank order is the most obvious in interactions between males and in horse groups where no stallion is present.[21,45,46]

In free-ranging mares, the social order can be difficult to determine and is considered inconsequential in many of the bands. The existence of a female social order has no known advantages. Reproductive success is not dependent on it, and aggression does not increase during food shortages. Rather than getting access to food by social ranking, getting to it first seems to be the method of choice, a behavior called *scrambling*.[33,47] In bands where rankings exist, they are established by the mares flattening their ears, not by actual biting or kicking.[33] Aggression can be costly; threats are safer. Mares rarely kick other mares—the opposite of what occurs with tamed horses.

The relative rank of individuals within harem groups does change, with that between mares changing as often as every few weeks.[33] In others, the position in rank increases with age as the older, higher-ranking horses leave or die off.[30,48] Foals find their rank within a band by approximately 6 months of age and retain that position as long as they remain in that band.[21]

Cooperative interactions between mares are rare, except for mutual grooming during fly season. The exception to this is when a new stallion takes over an existing harem group. Mares not being harassed by the stallion will come to the defense of those that are.[33]

Although membership in bachelor groups is fluid, changing several times throughout the year, the relative dominance-subordinate relationships are linear and well defined.[21,46] The social rankings are not related to age, weight, height, or aggressiveness. Not all young stallions associate with a bachelor group, but those that do go through an "initiation" during which they are sniffed, nipped, bitten, chased, and mounted before being accepted. Individuals that are at least 3½ years old do not seem to go through the initiation.[33] The highest ranking bachelor stallion takes on the role of leader when stressful situations occur. All the stallions show restlessness, but only the leader has a significant increase in stress hormones.[49]

Intergroup dominance hierarchies occur between individuals within the larger herd. This is regardless of the subgroup type in which the horse normally lives.[24]

Aggression

The role of adult horses in the harem band influences the development of aggressive behavior in juveniles.[50] While we tend to think that "boys will be boys," the presence of adult horses correlates with reduced amounts of aggressive behavior and increased amounts of social cohesion in the young.

Aggression comes with a high cost for horses. In addition to the risk of injuries, which can be life threatening, there is the expenditure of energy and disruption of normal behavior patterns.[30] As a result, most aggressive encounters between adult horses are not fierce. Depending on the context in which they occur, agonistic behaviors in free-ranging horses vary from subtle threats like a stare, to fierce contacts with teeth or feet.[23,37] In a free-ranging herd, bite threats are shown by higher-ranking horses toward lower-ranking ones; however, kicks tend to be defensive reactions of the submissive animals.[21,51] Stallions may show mutual olfactory exploration and fecal marking over each other's feces. This leads to the stallion moving his band away from the area.[23] Pinned back ears, an arched neck, and/or movement of the head toward the threat also represent minimal presentation. As intensity increases, the aggressor uses forceful contact with its head, neck, or shoulder. Chasing is another manifestation of the threat that usually is brief over short distances. At the extreme, the chase may last more than an hour and cover more than 1.75 mi (3 km).[37] Biting and kicking are the next advanced level of agonistic behavior, and serious injury is not common even when contact is made. Rearing does not make contact but is an elevated threat, positioning the aggressor to make contact by stomping or striking with its forelimbs. "Boxing" and "dancing" represent the most intense forms of aggression. Here the stallions are rearing and striking each other with their forelimbs (*boxing*). They also are close enough to each other to bite the head and neck of their opponent (*dancing*).

Aggression between stallions is variable. In some studies, higher-ranking stallions win more fights, but they do not show aggression more often than low-ranking ones. Conflicts are more likely to occur between the three lowest-ranking stallions than elsewhere in the social order.[52] In other studies, high-ranking horses are more aggressive, and contests are between higher to mid-level ranking animals.[53] Losers usually show submission by running away. Other behaviors that indicate submission can include the lowering of the head, ears, or hindquarters. Jaw chomping (snapping) is a submissive behavior most commonly shown by juveniles (see Figure 4-5 in Chapter 4).[37]

Within bachelor groups, aggressive interactions are not common. When they do occur, high- and low-ranking stallions are equally likely to be involved.[46] Agonistic interactions between stallions close in rank tend to be more severe than when the animals are farther apart in the rank order when a low-ranking horse submits sooner.

Aggression by a mare peaks soon after she reaches full size, correlating with the time the mare establishes her position within the harem group's social order. The frequency will progressively decrease as the mare gets older. Agonistic behaviors become more common as the number of mares increases, and they are more frequently directed toward subordinate mares with foals.[54]

Within the larger herd, harem groups frequently encounter other bands because of overlapping home ranges. When this happens, harem stallions keep their own group as a unit and move it farther from the other band. The stallions will also display posturing and threats toward each other. Approximately 50% of the time, stallions place themselves between the two groups to escalate their threat, approaching each other with heads and tails high. Defecation is followed by sniffing and/or covering of the feces with more feces. Nasonasal, nasogenital, and/or nasoanal sniffing occurs in about half of the stallion meetings and is often followed by biting or kicking.[24]

Movement as a Herd

Any group member can start the movement of a band in a different direction.[55,56] This is "movement by departure" in which one horse moves off in a different direction and is followed at first by other close associates and eventually by the rest of the group. In domestic herds, horses that initiate movement are more likely to be horses described as "bold" rather than "shy." These characteristics do not determine whether another free-ranging horse will follow or not, but the close relationship between the horses is most important.[55] Only the harem stallion will initiate "herding" in which he will drive the members of the band with a lowered head and stretched neck posture.[42,56]

Dispersal of Herd Members

Juveniles can sometimes be found within their own mixed-sex band for extended periods, but it is not until 2–4 years of age that juveniles permanently leave their maternal band. Approximately 80% of fillies will join existing harem groups, where they may encounter aggression from mares already incorporated into that group. When joining an established group, fillies prefer to move to a group with familiar mares and unfamiliar stallions.[57] The remaining 20% of fillies go with young stallions that are starting their own harem group.[58] Abduction of fillies from their maternal band by a stallion only occurs about one-quarter of the time. Emigrating fillies usually remain within visual contact of familiar terrain.

Dispersal is not forced, even for the colts. The exception to this occurs when a new stallion takes over a harem group, when approximately 40% of young males are forced out.[33] Foals not sired by the new stallion eventually disperse later than is typical (646 days vs. 426 days) and are in poor condition as yearlings.[59] Colts may remain solitary for months to years or they may join other young males in a bachelor group.[21,29]

Environmental conditions play a role in the stability of harem groups and emigration of mares to new groups. When food is relatively scarce, mares tend to change groups. Those that do are more likely to be leaving a band having a single stallion, rather than one with multiple males.[60]

Presence of Geldings Instead of Stallions

Geldings are not found under natural conditions and have introduced a new element to group dynamics. Without a stallion in the group, the geldings tend to become part of the juvenile subgroup, while the mares form their own higher-ranking subgroup.[61] Under free-ranging conditions, they would form their own bachelor group and not be allowed into any group having a mature stallion.[62]

SOCIAL ORGANIZATION OF TAMED HORSES

Groupings of horses under human care are artificially distorted. New individuals are constantly being added or removed, and the herd size and age structure is unlike in the wild. Mares with foals are kept together until the foals are weaned and then juveniles of the same sex stay together until they go into training. Stallions are isolated from mares except on a few breeding farms, and geldings are now part of pastured groups. Many horses are stalled with little direct contact with others. Separation from a group is stressful, as demonstrated by distress calls and increased agitation. A horse may normally spend about half its time out of visual contact with other horses in a pasture. However, if the horse is pastured alone, it is three times more active and spends 10% less time eating.[63] Visual or physical contact is often sufficient to negate the stress response. The need for physical contact with other horses has a stronger internal drive than that for exercise (Figure 5-2).[64] An example of the level of social bonding is shown by the young horse taken for a trail ride alone. Most become very distressed and will whinny and try to go back home. This is the reason that older, experienced horses are ridden with young ones until the youngster becomes familiar with places it will go and the types of things it will see. Most horses eventually learn to go alone, particularly if they come to trust their rider. The learning curve for a young horse trained with other horses present is not significantly different from one worked alone. However, the amount of stress experienced as measured by the heart rate is significantly lower if other horses are present.[65]

FIGURE 5-2 The need for social interaction in horses is underrated. Two mares in adjoining stalls at a horse show have managed to make contact by widening a small hole in the canvas stall divider.

Within tamed herds as well as in harem bands, subgroups tend to form based on *preferred associates*—typically two horses that spend the most time near each other.[42,66–68] Within a group, the degree of relatedness and proximity in rank between the mares strongly influences the development of these close bonds.[67,69–71] Close associations are more common between individuals in the middle social rankings than those that are either high or low in rank.[72] Foals of preferred associates preferentially associate with each other as well.[73] Although more aggression occurs between preferred associates, it is usually in the form of threats rather than contact.[42]

The relationship between preferred associates is important. In free-roaming horses, this relationship increases foal birth rates and survival, independent from all other factors such as habitat and age.[74] Even in domestic herds, individual horses will demonstrate the significance. Over half of grooming events between two horses will be interrupted if the partner is being groomed by an outsider.[75] With allogrooming and play, interventions and the preferred associates' relationship are considered to have rewarding properties essential in good welfare.[76]

Stallions, in particular, often have no physical contact with other horses. Their visual contact is often restricted too. Newer stall designs include adjoining outdoor runs or bars between adjoining stalls that allow some contact (Figure 5-3). This type of arrangement allows social interactions without serious aggression.[77] If isolated stallions are later put in pasture with other stallions, the stalled horses show more aggressive behavior than do ones that were raised in a group with other stallions.[78] When several adult stallions are introduced to each other in a large pasture setting, the frequency of aggression and threat displays decreases sharply over the first 3–4 days. Within 3 months a stable hierarchy is formed.[52,79]

Separating a horse from the group is stressful, as shown by dramatic changes in behavior, cortisol levels, and blood parameters.[80] Young horses that have been kept away from social interaction with other horses for several days will habituate to novel objects better.[81] However, horses kept in groups are more adaptable to training, have less objectionable behavior, are less reactive, and have a lower cortisol response compared to individually stalled ones.[79,82–85] Keeping young horses together provides the social experiences necessary for reduced aggression and appropriate group interactions as adults.[79,86] At the same time, it is important for the young to have adult horses present to learn other life lessons. Adult presence increases the number of new behaviors, promotes the formation of preferred partners, results in positive social interactions, and decreases the number of aggressive bouts.[86,87]

An additional benefit of social behavior is that a conspecific can have a calming effect on others. After a naïve horse watches an experienced demonstrator horse cross a novel surface like a tarp, the naïve animal has a lower heart rate when it crosses. The change in heart rate does not happen without the benefit of the demonstration.[88] This does not support observational learning in

FIGURE 5-3 Individually stalled stallions usually have minimal contact with other horses. Protected openings between stalls allows social contact for otherwise isolated individuals. Here two stallions exchange licks through an opening between adjoining stalls.

horses, however, because the number of trials to learn the behavior is not reduced. It does suggest that the calming effect might be useful in certain fearful situations. The lower heart rate is also shown in horses that are first separated from their group in pairs before undergoing solo separations.[65]

Social Orders

Social orders within groups of tamed horses are linear-tending, particularly among geldings.[53,72,89] As happens with free-roaming horses, shared positions and triangular social relationships can exist in tamed horses too, especially in larger groups.[21,90,91] Rank within a group appears to affect or be affected by body weight, but not by gender or age, although middle-aged horses are more likely to be dominant.[42,53,90,92,93] Body condition scores are higher in more dominant individuals.[92] Stallions are usually not the highest ranking horse in the domestic herd.[90] When no stallion is present in a group, social structure, herding, and defense are taken over by a dominant mare or group of mares.[21,61] The position of geldings correlates with age at the time of castration.[72]

Low-ranking horses usually defer to higher-ranking ones in association with eating, drinking, space, breeding, sequential rolling, and movement, with space being the most common.[21,94,95] A higher-ranking horse may displace a pasture mate when the two are about to enter a barn but defer to that same pasture mate when it comes to accessing a pile of hay. The perceived value of the situation can temporarily change the relationship.[21,69]

Foals tend to develop a rank similar to that of their dam. Those from higher-ranking mares are responsible for 80% of aggressive threats but receive fewer aggressions.[96,97] Birth order also has an effect on a foal's rank preweaning but not after.[21] The relative age of foals within the group does affect its rank.[98] Those from high-ranking mares suckle longer into the lactation than do foals of low-ranking mothers as well.[97] When a 2- or 3-year-old filly joins another herd, her rank generally correlates with the rank of her mother, particularly if the dam was high-ranking in the herd of origin.[90,98]

The introduction of new horses into a herd produces a disturbance until the new animal finds its place in the social order. There are several ways initial interactions determine social rank. As with free-ranging horses, the interactions can be very subtle. Increasingly severe aggression begins with visual threats and progresses from relatively mild physical contact involving head or neck bumping, to more severe biting, kicking, and striking. Resident horses are the most aggressive in these encounters.[99] Relative relationships are settled within a few days.[21] The amount of agonistic behavior occurring with new introductions is related to breed differences rather than to group composition, age, or sex. Some breeds are not highly reactive and have fewer associated injuries to horses entering their herd. Others are more likely to have injuries, usually minor, even a few weeks after introductions.[100]

Members of a herd may use the interactions between herdmates to aid their knowledge of social ranks. Horses can mirror social behaviors of dominant horses but will ignore those of subordinate ones.[101,102] Such lessons are easiest for young, low-ranking, and more exploratory horses. Older animals seem to have more trouble mirroring social behavior.[101] Once established, the social rankings remain relatively stable over time.

Perhaps as important as the relative dominance order is the "avoidance order"—where individuals avoid provoking higher-ranking ones.[103] Slightly over 10% of potentially aggressive encounters result in passive avoidance.[91] Ultimately, the social ranking, affiliations between individuals, and aversive relationships result in a complex social system.[21]

Aggression

While aggressive interactions are not common in free-ranging harem bands, they are in pastured horses. Even after being together 8 months, herdmates still have aggressive encounters. Most aggression (66%–80%) in domestic horse groups consists of threats. But the smaller the space, the higher is the amount of aggression.[84,104,105] Bites are the most common (74.7%), about half of which make contact. The second most common aggressive interaction is the head bump (8.3%), followed by kicks or kick threats (6.2%).[91] When turning horses out into a group after being stalled for part of the day, aggression can be reduced by providing access to grass or roughage such as hay or straw.[106]

Different levels of aggression between managed horse groups and free-ranging ones can be explained by group stability, dominance hierarchy, and learned appropriate social skills while young in roaming horses.[107] Domestic horses that stay within one group have a lower risk of injuries from agonistic interactions or play.[86,105,108] The amount of aggression in groups composed of mares and geldings is no different from that in groups of a single sex.[86,105]

Bachelor stallion groups of domestic horses are seldom discussed relative to behaviors, because group members usually remain together only from weaning until their training begins. An ethogram has been developed for these bachelors to pull together agonistic behaviors previously described by others.[109] Having a single reference can be helpful.

Potential for injury is a common reason owners keep horses separated. Introducing two horses carries the risk that one will be hurt. This is less likely when horses are first introduced in adjoining stalls and then in adjoining paddocks. Doing so does not reduce the total level of aggression, but contact aggressions, including bites, are lessened. Bite threats by stalled horses correlate with contact aggressions in a paddock.[110] If horses are being introduced in a paddock instead of in adjoining stalls, larger paddocks are associated with fewer agonistic interactions.[111]

Relations With Other Equids

Donkeys, mules, and horses are often discussed as if they were slight variations of each other. As previously discussed, donkeys and horses are distant relatives with significantly different genetic makeups, and the mule is the resulting hybrid. Mating behaviors of donkeys and horses are quite different as well, so the creation of a mule requires considerations different from breeding horses to horses or donkeys to donkeys. When pastured together, each equid subspecies will preferentially form their own affiliative group. Of the three, horses tend to be the highest ranking in a common social order, with mules second and donkeys third.[112]

PLAY

Play is a complex series of behaviors with a number of functions. There are locomotor and affiliative functions that prepare young horses for the social world they will live in and prepare their neuromuscular structures for adult life. Playful activities are also balanced between affiliative and dominance behaviors. It is likely there is also an "ethological need" for play expression.[76,113,114]

The significance of an ethological need is that the specific behavior is important to long-term welfare, and the animal will show signs of stress if it is prevented from occurring.

In any species, individual and social play behavior occur most in young animals. The number of playful interactions in horses tends to peak between 26 and 38 months and then decrease with age.[115] The various play patterns can be divided into three or four categories: locomotor, social or interactive (sometimes subdivided into play fighting and sexual play), and object play (Figure 5-4).[116,119]

Play is first seen as self-play and playful interactions with the dam. After the first few months, social play with other foals and yearlings is added.[117,120] Play accounts for most of the foal's exercise, including two-thirds of its running and 95% of the high-speed turns.[118]

As colts and fillies get older, there is no difference in the amount of running, bucking, and playing with inanimate objects between them. Both also show mounting play. There is a difference in that fillies show more solitary running and bucking, compared to interactive play bouts by colts.[116,117] Gradually, foals find preferred play partners that are about the same age and sex. Colts will engage in more play bouts than do fillies, and half of these bouts involve play fighting unrelated to social order (Figure 5-5).[38,96] Play fighting is common after mutual grooming between the peers.[121] By 2 years of age, a stallion-only group will have approximately one play bout per horse per hour. While play is associated with young individuals of the same age, it is still important for juveniles to have contact with adults in mixed groups to provide the other lessons needed in social learning.[86]

Play is not common in adult horses, especially free-ranging animals. When it does occur in pastured horses, it happens less in all female groups than in all gelding or combined mare-gelding groups.[86,105,122] The type of play is usually associated with motor activities, such as running and bucking.[116] When play is shown by adult horses, it is more common in individuals that fit a stressed profile.[122] Why this happens is unknown.

PROBLEMS RELATED TO SOCIAL BEHAVIOR

Horses are social and need access to conspecifics. When such contact is limited or prevented, behavior problems happen. The social behavior problems discussed here represent social structure gone awry.

Aggression

With the possible exception of harem stallion challenges, aggression in free-ranging horses is rarely serious. These horses are more likely to use subtle threats to get their message across. So why are owners concerned their horse will get hurt if it is in a pasture or paddock with other horses? The rate of injuries

Locomotor play
- Bucking (also called *kicking up, kick out, rear kicking*)
  - Weight is shifted to the front limbs while both rear legs come off the ground and are simultaneously extended backwards several times in rapid succession
- Cavorting (also called *frolicking, gamboling, capering, prop*)
  - Both front and rear legs are simultaneously come off the ground, with bucking, head shaking, and body twists. This may be followed by a sudden gallop
- Chasing (also called *charge, race*)
  - One foal pursues another in an apparent effort to catch or overtake the leader
- Jumping
  - Similar in appearance to a horse jumping over an object, the foal pushes off with both rear limbs in a forward trajectory
- Leaping
  - Fore and rear legs simultaneously propel the horse straight up off the ground in a stiff legged motion. Similar to the "stot" of antelope
- Prancing
  - Walking or trotting with an exaggerated knee action
  - Accompanied by arched neck, ears forward, and elevated tail
  - May snort. Often occurs at end of play bout
- Running (also called *galloping, exuberant galloping*)
  - Spontaneous cantering or galloping for no apparent reason
- To-and-from (also called *runs in opposite directions*)
  - Runs to and from a specific target horse or object

Social (interactive) play
- Play fighting
  - Evasive jumping
    - Part or all of the body is propelled off the ground away from the approaching horse's offensive move
  - Nipping/biting/grasping
    - Forelimb nipping/biting/grasping
      - Nips or bites are directed toward the forelimb of its playmate, which then may drop to its knees (called *kneeling* or *turtle posture*)
    - Head/neck/chest nipping and biting (also called *neck biting, face nip*)
      - Nips or bites are directed toward the playmate's head, neck, or chest

FIGURE 5-4 Play is a common behavior in young horses. The ethogram identifies three or four major categories with a number of behaviors under each.[96, 115–119]

Continued

serious enough to require veterinary care is not particularly high—about 1.7% are kick or bite injuries.[123] Of those, 18% are associated with changes in housing management. Breed makes a difference as shown by the risk of injuries, which is 4.3 times higher in Thoroughbreds, Arabians, and Warmbloods than in horses of other breeds.

Knowing what management practices are likely to result in aggression allows a person to make modifications to minimize the likelihood of injuries.

Hind leg nipping/biting/grasping
Nips or bites are directed toward the playmate's hind limbs
Rump nipping or biting
Nips and bites are directed at the rump of a cohort, with it often bucking
Kicking (also called *hindkicking*, *hindquarter threat*, *kick threats*, *kick out*)
The rump is turned toward the playmate and one leg raised as if to kick. The horse will simultaneously back toward the target (Kicking with both rear limbs usually ends the play session instead)
Neck grasping (also called *grip neck/mane*, *holding mane*, *neck grip*, *holding crest*, *grasp*, mane grip, *biting*)
Jaws are closed onto the dorsal aspect of a playmate's neck
Neck wrestling (also called *neck fencing*, *wrestle*)
Sparring with the head and neck; may involve pushing with the shoulder
Pushing (also called *bump against*, *push and bunt*)
Pressing of the upper body against a playmate in an apparent attempt to displace it
Rearing
Weight is supported on the hind legs with the front legs lifted off the ground so that the horse is nearly vertical
Stamping (also called *strike*, *front hoof beating*, *paw*, *stomp*)
One front leg is raised and sharply lowered to the ground
Swerving (also called *evasive balking*, *quick stop*)
As two horses approach each other, one abruptly stops and turns its head, neck, and chest to the side away from the approacher
Whirling (also called *evasive spinning*)
The body pivots approximately 180 degrees away from a playmate's offensive forward move

Sexual play

Marking
Voided urine/feces is sniffed and covered by urinating/defecating on top of it. The foal sniffs again and shows flehmen. The sequence is similar to that of adult stallions
Mounting (also called *sexual mounting*, *sex without coition*)
The foal rears and rests its chest or forelimbs on the back of another horse. Orientation can be from either side or rear. Sexual arousal is not apparent, even if there is an erection. An abbreviated form is just the resting of the head and chin on the other horse
Teasing

FIGURE 5-4, CONT'D

The smaller the space available, the greater is the incidence of aggression, so pastures are better than paddocks, which are better than corrals.[107] Resources like forage that simulate natural grazing conditions reduce aggression regardless of available space. Group composition also affects aggression. The presence of estrous mares, mares with foals, or new introductions into the herd

The juvenile sniffs/nuzzles another horse as a stallion would but not always in the same order. Flehmen usually occurs

Object play

Carrying (also called *drag*)
The horse pulls an object along as it moves while holding in its mouth
Chewing
The foal chews an object taken into its mouth. Head tossing may occur. The object usually falls from the mouth instead of being eaten. The dam's mane and tail are common targets
Circling (also called *running in loops*)
The foal moves in a circular pattern around an object, often the dam
Dropping or tossing
An object that has been picked up or mouthed while on the ground is released or tossed by an upward motion of the head
Kicking up
The rear is raised a few inches by a hopping motion while the rump is pointed toward and often touching the herdmate target. The hind legs do not extend backward
Mouthing (also called *manipulate by mouth*)
Part or all of an object is taken into the mouth. This is often followed by *lifting*, *dropping*, *tossing,* and/or *carrying*
Nibbling
The upper lip moves up and down the object, usually without biting it
Pawing
The horse moves a forelimb in a back and forth motion on or near an object. It may move the object. This is common in water and may initiate play
Lifting (also called *pick up*)
The object is held in the mouth and lifted off the ground
Pulling (also called *drag*)
The object is held in the mouth and dragged away from its original position
Resting rear
Similar to mounting from the side, the foal's chest and one or both forelimbs rest across the back of a playmate
Shaking (also called *swing head*, *wave about*, *scrape along ground*)
Once an object has been picked up by the mouth, it is moved in an up-and-down, side-to-side, or circular motion
Sniffing/licking
The object is sniffed or licked as if to take in the odor, texture, or taste

FIGURE 5-4, CONT'D

increases the potential for aggression. So do large groups. As an example, one group of 13 Przewalski's stallions showed an aggression rate of 1.46 per horse per hour, but the rate was 0.76 per horse per hour in a four-horse stallion group.[38,107]

Discussions about aggression are extremely variable, especially as they relate to categories. Interpretation of findings of various researchers without

FIGURE 5-5 Play fighting by yearling colts involves rearing and biting. The roughness of their play makes it potentially dangerous if they try to play with a human.

in-depth knowledge of the specific criteria they were studying can be difficult. The following classification system is based on the external drivers of the aggression and includes a description of what those criteria are.

Dominance Aggression

The use of the word "dominance" has recently gone out of favor with many behaviorists, being replaced by "status-related." Dominance aggression is that shown by a high-ranking individual toward a lower-ranking one in the context of a reminder that it is lower ranking. A pony gelding is pastured with an older gelding. The older horse is tolerant of the little pest as it frequently eats within inches of the old gelding and occasionally even swings its body to displace the older horse. Then one day, the older gelding grabs the pony's halter by the nose-piece and leads the squealing pony around the pasture for several minutes. Thereafter, the pony respects the personal space of the older gelding and yields to him in close situations.

Free-ranging stallions from different herds have a very ritualized type of aggression when they first meet.[30] As they leave their harem group, the approach is done with heads and tails held high. Defecation and sniffing of feces is ritualistic. In most encounters, one stallion will then turn away and dominance is settled. If the stallions are closely matched, body sniffing is the next phase, beginning nose to nose and progressing to naso-genital and naso-anal investigations. While side by side, they begin biting, and then striking, kicking, and rearing until a loser moves off.[30]

There are occasions when a lower-ranking animal will attempt to climb the social ladder, particularly if the intended victim is older and less likely to be able to fend off the challenge. This is most likely to happen in harem groups where an associated male finally drives off the harem stallion to take over the band, and in groups where the middle-aged mares are the highest ranking. These aggressions are related to the relative status between the two horses involved—thus the newer name "status-related."

There are situational differences is social rank, and thus differences in which horses will aggress toward others. Food access can have one ranking, while social dominance is different.[124] Determining social orders is complicated by disruptions, especially due to estrus, introduction of a new horse, or presence of a foal.

Equine dominance affects humans in a couple of different ways. First, people can be injured when interfering with dominance-related aggression between horses. More often, injuries occur while working with a horse that has a dominant personality and shows resistance to performing a previously learned behavior. Resistance can be shown by strikes, kicks, or bites. Occasionally it can be bucking, rearing, or bolting.[124] The biggest offenders of social order are yearling and two-year-old stallions. This is the age when they would normally try to find their place in a harem group's social order. Occasionally, older, trusted stallions will severely bite a person, doing serious damage.

Castration of a dominant, aggressive stallion reduces or stops the behavior.[124,125] Surgery is not always an option. In that case, it becomes very important that the offending horse learn appropriate equine manners. Techniques involve getting a mechanical advantage over the large animal using halters, bridles, and stud chains on lead ropes. Then appropriate and desired behavior is rewarded. Punishment of undesired behavior can be useful for some lessons, but can be problematic for others. Hitting with a fist or "swatting" often leads to the horse treating it as a game and becoming skillful at stealth bites. Excessive punishment can trigger pain-induced or self-protective aggression instead of stopping the problem.[124] A firm "thump" to the forehead with an open, flat hand is a reminder to "stop what you are doing and pay attention."

Fear-Induced Aggression

By definition, fear is the feeling of apprehension from the nearness of an object or situation. While we cannot ask a horse if it is afraid, body language strongly suggests that horses experience fear. As a prey species, escape is the primary method to respond to frightening stimuli. If escape is blocked, a frightened animal may show aggression until it can get past the blockage. A horse's usual reaction to a new object is alertness, followed by stopping to investigate the object.[124] If in doubt or if the object moves, the horse will shy or bolt and run. However, if given time, the horse is likely to hesitatingly approach and sniff the object. Only when there is a trusting bond between horse and rider can a rider's coaxing encourage forward progress. More often, a rider tries to hurry the investigative process, and this can cause the horse to panic and try to get away instead. The response could be pulling, rearing, kicking, or biting. Similarly, horses often do not want certain body parts touched, particularly their legs. While this may signal play for yearling colts, minimally handled adults are guarding their mechanism of escape. Using small rewards and gradually working down, then up, then down a little farther in a desensitization protocol is a useful approach for teaching them to allow human touch.

Idiopathic Aggression

Idiopathic means the cause is unknown. In many cases, the term is used because there is too little information about precipitating events to allow classification. This is particularly true when undiagnosed medical problems are involved. Other times, the picture of what happened is complex and difficult to sort out, so it is easier to call it idiopathic. Occasionally there are unique happenings. As an example, there are reports of horses that appear to deliberately go after, kill, and even eat small animals from baby chicks to coyotes, squirrels, calves, and a variety of other wild animals.[126,127] The behavior is difficult to classify, and it may relate to play, irritation, or even protection. Many episodes cannot be readily explained.

Intrasexual Aggression

Aggression between horses of the same gender that occurs only because of a dislike of others of that gender is classified as intrasexual aggression. This is most common between intact males in many species.

Intermale Aggression

Intermale aggression is the type of intrasexual aggression occurring between free-ranging stallions when the harem stallion reacts to approaching stallions. Intermale aggression is basically a behavior of "you look male, you smell male, you act male—I hate you." It is not related to territory or to protecting mares and foals. It is triggered by the physical presence of a male that the stallion does not want around. Stallion encounters tend to be ritualized and often end before serious aggression is reached. The pattern begins with a stare, and is then followed by body posturing, olfactory investigations, squeals and forequarter threats, and fecal displays.[124,128] Biting, rearing, and kicking follow if the early threats are insufficient to deter the intruder. Intermale aggression is driven by testosterone, so castration of one or both stallions significantly reduces the problems, although intermale aggression can occur between geldings too. Castration before fighting begins reduces the behavior expression in 70%–80% of horses. Castration after it begins is only 40% successful.[124,125]

Interfemale Aggression

Interfemale aggression can occur, although it is less common than between males. In this case, the mare aggresses toward another mare regardless of the location or estrous state.

Irritable Aggression

Pain is associated with increased irritability, reduced tolerance, and increased aggression.[129,130] This is true in humans, and it is true in horses and other animals too. Finding the source of the pain can be a diagnostic challenge. Lameness, sore backs, retained caps, abdominal discomfort, urinary infections, and reproductive issues have been implicated. Because irritable aggression can relate to many and diverse causes, a detailed history with videos of the horse's normal and abnormal behavior can be helpful in making an appropriate diagnosis.[131]

Irritable aggression is displayed by the "cranky" mare. Here the complaint may be difficulty in training, kicking, squealing, tail swishing, excessive urinations, hyperexcitability, and general crankiness. The behaviors are synchronized with the estrous cycle in approximately 50% of mares, so differentials include painful ovulations, vaginitis, cervicitis, and endometritis.[132] These mares usually show hypersensitivity in the flank, hindquarters, and abdomen. Performance misbehaviors include kicking (57%), bolting (29%), refusal to move forward (29%), tail swishing (36%), and back stiffness (43%).[132] Caslick surgery may need to be considered even when ovarian cycles are not implicated.[132,133]

Irritable aggression is not always associated with physically uncomfortable conditions. A horse that is around a dog that barks constantly can become irritated enough to attack the dog. Sharp noise is irritating to the sensitive ears of livestock species, and the horse is likely to try to attack the source of that irritation.

Learned Aggression

Aggression, or its milder form of not deferring, can be shown toward humans. While most horses do not intentionally try to dominate humans, their size can be intimidating and cause less confident people to yield to the horse. As rapid learners, horses that rear, kick, bite, or threaten will quickly figure out that doing so results in the person backing off and leaving the horse alone. While the initial behavior may have been accidental or fear-related, the resulting response was interpreted as negative reinforcement. Through trial and error, the horse learns that it can control the situation. As an example, the horse does something the rider does not like. Being mad, the rider spurs the horse to go forward but holds the reins tight to prevent it from moving forward. In the confusion, frustration, or fear, the horse responds by rearing. This frightens the rider, who then dismounts. The horse quickly learns that the rider will dismount if it rears.[124]

Some lessons of control are easily taught to a horse while leading it. Moving in a clockwise arc teaches the horse to yield to the person.[134] Teaching the horse to back makes it do a difficult, unnatural motion while yielding to the person in control. Eventually, mild jiggling of the lead rope should result in the horse backing and getting out of the human's personal space.

Biting at people and equipment while in crossties is a problem that horses use to get attention or perhaps to be taken back to their stall. Instead, the person should reinforce the desired behavior of standing quietly, not the biting, pawing, or other unwanted behavior. The horse can learn what is appropriate. A food treat of a piece of carrot, horse snack, or small clump of hay is given only when the horse stands quietly. Gradually, the duration of the quiet standing is lengthened before the reward is given.[135]

Horses that turn their rear to a person entering their stall are a special concern because of limited space to avoid danger. Teaching the horse to face the person can occur through two opposing protocols. In the first, a whip or broom is used to tap or swat the rear while the person stands out of reach if the horse kicks. The tapping/swatting is stopped when the horse shows the slightest turning of its head. When the horse turns back to the original position, the tapping resumes. Each turn in the positive direction gets negative reinforcement, first for the head, then front end, and finally the entire body. The concern here is the high possibility of abusive hitting, especially if the person is mad at the horse. Positive reinforcement uses human body language and is usually a better choice. The person begins a very slow approach toward the horse. As soon as the horse turns its head slightly, the forward motion stops and the person takes a small step backward. When the head returns to its starting position, the slow approach begins again. Each time the horse's motion progresses toward facing the approaching person, the horse is rewarded with a stop and step back. Once it faces the person, it should be gently rubbed and talked to before a halter or bridle is put on. The procedure needs to be repeated several times without catching the horse too.

Medically Related Aggression

While aggression associated with pain and irritability might be included in this category, medically related aggression usually involves a specific medical problem. Abnormal hormone levels such as with granulosa cell tumors prove the point. An estimated 31% of mares with this tumor show aggression.[136] Other causes of high testosterone levels in mares are also associated with aggression and stallion-like behavior.

Pain-Induced Aggression

Pain is something most animals try to avoid, although it is not always possible. The usual reaction to chronic pain is less interest in the surroundings and blocking out normal stimuli. Horses may show irritable aggression as a way to try to fend off interactions. One reason so many horses become needle-shy is the association of the procedure and resulting pain. Which detail the horse focuses on to make the connection varies by individuals. It can be an odor such as from an alcohol swab, or the three taps before the needle passes through the skin. It could be the sight of a syringe and needle or the color of the overalls worn by the veterinarian. Using a nose twitch only before painful events connects the appearance of a twitch and pain. Prolonged twitching also makes a pain-twitch connection. If used, the maximum time a horse should be twitched is 5 min.[137] It is also advisable to gently rub the upper lip before and after a twitch is applied.

Certain techniques can be useful to help prevent development of needle shyness or enable working with a horse that is shy of needles. Little things matter. Covering the eye on the side of the injection prevents the horse from seeing what is going on by its neck. Using smaller gauge needles when possible is helpful as well. Sharp needles will slide through the skin so stabbing them in is not necessary. Thumb pressure on the site of the injection will help desensitize the skin to the injection that immediately follows. Odors and tapping can alert the horse that something bad follows, so they should be avoided.[124]

Desensitization and counterconditioning work well to stop needle-shy horse reactions. The first step is to find a high-value food treat. In the following desensitization, the goal is to be able to do a jugular stick. Desensitization begins by touching the neck area with the flat of a hand. Calm is rewarded with the food treat for this and each of the following desensitization steps. This gradually progresses from the hand to a four-finger touch. Then it goes to three fingers, two fingers, and finally to a single one. Next is the progression from a light touch, to firm touch, to light stroking of the jugular groove, to light stroking while holding off the vein briefly. The vein is held off for longer periods. The process then goes back to the beginning, but this time a capped needle and syringe are used to touch the neck, progressing to a hand stroking the jugular groove followed by the capped needle touching the area. Eventually the uncapped needle is used for the venipuncture. Desensitization will require repeating the procedure several times, but if done correctly, each trial will go faster.

Play Aggression

Aggression during play is normal, especially for colts. This is part of normal physical development and learning for future fights as adults. People will sometimes encourage the behavior without considering the significance of what happens as the horse gets older. While horse-horse aggressive play is generally short in duration, humans can artificially extend the play period by encouragement. Even though the behavior is play, it can still do damage, particularly if it is directed toward something smaller than the horse, such as a dog or person. This is a good reason to keep similarly aged weanlings and yearlings together, so they can play fight with an animal of their own size (Figure 5-5).

Human Directed Aggressive Play

When horses try to play with people, they use equine-specific behaviors. Foals that have been handled before they were weaned are at a lower risk of aggressive play toward humans, but it does not totally stop the problem.[138] Colts, in particular, are the most likely to play aggressively, but fillies may as well. Aggressive play sessions commonly start if the colt's legs are touched. In this mock fight, the young stallion may drop to his knees and try to bite. Many people try to punish the behavior, which usually escalates the attempt to play rough. Instead, desensitization, perhaps with counterconditioning, will be much more successful than thwarting the play. The concept of desensitization is to begin rubbing or brushing areas of the body where the behavior is not elicited and gradually work into problem areas. The lesson typically does not happen in one day, but with consistent daily applications, most horses will quickly learn that good things happen if it stands still. For safety, the horse should be restrained so that it can only move about a foot (0.33 m). Brushing begins at the neck or shoulder, gradually moving closer to the legs, but stopping before there is a reaction. Then the grooming is gradually worked back to the starting place. The downward brushing is repeated, moving a little farther toward the legs each time. Each "advance" during which the horse stands still is rewarded with a small treat before going into the "retreat" phase of the lesson. The goal is for each successive advance to move slightly closer to the leg before the retreat phase. It is particularly important that each advance be small enough to not trigger the undesired moving or biting behavior. Because play biting has its own internal reward for the horse, triggering the unwanted behavior undermines the goal of desensitization.

Protective Aggression

Protection of things like food, foals, and self are normal, instinctive behaviors. They become problems when people fail to recognize what to expect from the horse under specific conditions. Assessing how horses are approached and managed is helpful to avoid aggression.

Material (Food) Protective Aggression (Resource Guarding)

Probably the most common type of aggression horses show relates to protecting a food source. This can happen when they have food nearby, when they are going from the pasture into the barn to eat, or when a person approaches with food.

If relatively unfamiliar horses are fed in a pasture or paddock, three factors influence the amount of aggression shown—the distance between horses, height of the feeder, and the relative number of feeders per horse. Significantly less ear pinning or other distance increasing signals are shown if horses are at least 33 ft (10 m) from each other, the feeders are on the ground, and there are at least 1.5

feeders/horse.[139] The animals can also be taught to go to specific buckets hung on a fence or feeders on the ground to help reduce their negative interactions. Distance protects the horse's individual (personal) space and reduces the need for vigilance. Extra feeders allow horses that are displaced by higher-ranking individuals to find another place to eat. Even for horses that have a well-established social order, the distance between horses remains a factor in the threats, but height and number of feeders do not.[139] When feeding horses in the pasture or field, it is good practice to watch the interaction of individuals to be sure each horse gets its share. Low-ranking horses may be so timid that they either need to be fed far from the others or in a stall or separate paddock to prevent bullying.

For horses that go from a pasture into a barn for their meals, individuals quickly develop a specific order in which they go through the gate or barn door. If one animal is removed from the group, especially if it was one of the first through the gate, disruption occurs while the group reestablishes a new order of passage. It may be necessary to hand lead specific individuals or force some back to prevent injuries.

Stall-fed horses often greet the person carrying their food with pinned ears. What does the person do next? They reward the bad behavior by dumping the food into the feeder. The behavior can be changed to an ears forward or neutral one by not rewarding the undesirable flattened ears. When approaching the stall, the person should stop and take one step back as soon as the horse pins its ears. When the ears come forward, the movement toward the stall resumes—stop and back when pinned, forward motion when not. If the ears are neutral or forward when the stall is reached, the horse receives approximately one cup of feed as its reward.[140] The reason for the limited amount of feed is that the training can be repeated several times each day until the horse "gets it." Overfeeding is also prevented.

Maternal Aggression

Mares are protective of their foals—a hormonally related behavior. The strongest manifestation of this occurs during the first several weeks postpartum. It then gradually wanes. Initially, the mare and foal stay close, but as the foal begins to spend more time away from its mother, the mare remains protective. She may simply position herself between the foal and an approaching horse, or she may show aggressive behaviors, particularly if the foal shows the jaw chomping (snapping) behavior.[141] The degree of maternal protection varies with individuals. The safest way to approach the mare is by keeping the foal between her and the person. Mares strongly avoid running over their own foal.[124] If other horses are in the same paddock or pasture, she might accidentally knock into her foal if the space is tight, so care must be taken to ensure enough space is available.

Self-protection

Foals depend on their dam for protection, but as they get older and humans start to work with them, some become aggressive to human touch. The ears are indicators of intent to bite or kick. Forward or neutral ears suggest the behavior is mainly associated with play. Flattened ears suggest the foal wants to be in charge.[142] Allowing people to pick the feet and brush the legs are important lessons. While the natural tendency is to say some bad words in a loud voice and jerk on the lead rope, this reaction teaches the foal not to trust humans. Instead, it is better to work in a safe position, wear appropriate protective clothing, and speak with a soft voice. The foal can be near but physically separated from its mother if she is comforting to the youngster, or they can be separated if she is bothersome. Training tolerance of nonharmful procedures is best done with the desensitization and counterconditioning using positive reinforcement.[134,142] Begin with touching the foal somewhere that it likes to be touched, gradually moving slightly away from that area. Initial lessons are short—just a few minutes—and they are repeated two or three times a day. As the foal complies, it is told "good" in a soft voice and the rubbing stopped. It is important to stop only when the foal is showing acceptable behavior, so only appropriate behavior is rewarded.[134] If biting is the problem, a muzzle can be used, or someone can distract the youngster to get the correct no-bite response for being touched. Sessions will gradually get longer and distances from the favored spot lengthened until the whole body can eventually be touched.

Redirected Aggression

Redirected aggression happens when a horse becomes agitated at something it cannot get to, and then redirects the irritation toward any animal or person close enough to be reached.[124] A horse at a livestock show becomes irritated because its schedule is changed, barn lights never go off, it is dusty, and people keep petting its nose through the stall bars. When it is not successful at biting those fingers and the flattened ears are insufficient to warn people off, the horse may redirect the irritation by biting the owner as she walks into the stall.

Orphan Foals

When a foal is orphaned, a number of problems are created. The obvious first consideration is how to meet nutritional needs. The concern after this is how to provide the behavioral lessons that will allow it to grow into a well-adjusted adult. Because of the heavy dependence on humans associated with bottle or pail feeding, foals may not properly socialize with their own species. They show a strong preference for human companionship to that of other horses because they regard and treat humans as their own kind. They may buck and play primarily when the person is near and follow them like a dog would. Foals with excessive human bonding do not react to human interactions like other horses.

They tend to be pushy and resistant to handling. The term "dull to discipline" is descriptive.[143] They also resist learning, and if pushed during training, they may become aggressive.

Foals that had early prolonged maternal separation also have problems with socialization. They remain excessively close to their mothers and do not develop the normal social relationships with their peers. As a result, social skills are impaired, and these foals show high levels of aggression toward others their own age.[4]

Nonnutritional sucking is a common problem with orphan foals. Suckling of the teat is the behavior programmed into newborn animals, and it is the default behavior until weaning occurs. Normal has been compromised for orphans, even for those presented with nurser bottles. As a result, orphan foals will suck on a number of objects: other foals, fences, clothes, or their own legs or tongue. Human pacifiers are sometimes used to direct the sucking to a more acceptable target.[1] Most foals eventually outgrow abnormal sucking behavior, but some do not and default to it in times of stress or relaxation. Creep feeding is helpful and should be started early. Fresh feces from a healthy adult horse needs to be provided for the normal foal coprophagy.[1]

The earlier precautions are taken to minimize the future problems for orphans, the fewer are future complications. Minimizing human interactions and maximizing equine contacts should be the goal. It is best to dissociate humans from nursing, so a foster mare or nanny goat is ideal. When that is not possible, providing a continuous supply of milk in a tub might be an alternative.[1] A technique that allows meal feeding rather than continuous milk is to deliver the milk into a bucket remotely through an angled PVC pipe or hose. Foals also need "horse time." This can be accomplished with kindergarten foal groups or by pairing the orphan with a foal-safe older mare, gelding, or a pony.

Older, poorly socialized horses are an even bigger challenge. They should be placed in a pasture with accepting horses for an extended period of time.[143] They still revert to the human preference when people are around, but eventually some of these horses learn how to be around other horses.

Separation Anxiety

Separation anxiety is not just a problem with dogs. As a social species, horses can show behavior changes related to the absence of a herdmate. The seriousness of this problem is dependent on the strength of the bond between the individuals, the dependency of one on the other, presence of other horses, and length of the separation. While fence walking or the increased expression of other stereotypies is a common response, some horses will loudly neigh, pace the stall or fence line, respond poorly to commands, and perhaps attempt to escape. Others can show aggression.[42,144] Aggression and inappetence are more likely to happen if the horses had been particularly close and the separation is for several days or longer.

REFERENCES

1. Grogan EH, McDonnell SM. Mare and foal bonding and problems. *Clin Tech Equine Pract* 2005;**4**(3):228–37.
2. Houpt KA. Formation and dissolution of the mare-foal bond. *Appl Anim Behav Sci* 2002;**8** (2–4):319–28.
3. Jensen P. Parental behavior. In: Keeling LJ, Gonyou HW, editors. *Social behaviour in farm animals*. New York: CABI Publishing; 2001. p. 59–81.
4. Henry S, Richard-Yris M-A, Tordjman S, Hausberger M. Neonatal handling affects durably bonding and social development. *PLoS ONE* 2009;**4**(4). https://doi.org/10.1371/journal.pone.0005216.
5. Krohn CC, Boivin X, Jago JG. The presence of the dam during handling prevents the socialization of young calves to humans. *Appl Anim Behav Sci* 2003;**80**(4):263–75.
6. Miller RM. *Imprint training of the newborn foal*. Colorado Springs: Western Horseman; 1991. p. 137.
7. Durier V, Henry S, Sankey C, Sizun J, Hausberger M. Locomotor inhibition in adult horses faced to stressors: a single postpartum experience may be enough!. *Front Psychol* 2012;**3**(Article 442):1–6.
8. Miller RM. *Understanding the ancient secrets of the horse's mind*. Neenah, WI: Russell Meerdink Co.; 1999. p. 138
9. Simpson BS. Neonatal foal handling. *Appl Anim Behav Sci* 2002;**78**(2–4):303–17.
10. Spier SJ, Pusterla JB, Villarroel A, Pusterla N. Outcome of tactile conditioning of neonates, or "imprint training" on selected handling measures in foals. *Vet J* 2004;**168**(3):252–8.
11. Scott JP, Fuller JL. *Genetics and the social behavior of the dog: the classic study*. Chicago: The University of Chicago Press; 1965. p. 468.
12. Beaver BV. *Feline behavior: a guide for veterinarians*. 2nd ed St. Louis: Saunders; 2003. p. 349.
13. Mal ME, McCall CA. The influence of handling during different ages on a halter training test in foals. *Appl Anim Behav Sci* 1996;**50**(2):115–20.
14. Rushen J, de Passillé AM, Munksgaard L, Tanida H. People as social actors in the world of farm animals. In: Keeling LJ, Gonyou HW, editors. *Social behaviour in farm animals*. New York: CABI Publishing; 2001. p. 353–72.
15. Makinodan M, Rosen KM, Ito S, Corfas G. A critical period for social experience—Dependent oligodendrocyte maturation and myelination. *Science* 2012;**337**:1357–60.
16. Christensen JW. Early-life object exposure with a habituated mother reduces fear reactions in foals. *Anim Cogn* 2016;**19**(1):171–9.
17. Henry S, Hemery D, Richard M-A, Hausberger M. Human-mare relationships and behavior of foals toward humans. *Appl Anim Behav Sci* 2005;**93**(3–4):341–62.
18. Søndergaard E, Halekoh U. Young horses' reactions to humans in relation to handling and social environment. *Appl Anim Behav Sci* 2003;**84**(94):265–80.
19. Ligout S, Bouissou M-F, Boivin X. Comparison of the effects of two different handling methods on the subsequent behavior of Anglo-Arabian foals toward humans and handling. *Appl Anim Behav Sci* 2008;**113**(1–3):175–88.
20. Jezierski T, Jaworski Z, Gorecka A. Effects of handling on behavior and heart rate in Konik horses: comparison of stable and forest reared youngstock. *Appl Anim Behav Sci* 1999;**62** (1):1–11.
21. Waring GH. *Horse behavior*. 2nd ed. Norwich, NY: Noyes Publications; 2003442.
22. King SRB. Home range and habitat use of free-ranging Przewalski horses at Hustai National Park, Mongolia. *Appl Anim Behav Sci* 2002;**78**(2–4):103–13.

23. Salter RE, Hudson RJ. Social organization of feral horses in western Canada. *Appl Anim Ethol* 1982;**8**(3):207–23.
24. McCort WD. Behavior of feral horses and ponies. *J Anim Sci* 1984;**58**(2):493–9.
25. Feist JD, McCullough DR. Behavior patterns and communication in feral horses. *Ethology* 1976;**41**(4):337–71.
26. Brubaker AS, Coss RG. Evolutionary constraints on equid domestication: comparison of flight initiation distances of wild horses (*Equus caballus ferus*) and plains zebras (*Equus quagga*). *J Comp Psychol* 2015;**129**(4):366–76.
27. Birke L, Hockenhull J, Creighton E, Pinno L, Mee J, Mills D. Horses' responses to variation in human approach. *Appl Anim Behav Sci* 2011;**134**(1–2):56–63.
28. Nicol CJ. The social transmission of information and behavior. *Appl Anim Behav Sci* 1995;**44** (2–4):79–98.
29. Feist JD, McCullough DR. Reproduction in feral horses. *J Reprod Fertil Suppl* 1975;**23**:13–8.
30. Keiper RR. Social structure. *Vet Clin N Am Equine Pract* 1986;**2**(3):465–84.
31. Keiper RR, Sambraus HH. The stability of equine dominance hierarchies and the effects of kinship, proximity and foaling status on hierarchy rank. *Appl Anim Behav Sci* 1986;**16** (2):121–30.
32. Kirkpatrick JF, Turner Jr JW. Comparative reproductive biology of North American feral horses. *Equine Vet Sci* 1986;**6**(5):224–30.
33. Berger J. *Wild horses of the Great Basin: social competition and population size*. Chicago: University of Chicago Press; 1986. p. 326.
34. Feh C. Alliances and reproductive success in Camargue stallions. *Anim Behav* 1999;**57** (3):705–13.
35. Nuñez CMV, Adelman JS, Mason C, Rubenstein DI. Immunocontraception decreases group fidelity in a feral horse population during the non-breeding season. *Appl Anim Behav Sci* 2009;**117**(1–2):74–83.
36. McDonnell S. Reproductive behavior of the stallion. *Vet Clin N Am Equine Pract* 1986; **2**(3):535–55.
37. Ransom JI, Cade BS. *Quantifying equid behavior—a research ethogram for free-roaming feral horses*. U.S. Geological Survey, https://pubs.usgs.gov/tm/02a09/pdf/TM2A9.pdf; 2009 [downloaded March 15, 2017].
38. Christensen JW, Zharkikh T, Ladewig J, Yasinetskaya N. Social behavior in stallion groups (*Equus przewalskii* and *Equus caballus*) kept under natural and domestic conditions. *Appl Anim Behav Sci* 2002;**76**(1):11–20.
39. Khalil AM, Miyazaki U. Early experience affects developmental behavior and timing of harem formation in Misaki horses. *Appl Anim Behav Sci* 1998;**59**(4):253–63.
40. Turner Jr JW, Kirkpatrick JF. Hormones and reproduction in feral horses. *Equine Vet Sci* 1986;**6**(5):250–8.
41. Price EO. *Principles & applications of domestic animal behavior*. Cambridge, MA: CAB International; 2008. p. 332.
42. Crowell-Davis SL. Social behavior of the horse and its consequences for domestic management. *Equine Vet Educ* 1993;**5**(3):148–50.
43. McGreevy P. *Equine behavior: a guide for veterinarians and equine scientists*. New York: Saunders; 2004369.
44. Houpt KA, Keiper R. The position of the stallion in the equine dominance hierarchy of feral and domestic ponies. *J Anim Sci* 1982;**54**(5):945–50.
45. Grandquist SM, Gudrun Thorhallsdottir A, Sigurjonsdottir H. The effect of stallions on social interactions in domestic and semi feral harems. *Appl Anim Behav Sci* 2012;**141**(1–2):49–56.

46. Heitor F, Vicente L. Dominance relationships and patterns of aggression in a bachelor group of Sorraia horses. *J Ethol* 2010;**28**(1):35–44.
47. Nicholson AJ. The self-adjustment of populations to change. *Cold Spring Harb Symp Quant Biol* 1957;**22**:153–73.
48. Heitor F, Vicente L. Affiliative relationships among Sorraia mares: influence of age, dominance, kinship and reproductive state. *J Ethol* 2010;**28**(1):133–40.
49. Wolter R, Pantel N, Stefanski V, Möstl E, Krueger K. The role of an alpha animal in changing environmental conditions. *Physiol Behav* 2014;**133**:236–43.
50. Bourjade M, de Boyer d, Roches A, Hausberger M. Adult-young ratio, a major factor regulating social behavior of young: a horse study. *PLoS ONE* 2009;**4**(3)https://doi.org/10.1371/journal.pone.0004888.
51. Wells SM, von Goldschmidt-Rothschild B. Social behavior and relationships in a herd of Camargue horses. *Z Tierpsychol* 1979;**49**(4):363–80.
52. Freymond SB, Briefer EF, Von Niederhäusern R, Bachmann I. Pattern of social interactions after group integration: a possibility to keep stallions in group. *PLoS ONE* 2013;**8**(1)https://doi.org/10.1371/journal.pone.0054688.
53. Vervaecke H, Stevens JMG, Vandemoortele H, Sigurjónsdóttir H, De Vires H. Aggression and dominance in matched groups of subadult Icelandic horses (*Equus caballus*). *J Ethol* 2007;**25**(3):239–48.
54. Rutberg AT, Greenberg SA. Dominance, aggression frequencies and modes of aggressive competition in feral pony mares. *Anim Behav* 1990;**40**(2):322–31.
55. Briard L, Dorn C, Petit O. Personality and affinities play a key role in the organization of collective movements in a group of domestic horses. *Ethology* 2015;**121**(9):888–902.
56. Krüger K, Flauger B, Farmer K, Hemelrijk C. Movement initiation in groups of feral horses. *Behav Process* 2014;**103**:91–101.
57. Monard A-M, Duncan P. Consequences of natal dispersal in female horses. *Anim Behav* 1996;**52**(3):565–79.
58. Monard A-M, Duncan P, Boy V. The proximate mechanisms of natal dispersal in female horses. *Behaviour* 1996;**133**(13–14):1095–124.
59. Cameron EZ, Linklater WL, Stafford KJ, Minot EO. Social grouping and maternal behavior in feral horses (*Equus caballus*): the influence of males on maternal protectiveness. *Behav Ecol Sociobiol* 2003;**53**(2):92–101.
60. Franke Stevens E. Instability of harems of feral horses in relation to season and presence of subordinate stallions. *Behaviour* 1990;**112**(3–4):149–61.
61. Sigurjónsdóttir H, van Dierendonck MC, Snorrason S, Thórhallsdóttir AG. Social relationships in a group of horses without a mature stallion. *Behaviour* 2003;**140**(6):783–804.
62. Kaseda Y. The structure of the groups of Misaki horses in Toi Cape. *Jpn J Zootech Sci* 1981;**5**(3):227–35.
63. Houpt KA, Houpt TR. Social and illumination preferences of mares. *J Anim Sci* 1988;**66**(9):2159–214.
64. Søndergaard E, Jensen MB, Nicol CJ. Motivation for social contact in horses measured by operant conditioning. *Appl Anim Behav Sci* 2011;**132**(3–4):131–7.
65. Hartmann E, Christensen JW, Keeling LJ. Training young horses to social separation: effect of a companion horse on training efficiency. *Equine Vet J* 2011;**43**(5):580–4.
66. Hauschildt V, Gerken M. Temporal stability of social structure and behavioural synchronization in Shetland pony mares (*Equus caballus*) kept on pasture. *Acta Agric Scand, Sect A Anim Sci* 2015;**65**(1):33–41.

67. Kimura R. Mutual grooming and preferred associate relationships in a band of free-ranging horses. *Appl Anim Behav Sci* 1998;**59**(4):265–76.
68. Van Dierendonck MC, Sigurjónsdóttir H, Colenbrander B, Thorhallsdóttir AG. Differences in social behavior between late pregnant, post-partum and barren mares in a herd of Icelandic horses. *Appl Anim Behav Sci* 2004;**89**(3–4):283–97.
69. Ellard M-E, Crowell-Davis SL. Evaluating equine dominance in draft mares. *Appl Anim Behav Sci* 1989;**24**(1):55–75.
70. Heitor F, do Mar Oom M, Vicente L. Social relationships in a herd of Sorraia horses. Part II. Factors affecting affiliative relationships and sexual behaviours. *Behav Process* 2006;**73** (3):231–9.
71. Roberts JM, Browning BA. Proximity and threats in highland ponies. *Soc Networks* 1998;**20** (3):227–38.
72. Van Dierendonck MC, de Vires H, Schilder MBH. An analysis of dominance, its behavioural parameters and possible determinants in a herd of Icelandic horses in captivity. *Netherlands J Zool* 1995;**45**(3–4):362–85.
73. Weeks JW, Crowell-Davis SL, Caudle AB, Heusner GL. Aggression and social spacing in light horse (*Equus caballus*) mares and foals. *Appl Anim Behav Sci* 2000;**68**(4):319–37.
74. Cameron EZ, Setsaas TH, Linklater WL. Social bonds between unrelated females increase reproductive success in feral horses. *Proc Natl Acad Sci USA* 2009;**106**(33):13850–3.
75. Van Dierendonck MC, de Vires H, Schilder MBH, Colenbrander B, þorhallsdóttir AG, Sigurjónsdóttir H. Interventions in social behavior in a herd of mares and geldings. *Appl Anim Behav Sci* 2009;**116**(1):67–73.
76. Van Dierendonck MC, Spruijt BM. Coping in groups of domestic horses—review from a social and neurobiological perspective. *Appl Anim Behav Sci* 2012;**138**(3–4):194–202.
77. Zollinger A, Wyss C, Bardou D, Ramseyer A, Bachmann I. The 'social box' offers stallions the possibility to have increased social interactions. *J Vet Behav* 2016;**15**:84.
78. Christensen JW, Ladewig J, Sondergaard E, Malmkvist J. Effects of individual versus group stabling on social behavior in domestic stallions. *Appl Anim Behav Sci* 2002;**75**(3):233–48.
79. Lesimple C, Fureix C, LeScolan N, Richard-Yris M-A, Hausberger M. Housing conditions and breed are associated with emotionality and cognitive abilities in riding school horses. *Appl Anim Behav Sci* 2011;**129**(2–4):92–9.
80. Strand SC, Tiefenbacher S, Haskell M, Hosmer T, McDonnell SM, Freeman DA. Behavior and physiologic responses of mares to short-term isolation. *Appl Anim Behav Sci* 2002;**78**(2o-4):145–57.
81. Lansade L, Neveux C, Levy F. A few days of social separation affects yearling horses' response to emotional reactivity tests and enhances learning performance. *Behav Process* 2012;**91**(1):94–102.
82. Losonci Z, Berry J, Paddison J. Do stabled horses show more undesirable behaviors during handling than field-kept ones? *J Vet Behav* 2016;**15**:93.
83. Rivera E, Benjamin S, Nielsen B, Shelle J, Zanella AJ. Behavioral and physiological responses of horses to initial training: the comparison between pastured versus stalled horses. *Appl Anim Behav Sci* 2002;**78**(2–4):235–52.
84. Yarnell K. A life less solitary. *Equine Vet Educ: Am Ed* 2016;**28**(12):659–60.
85. Yarnell K, Hall C, Royle C, Walker SL. Domesticated horses differ in their behavioural and physiological responses to isolated and group housing. *Physiol Behav* 2015;**143**:51–7.
86. Hartmann E, Søndergaard E, Keeling LJ. Keeping horses in groups: a review. *Appl Anim Behav Sci* 2012;**136**(2–4):77–87.

87. Bourjade M, Moulinot M, Henry S, Richard-Yris M-A, Hausberger M. Could adults be used to improve social skills of young horses, *Equus caballus*? *Dev Psychobiol* 2008;**50**(4):408–17.
88. Rørvang MV, Ahrendt LP, Christensen JW. A trained demonstrator has a calming effect on naïve horses when crossing a novel surface. *Appl Anim Behav Sci* 2015;**171**:117–20.
89. Estep DQ, Crowell-Davis SL, Earl-Costello S-A, Beatey SA. Changes in the social behavior of drafthorse (*Equus caballus*) mares coincident with foaling. *Appl Anim Behav Sci* 1993;**35**(3):199–213.
90. Houpt KA, Law K, Martinisi V. Dominance hierarchies in domestic horses. *Appl Anim Ethol* 1978;**4**(3):273–83.
91. Montgomery GG. Some aspects of the sociality of the domestic horse. *Trans Kans Acad Sci* 1957;**60**(4):419–24.
92. Giles SL, Nicol CJ, Harris PA, Rands SA. Dominance rank is associated with body condition in outdoor-living domestic horses (*Equus caballus*). *Appl Anim Behav Sci* 2015;**166**:71–9.
93. Yakan A, Akcay A, Durmaz S, Aksu T, Ozturk H. Paddock behaviors and dominance relationships of young male horses the first hours in the morning and again in the afternoon. *J Anim Vet Adv* 2012;**11**(19):3486–92.
94. Heitor F, do Mar Oom M, Vicente L. Social relationships in a herd of Sorraia horses. Part I. Correlates of social dominance and contexts of aggression. *Behav Process* 2006;**73**(2):170–7.
95. Krüger K, Flauger B. Social feeding decisions in horses (*Equus caballus*). *Behav Process* 2008;**78**(1):76–83.
96. Araba BD, Crowell-Davis SL. Dominance relationships and aggression of foals (*Equus caballus*). *Appl Anim Behav Sci* 1994;**41**(1–2):1–25.
97. Heitor F, Vicente L. Maternal care and foal social relationships in a herd of Sorraia horses: influence of maternal rank and experience. *Appl Anim Behav Sci* 2008;**113**(1–3):189–205.
98. Komárková M, Bartošová J, Dubcová J. Age and group residence but not maternal dominance affect dominance rank in young domestic horses. *J Anim Sci* 2014;**92**(11):5285–92.
99. Hartmann E, Keeling LJ, Rundgren M. Comparison of 3 methods for mixing unfamiliar horses (*Equus caballus*). *J Vet Behav* 2011;**6**(1):39–49.
100. Keeling LJ, Bøe KE, Christensen JW, Hyyppä S, Jansson H, Jørgensen GHM, Ladewig J, Mejdell CM, Särkijärvi S, Søndergaard E, Hartmann E. Injury incidence, reactivity and ease of handling of horses kept in groups: a matched case control study in four Nordic countries. *Appl Anim Behav Sci* 2016;**185**:59–65.
101. Krueger K, Farmer K, Heinze J. The effects of age, rank and neophobia on social learning in horses. *Anim Cogn* 2014;**17**(3):645–55.
102. Krueger K, Heinze J. Horse sense: social status of horses (*Equus caballus*) affects their likelihood of copying other horses' behavior. *Anim Cogn* 2008;**11**(3):431–9.
103. Lindberg AC. Group life. In: Keeling LJ, Gonyou HW, editors. *Social behaviour in farm animals*. New York: CABI Publishing; 2001. p. 37–58.
104. Christensen JW, Søndergaard E, Thodberg K, Halekoh U. Effects of repeated regrouping on horse behavior and injuries. *Appl Anim Behav Sci* 2011;**133**(3/4):199–206.
105. Jørgensen GHM, Borsheim L, Mejdell CM, Søndegaard E, Bøe KE. Grouping horses according to gender—effects on aggression, spacing and injuries. *Appl Anim Behav Sci* 2009;**120**(1–2):94–9.
106. Jørgensen GHM, Liestøl SH-O, Bøe KE. Effects of enrichment items on activity and social interactions in domestic horses (*Equus caballus*). *Appl Anim Behav Sci* 2011;**129**(2–4):100–10.

107. Fureix C, Bourjade M, Henry S, Sankey C, Hausberger M. Exploring aggression regulation in managed groups of horses *Equus caballus*. *Appl Anim Behav Sci* 2012;**138**(3/4):216–28.
108. Grogan EH, McDonnell SM. Injuries and blemishes in a semi-feral herd of ponies. *J Equine Vet* 2005;**25**(1):26–30.
109. McDonnell SM, Haviland JCS. Agonistic ethogram of the equid bachelor band. *Appl Anim Behav Sci* 1995;**43**(3):147–88.
110. Hartmann E, Christensen JW, Keeling LJ. Social interactions of unfamiliar horses during paired encounters: effect of pre-exposure on aggression level and so risk injury. *Appl Anim Behav Sci* 2009;**121**(3–4):214–21.
111. Majecka K, Klawe A. Influence of paddock size on social relationships in domestic horses. *J Appl Anim Welf Sci* 2018;**21**(1):8–16.
112. Proops L, Burden F, Osthaus B. Social relations in a mixed group of mules, ponies and donkeys reflect differences in equid type. *Behav Process* 2012;**90**(3):337–42.
113. Fagen R. Selective and evolutionary aspects of animal play. *Am Nat* 1974;**108**(964):850–8.
114. Fagen RM. Selection for optimal age-dependent schedules of play behavior. *Am Nat* 1977;**111**(979):395–414.
115. Hoffman R. On the development of social behavior in immature males of a feral horse population (*Equus przewalskii* f. caballus). *Z Säugetierkunde* 1985;**50**:302–14.
116. Crowell-Davis SL. Developmental behavior. *Vet Clin North Am Equine Pract* 1986;**2**(3):573–90.
117. Crowell-Davis SL, Houpt KA, Kane L. Play development in welsh pony (*Equus caballus*) foals. *Appl Anim Behav Sci* 1987;**18**(2):119–31.
118. Fagen RM, George TK. Play behavior and exercise in young ponies (*Equus caballus* L.). *Behav Ecol Sociobiol* 1977;**2**(3):267–9.
119. McDonnell SM, Poulin A. Equid play ethogram. *Appl Anim Behav Sci* 2002;**78**(2–4):263–90.
120. Tyler SJ. The behavior and social organization of the new Forest ponies. *Anim Behav Monogr* 1972;**5**(2):87–196.
121. Rho JR, Srygley RB, Choe JC. Sex preferences in Jeju pony foals (*Equus caballus*) for mutual grooming and play-fighting behaviors. *Zool Sci* 2007;**24**(8):769–73.
122. Hausberger M, Fureix C, Bourjade M, Wessel-Robert S, Richard-Yris M-A. On the significance of adult play: what does social play tell us about adult horse welfare? *Naturwissenschaften* 2012;**99**(4):291–302.
123. Knubben JM, Fürst A, Gygax L, Stauffacher M. Bite and kick injuries in horses: prevalence, risk factors and prevention. *Equine Vet J* 2008;**40**(3):219–23.
124. Beaver BV. Aggressive behavior problems. *Vet Clin N Am Equine Pract* 1986;**2**(3):635–44.
125. Line SW, Hart BL, Sanders L. Effect of prepubertal versus postpubertal castration on sexual and aggressive behavior in male horses. *J Am Vet Med Assoc* 1985;**16**(3):249–51.
126. McDonnell S. Follow-up: carnivorous horses. *The Horse* 2003;**XX**(5):67–9.
127. McDonnell S. Carnivorous horses. *The Horse* 2002;**XIX**(10). 61, 62 and 64.
128. Voith VL. Treatment of fear-induced aggression in a horse. *Mod Vet Pract* 1979;**60**(10):835–7.
129. Fureix C, Manugy H, Hausberger M. Partners with bad temper: reject or cure? A study of chronic pain and aggression in horses. *PLoS ONE* 2010;**5**(8). https://doi.org/10.1371/journal.pone.0012434.
130. Hausberger M, Roche H, Henry S, Visser EK. A review of the human-horse relationship. *Appl Anim Behav Sci* 2008;**109**(1):1–24.
131. Pryor P, Tibary A. Management of estrus in the performance mare. *Clin Tech Equine Pract* 2005;**4**(3):197–209.

132. Christoffersen M, Lehn-Jensen H, Bøgh IB. Referred vaginal pain: cause of hypersensitivity and performance problems in mares? A clinical case study. *J Equine Vet* 2007;**27**(1):32–6.
133. Bliss S. Options for managing estrus related behavior problems in performance mares. *The Remuda* 2011;**4**(2):13–6.
134. Houpt KA. *Behavior in horses. Notes from talk given at the southwest veterinary symposium.* Texas: Ft. Worth; 2010.
135. Fox AE, Bailey SR, Hall EG, St. Peter CC. Reduction of biting and chewing of horses using differential reinforcement of other behavior. *Behav Process* 2012;**91**(1):125–8.
136. Sherlock CE, Lott-Ellis K, Bergren A, Withers JM, Fews D, Mair TS. Granulosa cell tumours in the mare: a review of 52 cases. *Equine Vet Educ* 2016;**28**(2):75–82.
137. McDonnell S. Using the twitch properly. *The Horse* 2004;**XXI**(11):105–6.
138. Parker M, Goodwin D, Redhead ES. Survey of breeders' management of horses in Europe, North America and Australia: comparison of factors associated with the development of abnormal behaviour. *Appl Anim Behav Sci* 2008;**114**(1–2):206–15.
139. Luz MPF, Maia CM, Pantoja JCF, Neto MC, Filho JNPP. Feeding time and agonistic behavior in horses: influence of distance, proportion, and height of troughs. *J Equine Vet* 2015;**35**(10):843–8.
140. Houpt KA. Treatment of aggression in horses. *Equine Pract* 1984;**6**(6):8–10.
141. Crowell-Davis SL, Houpt KA. Maternal behavior. *Vet Clin N Am Equine Pract* 1986;**2**(3):557–71.
142. McDonnell SM. Retraining a rebellious colt. *The Horse* 2001;**XVIII**(9):49–50.
143. Marcella KL. Poor socialization can stem from a variety of circumstances. *DVM Newsmag* 2006;2E–5E.
144. Houpt KA. Aggression and intolerance of separation from a mare by an aged gelding. *Equine Vet Educ* 1993;**5**(3):140–1.

Chapter 6

Equine Reproductive Behavior

Survival of a species is dependent on successful reproductive strategies. The resulting behavior varies across equine species, but the focus here is with the domestic horse. Free-roaming horses serve as the model for what is normal behavior, but even there, human interference occurs. There are roundups and immunocontraception to try to manage numbers, so even in these instances, humans exert an influence on behavior. Management of tamed horses introduces a huge number of behavioral variations.

PRENATAL INFLUENCES

Harem groups constitute the social and reproductive units of free-ranging horses. Where these animals naturally roam, the ratio of one stallion to one mare found at birth is skewed more toward a 1:3 or 1:4 ratio in harem groups. Research is showing the sex of a foal is determined by more than pure chance. Mares older than 10 years are more likely to produce fillies than colts, resulting in 0.75 colts for every 1.0 fillies by the time the mare is 15. The same trend is true for stallions, especially between 15 and 20 years of age, when there are 0.84 colts born for each filly. Past 20 years, the ratio returns to 1:1. When mares and stallions are within 5 years of each other in age, the ratio of colts and fillies is equal, but when older stallions breed younger mares, the likelihood of colts being born increases. With a 15-year age gap, the ratio of colts to fillies reaches 1.53:1.[1]

Over generations, secondary sex characteristics have developed to assure greater reproductive success, especially in males. Across species, those that are polygynous, diurnal, and live in open habitats are more likely to develop marked differences between males and females.[2] With a sex ratio of 1:1 in free-roaming horses, strong sexual characteristics become important to a stallion for attracting mares and deterring rivals. Humans have influenced the degree of emphasis of the secondary sex characteristics by domestication and selective breeding. Quarter Horse colts develop heavy jowls, while the jaws of Arabian colts are much less prominent. Heavy, cresty necks are common in males because of androgen-facilitated muscle deposition patterns.[2] Another feature is the enlargement of the canine teeth. Over time, these features may change in a breed as the desired look in the show arena changes. Mares have

Equine Behavioral Medicine. https://doi.org/10.1016/B978-0-12-812106-1.00006-1

unique sexually dimorphic characteristics too. Their brains are larger; the pelvis is wider; and they tend to be slightly smaller than males of the same breed.[2]

MALE BEHAVIORS

Development of Male Behaviors

Components of adult sexual behavior are exhibited by young horses during play, but rarely in a complete sequence. Flehmen is usually the first of the sexually related behaviors to appear and can be shown as early as 2 days of age. It occurs with the greatest frequency during the first 4 weeks, approximately once every 1.2 h, and then decreases linearly so that by 17–20 weeks of ages it only occurs once every 12.5 h.[3] The event most associated with onset of the flehmen response is urination by another horse. The urine is closely investigated by the foal, which then flehmens. Fillies also show this behavior but at a reduced rate.

Colts as young as 1 month of age can have a full penile erection, although this is more common in the second or third month.[4,5] Mounting of fillies will occasionally occur in the context of play. Young colts will sniff the urine or perineum of an estrous mare and flehmen, but additional sexual behaviors do not follow. For weanlings, play is important for the development of motor skills and learning of techniques that become part of adult stallion behavior. This is best done with other foals of the same age. If the mother is the only other horse, she will become the play target. As adults, however, the solitary colts are generally awkward at first.[6]

Behaviors and breeding readiness do not necessarily coincide. Colts tend to play rougher than fillies do (Figure 6-1). As testosterone levels increase in the spring, behaviors become rougher and more stallion-like. Colts become touchier about their mouth, legs, flank, and genital regions. Touching the mouth is usually followed by rearing or biting. Touching the legs results in kneeling and biting, and touching the flank is followed by squealing and kicking.[7] Some youngsters nip or bite more than others do, a behavior likely to be retained as an adult. Repetition of touching in those sensitive areas should be used to teach compliance with human contact, and touch tolerance is easiest to teach weanlings before their spring surge of testosterone. Playing with other colts in a pasture also diminishes some of the "coltishness" so the youngster is more receptive to learning. Colts become increasingly inquisitive, which sometimes gets them into trouble (Figure 6-2). They also begin repeated defecations in the same place, followed by sniffing the urine and feces and then flehmen.[7]

Yearlings and 2-year-olds will breed if conditions are right, and the fillies are their choice to breed. Colts initially show quite a bit of interest in estrous mares, sniffing and nibbling them. At this time, though, their mating styles are still poorly developed. Some are shy acting, others seem very confident. Copulations usually are longer for the beginners than for an experienced stallion. Initial attempts at mounting may begin at the front, side, or rear of the mare.

FIGURE 6-1 Colts show increasingly rough play as they mature. Neck biting is common, and this sorrel colt was the most aggressively active of four yearling stallions pastured together.

FIGURE 6-2 Foals become increasingly inquisitive of their surroundings as they become more independent. The curiosity helps them learn about their environment, but it is a time when ensuring their safety is important. This foal managed to fall between two round hay bales that had to be moved for his release. *(Photo courtesy of Maggie Gratny Young and used with permission.)*

The pelvic thrusts of inexperienced colts tend to be uncoordinated, so they attempt several mounts and numerous thrusts before having a successful intromission. Even then, the intromission may be followed by an early dismount before ejaculation occurs. The first copulations will occur between 15 months and 2 years of age. This is also about the time spermatozoa can be identified in an ejaculate.[4,5] Although spermatozoa counts per milliliter of ejaculate in adult stallions range from 80,000,000/mL and up, researchers used a count of 500,000/mL to suggest sexual maturity occurs on average at 83 weeks (56–97 weeks).[8,9] Weanlings, particularly those born late the previous year, will occasionally sire foals even though their sperm counts are much lower than those born earlier in the prior season. The increasing length of daylight promotes increased testosterone production and speeds sexual function. For this reason, in domestic herds it is important to separate weanling colts from females to avoid unwanted pregnancies.

In free-ranging horses, young fillies periodically stray from their bands to mate with young bachelors. Breeding styles are awkward at first, but the fillies seem to tolerate this. Allowing colts to breed frequently accelerates maturation of behavior and hormone levels.[10]

Mating by Free-Ranging Stallions

In free-ranging harem groups, matings are seasonal in a period between late spring and midsummer. This typically begins in March and ends by August—peaking in May and early June.[11,12] Although stallion behavior is somewhat individualized, there are many common patterns. Attention to mares and libido are present year-round, but stallion-mare interactions and a stallion's testosterone levels, sex drive, reaction time, and number of mounts per ejaculation are greatest during the mating season.[5,7,13] At the beginning of the season, stallions are highly responsive to signs of estrus in mares and stay close to receptive mares for several days. By the end of the season, they may show little interest until the mare is in full estrus.[4]

Reproductive behavior is polygynous, with the harem stallion being responsible for slightly less than 90% of all matings.[14,15] These stallions are also in the age groups that produce the highest percentage of foals—7–10 years, followed by 11–13 years.[16] In multiple stallion groups, variations exist in which stallions do most of the mating. It usually is the dominant harem stallion, but in some groups, several males will breed a mare serially.[15]

During breeding season, scent marking is common. Stallions start marking about a month before matings begin and continue for a month past the peak season.[12] Feces marks the feces of other stallions, and urine covers the feces of mares in the harem band. Scent investigations include interest in communal fecal piles (also called dung heaps), eliminations of others in the harem band, and defecations of other males.[17,18] Stallions bring their nostrils close to the excreta without actually contacting it. Their head may move slowly side to side over the urine or feces. The nostrils dilate and respirations become more vigorous. The upper lip may quickly be raised and lowered again, which is then repeated several times in rapid succession, or the stallion may flehmen.[19] Not only can the stallion smell the excreta, the behaviors probably allow odors to go to the vomeronasal organ. The stallion will investigate another stallion's feces, walk forward a few steps, defecate a small amount on top of the pile, and step back to sniff it again.[4] Interest in and the rate of marking of mare eliminations is approximately 90% in May and June, but drops to 1% in November to March.[20] During the breeding season, stallion urine contains high levels of cresols, suggesting that one role for marking may be to mask the manure odors produced by the mares.[21] The flehmen behavior more often precedes marking rather than breeding, probably serving to help prime the stallion for mating.[22,23]

In the spring, stallions begin showing interest in mature mares in their harem groups before they are in full estrus. The stallion becomes more protective of the group and will tend the mares by staying close and moving as they move.[12,24] While protective of young mares, the stallion does not guard them as much when they are in estrus.[4,13] Stallions search for mares nearing estrus and approach cautiously to avoid aggressive rejection. Stallions may transition from grazing to sexual arousal rapidly, as shown by stomping, pawing, and nickering or whinnying. The approach for mares in full estrus is a high stepping, prancing gait with the head and tail arched. It is accompanied by loud nickers and whinnies. The approach is aggressive, involving nipping at the mare's neck, shoulder, or flank. Kicking and striking are also common.[5,25] If the mare is not receptive, aggression by both horses continues, and the stallion will eventually move away. Other things that can reduce the stallion's arousal include withdrawal by the mare, unsuccessful mounts, and prolonged and repeated flehmen.[5] There are many interactions by the stallion with the mares, even before they are in full estrus. As a result, the actual mating sequence can be relatively rapid, often less than 1 min.[25] Stallions that take over a harem group will commonly force copulations on the unfamiliar females, absent typical courtship behaviors.[16]

Dominant mares tend to be preferentially favored by stallions, even though the stallion does not spend more time with them than with others.[26,27] However, he will spend more time investigating their eliminations.[27] Males may reject certain mares. While reasons are not always obvious, the mare's coat color seems important. Harem stallions have been known to gather only buckskin mares, and others only sorrels and bays.[5] They also are more attracted to adult mares as compared to younger ones.[4]

If the mare is in estrus, the amount of aggression diminishes, and the stallion begins the precopulatory behavior. He is attracted to a number of odors and behaviors, including the mare's urination-like stance.[5] Most of the breedings that end with ejaculation begin with the mare approaching the stallion. When the two horses are near each other, he begins sniffing the mare. Erection begins gradually. Vocalizations become quieter, as soft nickers. The stallion continues with nuzzling, licking, nibbling, and/or nipping the mare, usually starting at her head and working back toward the perineum. He will rub his head in her flank or rest his chin on her back, and the penis will be extended from the prepuce.[24] These courtship behaviors are necessary because the erectile tissue within the penis takes some time to become fully engorged. The erection is usually complete prior to the stallion mounting the mare.[6,25]

Mounting is the first behavior of the copulatory sequence. Experienced stallions usually mount the rear of the mare, but younger stallions may start farther forward and work their way back. Stallions copulate just over 60% of the time when they mount fully receptive mares, and of those, 45% involve more than a single mount.[4] Multiple mounts are considered normal. A few of these mounts may also occur without an erection in both younger and mature horses.[13] There is a sudden rearing motion with a shuffling of the hindlimbs to move forward. The forelimbs are positioned with the knees just ahead of the mare's pelvis. The stallion's head rests against her back, near or just behind the withers. He may grasp the neck or mane with his teeth. One or two pelvic thrusts are usually sufficient for successful intromission. Once this occurs, the stallion plants his hind feet and "couples up" closely to the mare.[25] Pelvic thrusts intensify, and ejaculation begins within 16 s after full intromission. Externally, the stallion's facial muscles relax, and his head and ears droop.[4,5] There are rhythmic muscle contractions in the rear limbs and an increased respiratory rate. The up and down, jerky swish of the tail, called *flagging* or *tail flips*, is apparently related to the contractions of the urethral muscles as the ejaculate is spurted out, but there can be flagging in the absence of ejaculation.[5,25] A well-trained stallion, especially with semen collection for artificial insemination, will tolerate slight hand pressure on the ventral penis during a collection. If 6–10 distinct pulses are felt, that is confirmation that a complete ejaculation has occurred.[28] This permits the collector to keep full attention on the artificial vagina, the stallion's penis, location of his hind feet, and safety, without any need to turn his/her head toward the coned end of the artificial vagina to look at the collection bag and/or bottle for collected semen.

Postcopulatory behaviors begin with a brief relaxation while still mounted, and it is followed by a dismount approximately 10–15 s after ejaculation.[4,13,25] The relaxation is associated with a large release of beta endorphins. Initially the stallion may seem dazed, and first-mating stallions have been known to "faint" immediately after breeding. The dismount happens as the mare moves forward rather than by an action on the part of the stallion. The penis becomes flaccid and is retracted into the prepuce within a minute.[5] During the dismount, there is a short, soft squeal. Immediately after dismounting, the stallion will usually stand quietly behind the mare for a short period of time. The stallion usually sniffs spilled ejaculate or vaginal discharges and shows a flehmen response. He may urinate or defecate over those odors too. Eventually the mare and stallion separate, with the stallion moving away 26% of the time and the mare moving first 60% of the time.[4,5] Compared with mounting a phantom, natural cover or a tease (live) mare permits a more natural dismount for the stallion after ejaculation. Ejaculation requires great exertion by a stallion and he routinely needs time to recover before being asked to back away from a fixed mount. Forced too quickly, many stallions will stumble and have to catch themselves from falling.

Following successful mating, there is a refractory period before a male becomes interested in the same or another female again. The length of this time is variable, depending on the number of previous breedings occurring within a tight timeframe and the number of females in estrus. In free-ranging horses, the refractory period appears to be short, perhaps less than 30 min.[29] Within a 6-h period, one stallion is reported to have successfully mated twice each with three different mares.[4] The greater the number of mares serviced within a short time, the longer the refractory period becomes. There is individual variation as to the number of copulations a stallion will perform in a day. The average is 2.9 ejaculates, but some horses may go as high as 10 before reaching sexual satiation.[5] Harem stallions spend more time with low-ranking mares but mate all mares with equal frequency regardless of their social rank.[30] The exception to this is when the number of estrous mares is high; then stallions preferentially tend to copulate with the higher-ranking mares.[5]

Bachelor stallions or other lower-ranking stallions in the group may be successful in breeding mares. Reproductive success is lower for individual males if there are multiple stallions and for mares that leave their bands, suggesting an advantage for single stallion harem groups.[31] The former is somewhat counterbalanced because foal survival in multistallion bands is greater.[32] Incestuous breeding is not common because of the dispersion of the offspring, but it can occur when fillies do not leave their natal band. For those fillies, the foaling percentage is significantly lower than for mares bred to unrelated stallions (22.7% vs. 61.8%). Breeding between half siblings is also associated with poor reproductive success (36.8%).[14]

Mating by Domestically Managed Stallions

The most natural mating system is for the stallion to be pastured with the mares he is to breed. The behaviors shown by these stallions are closer to the full range shown by free-ranging horses, except that there may or may not be the equivalent of harem groups. If multiple stallions are kept within the herd, mares may be bred by more than one stallion. Stallions can breed as often as once every hour or two throughout a 24-h period without a decrease in fertility.[13] A young stallion is introduced in a large paddock to a patient, experienced mare, giving him time to learn his technique in a safe environment. He can then be introduced to the pasture containing other mares.

More common than pasture breeding is the controlled environment where stallions are brought to a breeding shed with an already restrained, estrous mare. While the general behaviors are the same as those of free-ranging horses, the investigatory and precopulatory behaviors are significantly abbreviated or absent. Stimuli involved with stallion sexual behavior are complex, particularly those of vision and olfaction.[5,6] For the first few visits to the breeding shed, stallions will require time to learn what is expected. Allowing him time to tease the mare and develop an erection lessens his stress. Over time, though, the mere walk from the stall to the shed can be sufficient to prepare him to mount and copulate. Stallions quickly learn the connection between prebreeding and breeding routines and locations. Experienced stallions can achieve an erection in just over 60 s and mount for the first time within 100 s of entering the breeding shed. About 70% of the time, a stallion will require only a single mount for ejaculation. If allowed, 20% of stallions will mount before achieving a full erection. This mimics what occurs in free-ranging horses and should not be discouraged. It is his insurance to protect himself from injury until he confirms a mare is in proper standing heat. The average stallion will thrust seven times and remain mounted 27 s with an ejaculation.[25] The ejaculations occur approximately 15 s after intromission.[6] Stallions with high libido may quickly mount a collection dummy and even develop a preference to it over a mare.[33] The dismount will involve the mare stepping forward when mating occurs under pasture conditions, but when done in-hand, the stallion will actively have to dismount on his own. This increases the importance of having safe footing and good muscle fitness for the horse.

Even though horses are not territorial, location can be important. In a novel location, even an experienced breeding stallion might require more time before successfully mating.[34]

From time to time, discussions occur about the potential for overuse of a breeding stallion. This is less of a concern in breed associations that allow artificial insemination, but not all do. Studies that look at the number of foals born compared to the number of mares bred by live cover have downplayed this concern. Previously, a 40-mare book was considered an appropriate number by the Thoroughbred industry. When comparing live foals from stallions that booked 40 or less, 41–80, 81–120, and greater than 120 mares, the larger the number of mares, the greater was the live foal rate.[35]

In addition to being sexually active, domestic stallions are often being shown or raced. Keeping them active does reduce libido, but it may not be enough to totally control stallion-like behaviors.[36] Working stallions are strongly discouraged from showing reproductive behaviors at times and locations away from the breeding shed. Stallion rings, a rigid band, are placed on the penis caudal to the corona glandis. The intent is for it to fit comfortably around the nonerect penile shaft but cause discomfort when an erection begins to develop. Alternatively, a stiff-bristled brush attached to a bellyband is strapped around the ventral abdomen such that the brush rests just in front of the preputial opening. As the stallion's penis emerges from the prepuce, it comes in contact with the bristles. These are used to discourage erection and masturbation. Punishment for erection and negative reinforcement, to elicit penile withdrawal back into the prepuce, quickly suppresses sexual arousal, but they can result in serious sexual dysfunction.[25] This raises the question of whether it is better for a stallion to be allowed to breed mares while still actively performing or to wait until the showing or racing career is over. Both options are viable as most stallions quickly learn the difference between opportunities to breed and nonsexual activities associated with work. It may be easier for a stallion to learn "not now" than "never."[33,37] Classical conditioning can be used to pair a specific cue with imminent breeding behavior. Within the horse community, this is accomplished in a number of different ways. Specific halters, halters with a bell attached, a unique work cue, the chain of a stud shank over vs. under the stallion's chin, and the presence of estrous mare urine are examples. A specific breeding site or handler that is unique to collection are other cues that stallions can quickly learn.

The selective breeding can also affect the fertility and semen quality of the stallion, especially by breed and age.[2,38] Even within a breed, semen quality can vary. As an example, Warmblood stallions of the carriage-type have poorer quality semen than do those used for riding. There was also a significant sire effect, indicating differences in heritability.[2,39]

Conflicting information makes it difficult to evaluate if the presence of multiple stallions affects semen. It has been suggested that stallions exposed to other stallions will have higher sperm numbers in their ejaculate than will those that are exposed only to mares.[40] In contrast, being within a bachelor group reduces androgen levels, sexual and aggressive behavior, accessory sex gland and testicular size, and semen quality.[13] These changes may occur in barns where several stallions are housed together as well. Moving a stallion to a barn with mares should reverse these effects, thereby increasing an individual stallion's libido and breeding success.

Masturbation is a normal behavior in stallions from newborn through old age (Figure 6-3). While standing at rest and when mutually grooming other horses, stallions will have a full or partial penile erection. It is repeatedly moved up toward contact with the ventral abdomen and down by rhythmical contractions of the ischiocavernosus muscle. The erect penis may also be rubbed against the abdominal wall. Masturbation generally lasts 1–5 min and may recur every 2–17 h, but ejaculation is rare.[5,13,41] On average, an erection occurs 7.4 times in 24 h. The total erection time is greater in the summer and in the morning. Over half the time of a full erection is spent masturbating.[5] Forty percent of times when lateral recumbency resting is initiated, masturbation follows within 5 min.[42] Devices designed to discourage the behavior rarely work and may increase the frequency and duration of the episodes.[13]

Gelding Behavior

Castration changes stallion-like behavior in different ways. If surgery is done on an adult, the horse may maintain a normal sexual drive for an average of 5–16 days postsurgery.[5] If gelded early in life instead, the drive is greatly reduced. Between 20% and 30% of horses gelded before 2 years of age retain

FIGURE 6-3 Masterbation is common in stallions of all ages as seen here in (A) a suckling foal, (B) a yearling, and (C) a 3-year-old.

stallion-like behaviors. This number is similar to that for horses gelded later in life.[43] Retained behaviors can include sniffing with flehmen, herding other horses, arching the neck, aggression, mounting with pelvic thrusts, and penile intromission. Geldings may only show the behaviors in pasture when uninhibited by humans, or the behavior may be directed toward a specific mare.[44]

The frequency, duration, and intensity of masturbation is significantly lower for geldings than for intact stallions. Following castration, frequency and duration are gradually reduced to about half of stallion levels over a 1-month period.[41]

FEMALE BEHAVIORS

The reproductive behaviors of mares developed over the millennia to allow the continued survival of the species. Well-preserved, pregnant *Eurohippus messelensis* mares and fetuses have been described. From these, it has been possible to examine what were soft tissue structures and determine that placental and fetal development have not changed in millions of years. Even then, mares could become pregnant before reaching full adulthood.[45]

Development of Female Behaviors

Flehmen is the first behavior a filly will show that will eventually be associated with sexual behavior. It can be seen as early as day one and reaches a frequency of approximately 25/h by week 10.[3] Both the rate and time spent showing flehmen is less than for colts. As the animals get older, there is great variability in the frequency of this behavior.

Puberty for fillies will occur between 10 and 24 months of age, with the average being about 18 months. Under open-range conditions, it is rare for yearling fillies, or even for 2-year-olds, to become pregnant. Just 0.9% of free-ranging mares foal when they are 2 years old, and 13.5% foal as 3-year-olds.[4,5] When fillies are well fed, as occurs in domestic and occasional range environments, a higher percentage of yearlings can become pregnant. As food quality decreases, there is a proportional reduction in pregnancies of young horses.[16]

Fillies disperse from their native harem band, but not in a random pattern. They usually join groups having familiar mares but in which the stallion is not known.[46] This significantly reduces the chances of inbreeding. Even with the stress of becoming established into a new group, the fillies show no delay in the age of first reproduction or the survival of the resulting foals.

Behavior in Free-Ranging Mares

Mares are long-day, seasonally polyestrous. Their ovaries cycle several times, but only during a specific time of year when daylight hours lengthen, corresponding to the natural and abundant availability of food in the spring and early summer. The estrous cycle ranges from 19 to 25 days, averaging 21 days. The length of the cycle is slightly longer in nonlactating mares, certain horse breeds, and ponies.[47] There is some variation in the follicular phase of the cycle, with the normal estrus averaging 5.5 days but ranging from 2 to 9 days. Extremely long periods do occur, particularly in young and elderly mares.[4,47] Compared to other large mammals, the length of estrus in mares is relatively long. It is theorized that this developed so that lower-ranking mares will have a better chance of conception.[26,48] Because aggression by mares toward mares is disproportionately directed toward subordinate mares with foals, associated stress and access to the harem stallion may delay breeding.[48]

Foal heat, the first estrus after foaling, occurs between 4 and 18 days after the mare has given birth. The average time is 8 days postpartum.[5,49] Free-ranging mares that foal early in the season have a longer period before the foal heat occurs. Those foaling near the end of the season, however, have a much shorter time before they cycle again.[16] Mares have been seen breeding within hours after foaling, but associated conception has not been reported.[5] Successful breeding during foal heat is common in some herds and is a time when mares are apparently highly fertile.[13,16] In other herds, conception is more common in the second heat.[4,16] Pregnancies in free-ranging mares rarely occur after the second estrus following parturition.[4,29]

The luteal phase of the cycle (diestrus) is a relative constant 14–15 days.[50] There is a long anestrus period throughout the fall and winter months. While in anestrus and diestrus, mares are not receptive to stallions. Approaches are met with flattened ears, restlessness, tail swishing, and attempts to move away. If a stallion persists, his advances may be perceived as threats and will be met with squealing, biting, striking, or kicking by mares. Older mares do not tolerate young colts and will drive them away.[4]

Female estrous behaviors necessary for successful mating have been divided into three phases.[29,51] The first is *attractivity*, which can be defined as the stimulus value of the mare that evokes an interest by the stallion. Next comes *proceptivity*—the mare's active solicitation of contact and mating by the stallion. Last is *receptivity*. This includes the mare's responses that are necessary and sufficient for a fertile copulation.

Estrus usually comes on gradually, paralleling the growth of the ovarian follicle and the changes in the mare's receptivity to the stallion.[52] Initially, she allows him to sniff and nuzzle her, but she will still walk away if he tries to mount. In estrus, the mare is increasing alert and more active. There is a unique facial expression (*mating face*) characterized by ears turned back but not flattened, loose lips, and relaxed facial muscles.[53] The mare approaches the stallion and presents her hindquarters to him, followed by a slightly spread-legged, squatting posture. The mare urinates small amounts frequently and retains the urinary posture for a prolonged time. The tail is slightly elevated and deviated to one side (Figure 6-4). At the same there is opening and closing of the labia, exposing the clitoris. This is been called *winking* or *clitoral winking*. Because the clitoris is lined with sebaceous glands, it is speculated that the winking behavior aids in dispersion of scents from the smegma that accumulates in the fossa.[26] A young mare coming into estrus occasionally spurts urine and swishes her tail as she moves away from a stallion.[4,29] Pony mares may elevate the tail away from the perineum instead of flagging it to the side.[4,29,54] The tail can be raised slightly or into an almost vertical position.[26]

FIGURE 6-4 Signs of estrus vary but usually include a deviation of the tail to one side and a "winking" of the labia, with or without accompanying urination.

The stallion usually ignores young mares in the harem group, so the filly often wanders off. By doing so, she can associate with young stallions. Mares in a group without a stallion typically wander off when they are in estrus as well. Even when in estrus, the young mares may display estrous behaviors near the stallion, but move away and exhibit submissive behaviors such as jaw chomping (snapping).[4,5,54] On rare occasions, older mares may show jaw chomping too.[55] While fillies may remain solitary or become part of a stallion's newly forming harem, older mares usually return to their female social group.[5]

Mating

In full estrus, the mare shows a passiveness toward the stallion's advances. If he does not actively court her, she will seek him out and display estrous behaviors near him. If he still fails to respond, she will graze a while and then display again.[4,5,26]

As the stallion mounts the estrous mare, she remains immobile or leans slightly backward. She may also move slightly side-to-side to help position his penis for intromission. During ejaculation, some mares look back toward the stallion.[29] A short time after mating is complete, the mare slowly moves forward, allowing the stallion to dismount. It is then that some ejaculate and seminal fluids may spill from the vulva. The mare frequently urinates a small amount shortly after the stallion dismounts and he will smell both fluids.[29]

The estrous mare will mate several times in a 2- or 3-day period. She may also mate with different stallions if conditions are right.[4]

Pregnancy

Although most mares in a harem band will breed during the spring season, a few do not. This suggests some may vary in their cycle, or they may have been bred but suffered embryonic loss. In free-ranging herds, the foaling rate is usually less than 60% and often less than 50%.[11,56] Mares that are 4–15 years of age have the highest pregnancy rates.[12] However, dominant mares are more likely to produce foals than lower ranking ones.[27] Mares that have formed social bonds with other mares have better foaling rates and foal survival than do solitary mares.[57] About 30% of mares foal in consecutive years. Forty-five percent will foal twice during a 3-year period.[11] Those that have had to protect their foals from stallion aggression are less likely to become pregnant.[58] Higher foaling rates may be the result of lower stress levels from less stallion harassment, favorable environmental conditions, and no prolonged lactation periods.[11] Mares in multistallion bands tend to be in significantly poorer condition in the month before foal birth and are significantly more likely to have female foals.[58]

Gestation lasts about 340 days, ranging from 315 to 365 days.[12,59] It is hard to have exact expectations for free-ranging mares to foal, but situations are known that can alter expected foaling dates. When conception occurs near the end of the breeding season, the length of gestation tends to be shorter.[59,60] Fetuses are most susceptible to loss during early pregnancy (less than 50 days), and much less likely to be lost later in gestation. If a new stallion takes over a harem band, mares that have carried foals for at least 6 months are not affected by the takeover, but pregnant mares might lose an early pregnancy.[16] Estrous behavior during pregnancy is observed in a small percentage of both free-ranging and domestic mares.[61] While not rare, it concerns owners and veterinarians.

Foaling

In free-ranging herds, most foals are born between April and June, with higher-ranking mares foaling first.[15,26] It remains unclear if this early foaling is due to preferential breeding by the harem stallion or to complex neuroendocrine factors.[26] Foaling occurs at night or near dawn in 85% of horses. When foaling occurs during the day, it is usually between 9:00 am and 4:00 pm, and 80% of daytime foaling involves nulliparous mares.[16]

The mare usually goes off by herself as parturition nears, perhaps nature's way of ensuring that mare-foal bonding occurs without interference.[62] The terrain of birthing sites is quite variable. While it may be presumed that flat areas with a great deal of cover would be ideal, free-roaming mares show a slight preference for a ravine or slope. Rocky areas are also common. Even with the potential hazards associated with rocks and slopes, survival rates between these sites does not vary.[16]

Behavior in Domestically Managed Mares

Managed mares usually do not have the social connections of other mares as occurs in a harem band. Continuous exposure to the presence of a stallion will cause mares to cycle earlier and more often than those not exposed.[63] It also tends to synchronize the cycles of these mares.[30] At the beginning of estrus, free-ranging mares play a much more active role in stimulating the stallion than happens in the more managed environments, because she approaches him. Of the precopulatory interactions that ultimately lead to successful breeding, 88% are initiated by the mare.[13] Managed mares rarely have prolonged exposure to a stallion, and with artificial insemination and embryo transfer, some have no exposure at all.

Most of the sexual behaviors managed mares show are similar to those of their free-roaming counterparts, although they can also be abbreviated or varied. Both are seasonally polyestrus and have similar estrous cycles. During the 2 months having the shortest daylight hours, 85%–95% of mares are anovulatory, with quiescent hormone circulation.[64] While larger mares may have a prolonged cycling season and have breed and individual variations, ponies, in contrast, have the most delineated ovulatory season.[12,26,29] Behavioral estrus correlates with ovarian activity in some mares, but for others, especially early in the season during vernal transition, the ovaries are quiescent and estrus is solely behavioral (*psychic heat*). This behavior is associated with hormones from the adrenal cortex.[64–66] Behaviors of ovulatory and nonovulatory cycles are indistinguishable.[11,24,53,54] The display of estrous behaviors during anovulatory periods is thought to facilitate maintenance of the horse's social structure.[67] Health issues such as granulosa cell tumors should be considered when estrous or anestrous periods are excessively long.[68,69] The industry's desire to breed mares for January and February foals is resulting in selection of mares that are cycling earlier than normal. This pushes the trend toward developing true polyestrous mares that ovulate throughout the entire year.

Intensity of estrus varies considerably and does not always correlate with the actual reproductive status or estrogen levels of the mare.[11,50] It can be very strong, associated with a typical estrous behavior; be very weak and difficult to recognize; or fluctuate in strength during the estrous period. Even so, the intensity in managed mares is less than in those closest to being truly feral.[52]

The behaviors of estrus are typically strongest in the 3 days before ovulation, which usually happens in the 3 days before the end of estrus. Sixty-nine percent of the time the oocyte is released in the final 48 h (average is 36 h).[5,26,70] A split estrus can occur during which the mare shows estrous behavior, stops showing it for a day or so, and then demonstrates it again. The frequency of a split estrus occurs in 4.9%–12% of mares, primarily in the spring. The frequency of this splitting in any specific mare is not known.[5,53] Prolonged estrus is most likely to occur in first estrus mares and near the start of the breeding season.[4] It may relate to poor nutrition or other physiological disruptions.[5] With both prolonged and split estrus, the mare's ovaries may be active, but the follicles do not mature.[53] As the season progresses and days lengthen, the cycles usually stabilize. Shortening of the estrous period is normal toward the end of the breeding season.[5] It can also be associated with stress, including that from palpation.[71]

Breeding programs for domestic mares are often highly managed with palpation, ultrasound, or teaser stallions being used for heat detection. By detecting follicular development artificially, there is the potential that a higher percentage of mares will show negative behaviors toward the servicing stallion if the individual mare is not yet in estrus or if she is restrained during exposure to the stallion and unable to separate herself from his advances. Mares that are difficult to handle in an artificial breeding environment and are aggressive to the stallion may be accepting or less aggressive in a paddock environment.[29] When using a teaser stallion, the most reliable indicators of estrus are ears forward, leaning toward the teaser stallion, unwillingness to move from him (*standing estrus*), deviated tail, clitoral winking, the urinary stance, and the absence of kicking.[28,29] Estrous mares commonly urinate when the stallion is presented, and this is often followed by the stallion's flehmen response. Stallions respond primarily to the immobile stance and the head-to-head approach, followed by the mare moving her hips toward his head.[13] Just like stallions, mares can show preference or rejection behaviors in breeding partners if given the opportunity.[5, 26, 47, 72]

The detection of estrus without a stallion can be difficult. While some mares display signs when exposed to other horses, most do not. Mares in heat may mount other mares (Figure 6-5).[29]

FIGURE 6-5 A mare in heat is mounting another mare in the herd. This is common in cattle but uncommon in horses.

Mating

One of the biggest concerns during live-cover breeding arises if the mare has a foal at her side. When separated from the mare, foals whinny, are agitated, and can become very disruptive. If confined in a padded box, they keep trying to escape. They can become very panicky just as the stallion begins to mount the mare, and this tends to upset the mare too. In free-ranging horses, foals tend to be almost oblivious to mating behavior, and if upset, go toward the mare's front, standing parallel to her chest.[73] A "foal-proof" breeding shed is large enough so the foal can be free to explore the inside. This allows the foal to go to the front of the mare if it becomes stressed. Assigning one person to be in charge of guiding the foal if necessary is precautionary. An alternative is to have the padded foal box located close to the mare's head so that she can reassure the foal as necessary.[73]

The sex of the foal is generally attributed to the success of a Y chromosome. It turns out this is not the whole story. The mare plays a small role in determining the foal's sex. Parous mares are more likely to have colts (52.9%) than are nulliparous mares (34.9%). Additionally, 57.4% of foals that develop from oocytes ovulated by the left ovary are male, compared to 35.3% when the oocyte is from the right ovary.[74]

Pregnancy

Under managed conditions, stress during conception and pregnancy affects the number of foals born. This is particularly critical for barren mares. Reducing the amount of teasing by stallions, minimizing stall confinement, and maintaining mares in a small, stable group during the breeding season do not change the initial pregnancy rate. Those same factors, however, significantly lower embryonic death rates so more foals are born.[75]

Fetal growth is related to the size of the mare carrying the foal. Her size interacts with the genotypes to influence the gross area of the allantochorion and the density and complexity of the microcotyledons on the placental surface.[76] In studies of foal size when the fetus was carried in pony, horse, and draft horse mares, the neonate's size highly correlates with the size of the dam and much less with the foal's genetic makeup. Pony embryos in draft and horse mares result in foals that are larger than pony embryos carried in ponies. Conversely, draft and horse foals carried in pony mares are smaller than were draft or horse controls.[76–79] Insulin sensitivity and resistance in these foals were also different.[77,78]

The length of pregnancy is influenced by internal and external factors. Determining when the mare will foal is difficult. Unfortunately, mares can get into serious trouble rapidly and develop dystocia. In domestic mares, most pregnancies last 335–342 days, with a range of 305 to more than 400 days.[80] About 3–4 weeks before foaling, the size of the mare's udder will begin to increase. The teats begin to enlarge within the last few days. In about 60%

of mares, a waxy substance (accumulated serum) appears at the openings of each teat. When present, it serves as an indicator of imminent parturition.[81] During the 3 days before foaling, the mare will show a stepwise increase in activities of walking, feeding, drinking, grooming, and small movements during sleep.

Pregnancies shorter than 325 days are associated with incomplete ossification of carpal and tarsal bones in the neonates, and that can be associated with poor performance in later life.[82] Genetics plays a role, in that the heritability of gestational length is 36%. Male fetuses are carried 4 days longer than females.[83–85] The time of year when the mare was bred also contributes to the length of gestation and accounts for 44% of the variance in length. Pregnancies from breedings done in December through May average 10 days longer than those occurring at other times of the year.[84,86] Gestation length is independent of nutritional influences, but postpartum foal maturation is dependent. Testicular maturation of yearling colts is delayed and insulin resistance is increased during the first 2 years if the dam was on a forage-only diet.[87]

Foaling

Managed mares foal at night and in the early morning 80%–86% of the time, with spring foaling more likely to occur at night compared to winter foaling.[60,81,84] Variations can occur. Light seems to be the primary environmental factor that controls the time of foaling in both free-ranging and owned horses.[84] The constant presence of humans may also change the frequency of nocturnal birthing to daylight hours.

Foaling mares are almost always recumbent and remain so for up to 40 min after delivery.[88] Once parturition begins, it is typically staged into three parts. The first stage is associated with relaxation and dilatation of the cervix and with the onset of uterine contractions. Behaviorally, the mare becomes restless in association with uterine contractions. Stage one can be harder to identify in multiparous and older mares, because they can mask that early discomfort. The mare may pace around her stall or paddock, paw, sweat in her flank and neck regions, look at her flank, show mild signs of colic, flehmen, swish her tail, and/or drip milk. She may also lie down and get up repeatedly. It is during this time that the fetus is moving into a dorsosacral position with extended head and forelimbs. The wedge formed by the front feet and head, which is between the front legs, stimulates cervical dilation.[80] Disturbances during this early stage can abate the mare's progression to foaling, delaying it for several hours or even for days.[59,89] Stage one usually lasts 1–4 h (range 10–330 min). It ends with rupture of the chorioallantois at the cervical star of the placenta and release of the large volume of allantoic fluid.[80,81] The lay name for that event is *breaking water*.

Stage two begins, during which the fetus passes through the birth canal. Toward the start of this stage, the mare may lie down, get up, and circle before lying down again. She may also roll in an effort to correct fetal positioning. On rare occasions, a mare may deliver the foal while standing.[89] Following rupture of the chorioallantois, abdominal contractions are added to ongoing uterine contractions. Stage two abdominal contractions of the mare are quite forceful to bring the fetus up to and through the birth canal. Commonly, the mare is lying on her side. With each expulsive effort, her limbs will extend straight out, and she often grunts. Within 5–10 min of the chorioallantotois rupturing, the whitish-blue amniotic membrane comes into view through the vulvar lips.[80] The amnion is attached only at the umbilicus, and its presence is an indicator that the fetus is entering the birth canal. The fetus' hooves, whether in anterior or posterior presentation, the legs, and finally its torso continue to dilate the cervix and stimulate ongoing, extremely strong contractions (*Ferguson's reflex*). The foal normally presents in the dorsosacral and head first, anterior presentation with one front foot being a few inches ahead of the second one. This leg placement narrows the width of shoulders as they pass through the pelvis, easing delivery. Harsh contractions stop once the foal's hips pass through the birth canal. The amnion usually tears as the foal emerges from the birth canal or by its movements after delivery. If it does not tear by the time the foal has been delivered, immediate human intervention is necessary to prevent suffocation of the newborn. This is especially true if the foal is not vigorous in its actions or the mare does not lick the membrane free from the neonate's face within the first few minutes following delivery. The average length of stage two is 15–47 min.[80,81,88,89] Fetal viability is unlikely if stage two exceeds 60 min.[28] If transit time through the birth canal is too short, "dummy" foals can result. Recently, it was discovered that a rope used around a foal's chest for several minutes will simulate being squeezed in the canal and thereby "arouse" many of these foals as the pressure is released.[90,91]

Stage three involves the expulsion of the fetal membranes and the involution of the uterus during the 30 days postpartum. Over 80% of mares will begin expelling the fetal membranes within 60 min of parturition.[60,92] Regardless, the placental passage should occur within 3 h of delivery. This is the earliest stage of involution, and signs that accompany the myometrial contractions range from mild discomfort to abject signs approximating colic.[28,80,81] Unlike a number of other species, mares rarely eat the placenta after it passes. It only happens after 1.4% of births if the opportunity presents.[93] Because normal equine neonates stand shortly after foaling and quickly "find their legs," the dam's need to eat the placenta to mask evidence of a recent birth and hide a nonmobile offspring from predators is less important. Placental ingestion is also not needed as a nutritional supply for the dam because her food intake over the next several days is not limited by maternal duties.

The umbilical cord remains intact for up to 30 min, eventually breaking as the foal awkwardly tries to stand.[88] It used to be thought that the detachment delay allowed significant amounts of blood to pass from the placenta to the neonate. It is now known that less than 100 mL of blood is transferred via the umbilical cord following birth. Weak neonates are more likely to be in poor health compromised by a low-grade placentitis or in utero acquired septicemia or viremia.[28,94] Because of the extreme exertion of parturition, the mare remains lying for 1–30 min.[81,89,95] Occasionally while she is lying down and most commonly after she stands, the mare will nuzzle the foal and smell and lick the birth fluids. She may flehmen in response to the fluids. A similar response is very pronounced in any horse, including geldings.[22] The mare contacts the fetal membranes an average of 16 times, mainly with the first 10 min postpartum. She will also direct behaviors toward the foal an average of 81 times in the first 30 min.[95] In both cases, the behaviors gradually lessen in frequency. The dam vigorously licks the neonate to help stimulate its activity. By the end of the first day postpartum, the mare's activity level is similar to what it had been 3 days earlier. It then it continues to normalize during the following 2 days.[96]

PARENTAL BEHAVIORS

Maternal Behavior

Maternal behavior is all about foal survival through nourishment, life lessons, and protection. The mare's high investment in energy highlights the importance of a next generation to ensure genetic continuance. Foal survival is enhanced when the mare is a member of a harem group, but without an associated stallion, foals are less likely to survive. Stallions are protective of, and not aggressive to, their own foals. On the other hand, the stallion is significantly more aggressive toward foals that were not sired by him. The dams of these foals are more highly protective. Mares are also more protective of their young in multistallion bands and if the stallion has previously aggressed against the foal.[58,97]

During the first day, the mare will continue to sniff and lick her foal, particularly around its head and perineum. This apparently helps her bond to this foal. The mare's attachment to the foal happens within the first 30 min postpartum. It takes the foal a little longer, with imprinting to the mare developing gradually over the first 2 h.[5] By then, separation results in extreme excitement by the mare and disorientation in the foal. It will try to stay close to and nuzzle the dam.[5,98] The foal's distress at separation continues to increase over the first 2 weeks, suggesting that the strength of bonding continues to evolve.[95] The foal's attentiveness to its mother can begin before stage three is complete, as it circles and nuzzles its dam, standing near her as she lies down to pass the placenta.[5] By the end of the first day, the mare resists attempts of other foals to nurse. Foals increasingly avoid strangers and have a significantly decreased tendency to follow large moving objects.[5,59]

Nursing the foal is essential for its survival. While this occurs, the mare may show different responses. Most mares will stand still with her weight distributed equally (see Figure 7.1A in Chapter 7), but she may also extend the hind limbs slightly. At other times, she may flex the hindlimb on the side opposite from where the foal is nursing. A few mares will continue at a very slow walk while grazing and allow her foal to suckle. In multistallion bands, mothers terminate nursing bouts more often than when there is a single or no stallion present.[58] Nursing frequency is highest during the first 2 weeks of life, and mares terminate about half of those nursing bouts by walking away or, occasionally, by raising a rear limb to block access to the udder.[59,99] After that, the foal ends 70%–80% of the bouts.[100] The duration of a nursing bout is only slightly shorter when ended by the mare than it would be if ended by the foal.

The senses play an important role in mare-foal recognition, especially as the youngster becomes more independent. At great distances, the mare orients and focuses her attention toward the whinnies of her foal, but foals do not seem to recognize the calls of their mothers.[101] Visual and olfactory cues are the most important, especially as the two approach each other. Coat color is one part of this recognition process.[59] As the approach continues, the foal uses the mare's reaction to its approach as the clue for acceptance and responds with the jaw chomping behavior. Soft nickers by both dam and foal are common as they near each other and confirm their identities.[101]

Some aggression toward the foal is considered normal.[4,99,100,102] It is rare during the first 4 weeks, happening in only 0.6%–5% of nursing attempts primarily during the prenursing stage.[99,100] The frequency of aggressive interactions peaks at just over 30% when the foal is between 3 and 5 months of age.[59,99] When aggression occurs, it most likely involves a nulliparous mare. Other than nursing, a mild rebuke may result from the neonate nuzzling and bunting the udder, suggesting it might be painful for the dam.[99] Later, the aggressive behaviors may be a response to pain, such as when a teat is bitten. This is more common when the mare is newly pregnant or in estrus.[99] It also coincides with the time when the foal's primary source of nutrition shifts from milk to grass. In this way, aggression may serve to encourage the foal's transition to grazing.[99]

The mare's aggression seldom lasts for long, and there is a range of agonistic behavior she may show. It varies from the mild flattened ears, squeals, and swishing tails to serious biting or kicking the foal.[59,99,100] One behavior, called *smacking*, is unique to the nursing mare.[59] The ears are flattened. The mare's head is turned toward the foal, and she abruptly opens her mouth making in a loud, smacking sound. Foals usually do not respond at all.[99] If they do, it is to pause or stop the suckling, kick or threaten to kick the mare, or move its rump while continuing to nurse.

The mare-foal duo spend significant time in close proximity with each other. Rather than bedding down the foal while the mare goes off to graze as do cattle and deer (calves and fawns are called *hiders* because of this behavior), the mare and foal stay close to each other (foals are called *followers*). On day one, mares and foals will be within 3 ft. (1 m) of each other over 90% of the time. In subsequent days two through seven, they will spend over 80% of the time within 3 ft. (1 m) of each other and over 90% of the time within 15 ft. (5 m) (Figure 6-6).[59,98–100,103] As the pair moves around the pasture, the mare will stop her normal grazing pattern if the foal lies down—called the *recumbency response* (Figure 6-7).[99,103] She may just stand over the foal or graze in a tight

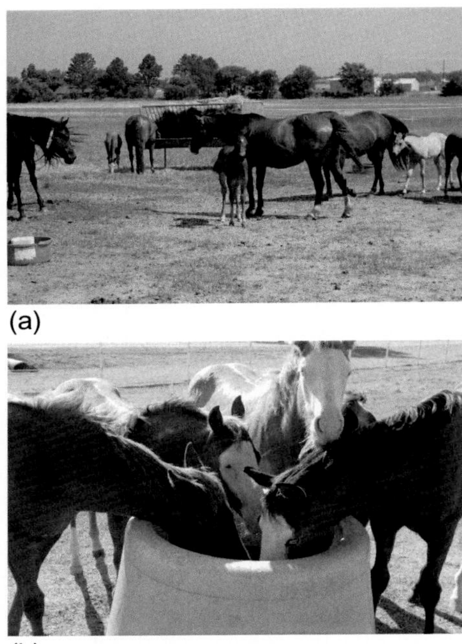

FIGURE 6-6 Social interactions for foals vary with age. (A) Young foals stay close to their mothers, but as they gradually become more independent and (B) the youngsters spend increasing amounts of time with others of their own age.

FIGURE 6-7 Mares stand very close to their sleeping neonates, a behavior called the *recumbency response*. If the mare moves, she remains in a tight circle around the foal or will position herself between her foal and other approaching horses.

circle around it. This is mainly the result of the mare following her foal. The mare-foal distance increases as the foal gets older, and by 21 weeks, the two spend less than half their time within the 15 ft. (5 m) distance.[99,103] By now, the foals are interacting with each other socially and in play. Their usual playmates are foals of mares that associate with the dam.[98] When close to its dam, the foal is more likely to be resting than playing. As the foal continues to grow, it will gradually develop a relationship with its older maternally related siblings.[98]

Mares show a range of social behaviors to other horses immediately before and after foaling. Any yearling that is still nursing will be driven off. This subadult yearling and the mare's 2-year-old offspring will remain her next closest associates in a harem band.[59,98] Sometimes, the only changes shown are less affiliative behavior with other mares and spacing herself and foal farther away.[104] Other mares become increasingly aggressive as parturition nears and then gradually decrease the aggressive behavior as the foal gets older. There is a high energy requirement for nursing the growing foal. A mare will produce between 1800 lbs. (823 kg) and 4000 lbs. (1845 kg) of milk for this one foal before it is weaned.[16] This requires that free-ranging mares spend more time grazing and less time interacting with others in the harem band.[105] The new mother-foal pair tends to separate off from other mare and foal pairs, interacting less with the mares than is done when barren or pregnant.[106,107]

Mares remain protective of their foals against the approach of another horse. This is shown primarily by her positioning herself between the approacher and her foal. If her foal begins to show submissive jaw chomping behavior toward another horse, the mother will actively aggress toward the intruder.[108] Higher-ranking, older mares are less aggressive, especially when compared to mares less than 5 years old.[109]

In some species, mothers will establish differential investments in offspring of one sex. This evaluation is typically initiated while the offspring is very young and still dependent on nourishment from its mother. The behavior can be expressed in ways such as longer or more frequent nursing bouts, later age at weaning, and increased tolerance for neonatal pranks directed toward the dam. This type of preferential treatment forward nursing foals of one sex has not been shown in horses; however, there is evidence that the dam-colt bond for yearling and 2-year-olds is stronger than that for fillies. Dams and subadult sons maintain closer proximity, have higher rates of affiliative interactions, and share more frequent suckling bouts, compared to interactions with her subadult fillies.[110]

If the foal is stillborn, the mare will nuzzle or paw it for several hours. Eventually she will either leave or begin grazing, slowly moving in circles farther away from the body.[59]

Foster Mares

Some mares, particularly those close to parturition, will steal another's newborn foal. Then when she goes into labor, the stolen foal is abandoned. Because the

FIGURE 6-8 A mare that will accept another mare's foal is valuable for raising orphans. This mare has her own foal on the left, but quietly stands while another foal nurses. *(Photo courtesy of Tami Nelson and used with permission.)*

foal and its biological dam failed to adequately bond, the foal is often rejected by the dam too. For orphaned foals, the ideal solution is to have a foster mare. Adopting another mare's foal by a surrogate can be difficult, so the rare mare that willingly accepts such foals is priceless (Figure 6-8). The best candidate is a mare that just lost her own foal. Mothering instincts are lost within 3 or 4 days and is coupled with a regression in udder development and milk supply.[89] Foal acceptance is more successful if the foal was recently born. If the surrogate's placenta was kept, it can be draped over the orphan. Likewise, if available, some of the surrogate's fetal fluids can be smeared on the foal. The older the foal, the less likely the mare is to accept it, but a foal will continue to show a willingness to nurse from any mare for its first few months.[89]

Even if a foster mare does not accept an orphan foal, it is very important for good species socialization that the foal be raised around other horses. This prevents a hyperattachment to the wrong species, and it allows the foal to learn correct equine social signals and manners.[5,111]

Fraternal Behavior

Stallions in free-roaming herds are around mares as they foal. In domestic situations, stallions can also be kept with foaling mares. It is better to keep the stallion with the mares throughout the year rather than to remove him prior to foal arriving and try to reintroduce him afterward.[112] Problems are rare when there is familiarity. The stallion might become more protective, especially as the mare goes into her foal heat, but keeping them together prevents overly protective behaviors. When stallions are kept with mares, it is advisable to have a small catch pen within or adjacent to the pasture so that a mare can be separated from the band should the need arise.

Stallions are able to identify their own offspring and are very protective of these foals. Youngsters that stray too far from the mare will be guided back. Owners do not always realize that is what is happening because the stallion's head is lowered and ears back, similar to the posture used by stallions to drive their harem band including the foals.[112] This protectiveness also makes it difficult to introduce new horses into the herd. Both mares and stallions will be aggressive toward the introduction, which can be particularly problematic in managed herds.

Around 3 weeks of age, the foal becomes increasingly independent from its dam. Soon after, it will start showing playful interest in the stallion by nibbling and even biting him. Stallions tolerate these playful antics.[5]

Stallions show more aggression toward the yearlings and 2-year-olds in the harem band, which is part of the reason that the colts and fillies ultimately leave. His aggression is directed at the colts significantly more than the fillies.[110]

FOAL BEHAVIORS

During the third month of gestation, the fetal horse will begin to show two simple, singular movements approximately every 10 min. This movement peaks at around 16 movements/10 min from the fourth to ninth month of gestation. By the ninth month, the singular movements become biphasic, and gradually increase in complexity with multiple components. The frequency of these complex movements increases to approximately 20/h during the last month of pregnancy.[98] In the last 3 days before birth, the movements cluster into continuous 10-min segments, probably associated with fetal repositioning for passage through the birth canal.

A fetus remains nonresponsive to external stimuli until its pelvis has passed through the birth canal. Then, the foal may take a few gasping respirations, quickly establishing a respiratory rhythm. It can lift its head and then right itself within a few minutes. Pupil constriction is present within 10 min of delivery.[92,98,113]

Once born, it is at least 15 min before the foal tries to stand, then taking many tipsy tries before being successful. This happens, on average, in about 32 min after birth in ponies (see Figure 2.8 in Chapter 2).[81] For Arabian, Anglo-Arabian, Warmblood, and Thoroughbred foals, the time until standing is longer.[60,85,88,92,113] Warmblood fillies average 43 min; Thoroughbred fillies average 56 min. Colts take longer: Warmbloods take 51 min and Thoroughbreds average 70 min. Foals that take longer than 130 min to stand should be considered abnormal.[88,98] Even with an unsteady stance, the foal will start to hunt for the udder (*thigmotaxis*). The first movements are an extending of the head and neck, accompanied by the slow opening and closing of the mouth. Its tongue is extended and it shows sucking behaviors toward the air.[98] Most foals nurse within the first 2h after birth.[92] Although ponies will average closer to a half hour, Saddlebred foals take about 1.5h and Thoroughbreds typically take closer to 2h.[81,98,113] Foals that nurse spontaneously are less likely to develop social problems later. The one that must be taken to the teat or is separated from the mare for up to an hour will remain closer to its mother and play less with other foals by a month of age.[62,114] Meconium is passed within 4h.[81] The first urination occurs in approximately 8h (range is 2.75–15h) after birth, with colts showing the behavior around 6h after birth and fillies first urinating 4h later than the colts.[81,113]

Sleeping is second to eating for neonates. During the first week, foals will be recumbent and resting up to 40% of the time. By week 5, this is reduced to just under 30%, and by week 14, it is about 5%.[99] Foals initiate many of these resting bouts in association with times the dam is already resting in a standing position. This foal-dam resting association begins during the first week with 25% of the foal's recumbent bouts and peaks as high as 60% of the time by week 17.[99]

Locomotor activities initially are helpful for establishing the mare-foal bond.[59] Neonates of many animal species have a tendency to follow and imprint on the first large, moving object they encounter. Proper mare-foal bonding makes it important that that first object be the dam.[4] The newborn foal will be suckling, and the mare terminates the bout by moving away. The foal will try to catch her to be able to nurse again—a reward for following that large object. The repetition of this over the first few days ultimately helps the foal imprint on its mother.

There are other lessons to be learned during the first few days for which the mare's behavior is important. When the mare is comfortable around people and allows humans to brush and feed her by hand, the foal will be more accepting of human contact. The foal will stand closer to people, is more inquisitive, initiates sniffing and touching contacts, and has a significantly reduced flight response compared to foals that were not around people or had mothers that sheltered them.[115,116] By 6months of age, a naïve foal is less predictable in how it will react to human exposure than it was at a few weeks of age.[117]

As foals get older, they gradually become more independent. During the first 3weeks, foals spend most of their time near their mother, and 70% of their playtime is also with their mother.[118] Interactions with other foals start shortly after the third week as the foal becomes more adventuresome. They start with investigative approaches and progress to play. By 2months, they begin to interact with older members of the herd.[119] The mare's rank within the herd is relatively insignificant to the foal's maturation. It does not affect maternal protectiveness, foal independence, or the development of relationships with other foals.[120] While the progression of independence does not vary by sex, it does seem to be consistent among all foals from a particular mare.[59]

Foals begin to explore the world visually and orally (Figure 6-9). Oral exploration extends to social interactions as well. Mutual grooming begins as close

FIGURE 6-9 Young horses explore their environment in several ways, but using their mouth to grasp and pick up objects is particularly common. Most objects are fixed or large enough that the foal cannot swallow them.

bonds form between two foals. Individuals that become the closest are generally of the same age and sex, related, and similar in social ranking.[5] When the pair is composed of two colts, they interact with mutual grooming and long bouts of play fighting. Colt-filly and two filly pairs interact almost exclusively with mutual grooming.[5]

Orphan foals raised with other foals show many of the same developmental behaviors as do foals raised with their dams. One difference is in the frequency of self-grooming. Orphans scratch approximately nine times an hour, while mothered foals do this only five times an hour.[121] Orphans also exhibit less stress behavior when placed alone in novel environments. Over a 5-min period, mothered foals will whinny 19 times, take 508 steps, and defecate twice. Orphans whinny five times, take 149 steps, and do not defecate.[121]

NEUROENDOCRINE RELATIONS TO BEHAVIOR

The relationship between hormones and sexual behavior is complicated, particularly in mares, as are the interrelationships between hormones. The following is a general, simplified view of the relationship of sex hormones to male and female sexual behaviors.

Both stallion and mares show seasonal differences in sexual behavior and hormone circulation. The factors that drive seasonality are complex. In theory, the increased length of daylight triggers the pineal gland to reduce melatonin production, which in turn affects the hypothalamus's steroid production. In part, this is true; however, it is not the total picture.[122] Removal of the pineal gland does not stop seasonal cycles, melatonin levels do not always reflect the photoperiod, and horses kept in constant long-day photoperiods still cycle.[123] Colder than normal environmental temperatures have been associated with a delayed onset, and hot temperatures accompany the shortening photoperiods to signal onset of fall.[70,124] Dietary supplementation speeds up ovulation under natural and artificial light environments.[123] Mares with low body fat will begin cycling later than those with normal to excess fat, but high-energy diets can aid the former.[125]

Stallions

Testosterone and its metabolite, estradiol, are the primary hormones associated with stallion behavior, but their relationship to behavior is complicated. A surge of testosterone around the time of birth masculinizes the fetal brain.[6] Yearlings do show a rise in sex steroids during the normal adult breeding season, but estrogens are the major steroids in immature males.[126] Androgenic hormonal activity is relatively quiescent until puberty at about 83 weeks, when endogenous and exogenous stimuli begin to influence the horse. As melatonin is reduced and other factors indicate the onset of the breeding season, gonadotropin releasing hormone (GnRH) is released by the hypothalamus and transported to the anterior pituitary. This induces the release of luteinizing hormone (LH) which travels to the testes, acting on the testicular Leydig cells to produce the steroid hormones, primarily testosterone. GnRH also stimulates the testes to produce follicle stimulating hormone (FSH).[5,127] Together, testosterone and FSH act on the Sertoli cells to stimulate spermatogenesis.[127] FSH and LH increase in parallel in yearlings during weeks 36–40, then decrease again until weeks 68–80 when both levels rise slowly again until puberty.[9] The levels then increase rapidly for another 12 weeks.

Photoperiods can affect the sexual cycle of the stallion. Artificially creating 16 h of daylight will increase sperm count within a few months.[127,128] Hormone levels not only fluctuate by season, but they vary throughout the day too—higher in summer, the afternoon, and around midnight. Spermatozoa levels do not exactly parallel testosterone production.[5] Ultimately, testosterone is not totally responsible for the increase in sexual behavior that accompanies the breeding season. Estradiol-17β and cortisol have a more parallel relationship.[128,129]

Stallion sexual behavior has a seasonal cycle, and just as with sperm, it is regulated by more than hormones. Experience has a direct connection with stallions' sexual behavior, both in showing or inhibiting it. The number of mares in the harem and the genetic makeup of the mare also play a role.[40,130] Measurements of testosterone levels in different housing and breeding situations suggest that hormone levels are influenced by social context more than sexual activity.[40] Harem stallions have significantly higher testosterone concentrations than do bachelor stallions. If the harem stallion is displaced, his testosterone levels drop sharply. Conversely, when a stallion assumes the status of harem stallion, his testosterone increases sharply.[40,131]

Mares

The interplay of hormones throughout the reproductive cycle of the mare is complex (Figure 6-10) and somewhat variable by season of the year and breed of the horse. As a result, the following discussion is a general picture of the interactions and not a complete physiological description. As in the stallion, the hypothalamus releases GnRH, which is necessary for the anterior pituitary to release FSH and LH, both of which regulate ovarian activity.[64] The divergent pattern of the release of these two hormones is more pronounced in mares than in females of other species.[47] During anestrus, GnRH is barely detectable, but during the mating season GnRH pulses increase to twice an hour during diestrus and hourly during estrus.[70] Because of the bimodal pattern, there is a surge in FSH secretion resulting in follicular growth, maturation, and estrogen synthesis occurring in two ways.[64,70] Ponies, Quarter Horses, and some other breeds have a single follicular wave, while Thoroughbreds, Warmbloods, and other closely related breeds are more likely to have two waves of follicles per estrous cycle.[47,70] If the FSH wave is insufficient, follicles regress. When a dominant follicle develops, it suppresses further FSH as it grows and the dominant follicle

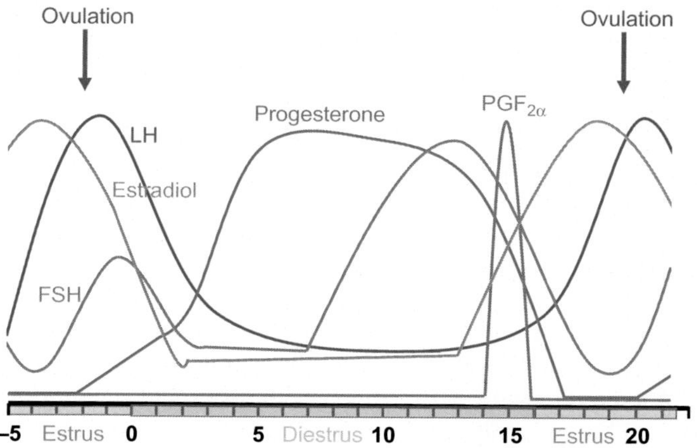

FIGURE 6-10 The relative relationships of sexual hormones during the mare's estrous cycle are complex.[47, 53, 70, 132–135]

produces estradiol beginning in estrus and continuing through to ovulation.[47] Spontaneous double ovulation does occur at a rate of about 2% in ponies, 25% in Thoroughbreds, and 40% of the time when two dominant follicles happen to develop simultaneously.[47] FSH waves are relatively constant throughout the year, unrelated to seasonal cyclicity.

LH works in harmony with FSH for the final maturation of the follicle and ovulation.[64] LH is involved in the development and maintenance of the corpus luteum following ovulation, so its level parallels, but lags behind, that of estradiol. Seasonally, LH is almost undetectable during the anovulatory season, and returns to higher levels during the breeding season.[47,133]

Corpus luteum function is controlled by LH to induce ovulation and prostaglandin $F_{2\alpha}$ ($PGF_{2\alpha}$) to terminate it.[28] Progesterone is secreted by the corpus luteum and is highest during the diestrus and the first 90 days of gestation.[70,135] It inhibits estrus. Progesterone concentration rises immediately following ovulation and reaches a maximum 5–8 days later. It declines rapidly day 17 of the cycle following a spike of $PGF_{2\alpha}$, except in pregnant mares in which it continues.[135,136] There is a slight increase in baseline progesterone levels during the breeding season but not enough to be statistically significant.[133]

Estradiol is the most active form of estrogen in nonpregnant mares, and estrone sulfate, a product of the fetus and placenta, has that role during pregnancy.[134] Estradiol is secreted by the developing follicle, and reaches its highest level about 2 days before ovulation. This is the hormone responsible for the estrous behaviors. Estrone sulfate rises rapidly from day 60 of gestation and remains elevated throughout gestation. At parturition when the fetus and placenta are gone, it drops to zero.[28] The equine pregnancy transitions from ovarian to placental maintenance between 70 and 90 days.[28] The placenta is a significant hormone factory. Both estrone sulfate and total estrogens are endocrine measures used to confirm pregnancy in a mare that cannot be palpated, such as small miniature mares, mustangs, and others difficult to examine safely.

Prostaglandin $F_{2\alpha}$ also plays a role in the cycle by initiating luteolysis in the late luteal phase 4 days prior to the next estrus.[47,64,134] As a result, progesterone levels drop precipitously and estrus can begin.

Corticosteroids and stress can either inhibit or facilitate reproduction in mares, depending on the length of exposure and amount of estrogen priming.[137] Stress can inhibit the release of GnRH and FSH to negatively impact sexual behavior. If, however, the mare has estrogen priming, steroids can play a role in the initiation and synchronization of the preovulatory LH and FSH surge.

The major histocompatibility complex (MHC) is a group of genes that are known to play an important role in the adaptive immune response, social signaling between parent and progeny, and kin recognition in horses and other vertebrates.[138] Reproductively, MHC affects both mares and stallions. Mares are more likely to become pregnant if exposed to an MHC-dissimilar male, with a pregnancy rate of approximating 50% compared to less than 40% if there are shared MHC antigens.[139] For stallions, exposure to MHC-dissimilar mares results in higher testosterone levels and mean sperm numbers per ejaculate than if that stallion is near MHC-similar mares.[138]

PROBLEMS RELATED TO REPRODUCTIVE BEHAVIORS

Certain behavior problems are sex-specific. These relate to normal reproductive and neonatal care behaviors but have become excessive or reduced.

Problems in Male Sexual Behavior

Stallions can show a number of behavior problems that may or may not relate to their physiological ability to breed a mare. Behavioral abnormalities can appear to be similar even though underlying causes can differ. Many excellent articles and books exist describing specifics of training stallion handlers, building appropriate facilities, and working with horses during collection to minimize problems. The intent of the following section is to describe the types of behavior problems that can occur and possible solutions.

Shy Breeders

Shy (slow) breeders are stallions that fail to show normal libido around mares. When compared to normal stallions, this can be manifested by taking significantly longer times to mount, latency to achieve an erection, and/or an increased number of mounts before finally collecting an ejaculate.[129] Several things can be causative. Some genetic lines are quiet, cautious breeders.[34] Weanlings without same-age play partners can become shy breeders, particularly during the first few times they are used.[6] This is somewhat similar to behavior observed in horses that are bonded more strongly to humans than to other horses.[25] Stallions with a history of being repeatedly reprimanded for showing normal sexual behaviors when exposed to estrous mares, such as with retired successful race and performance horses, can be shy breeders when they transition to stud work. The stallion needs to be retrained, to learn that expressing libido, interest in mares, and mounting to breed are normal and acceptable behaviors. This is not a license for bad behavior, but with patience and encouragement he can accept what is appropriate. In some cases, shy breeders may have had a bad experience associated with a previous breeding, including physical or psychological injury.[5]

A shy, novice breeder can benefit from being around patient, usually older, mares to gain confidence and help increase libido from increasing testosterone.[13,34,40] It is helpful when he is exposed to an estrous mare that she is tolerant of him and she is allowed some natural movements. Walking the mare toward the stallion, allowing a head-to-head approach, and letting the stallion sniff the mare are examples of positive interactions.[13] Some stallions have favorite mares or colors, which can also help increase libido.[6,140]

Overuse as a breeding stallion, particularly young horses, can result in low libido, impotence, control difficulties, or aggression. Excessive aggressiveness occurs more often when the stallion is breeding outside the normal season.[6] Once the stallion exhibits normal libido but aggressiveness continues, it may be necessary to use a muzzle or withers shroud to protect the mare and prevent his savaging her during breeding.[6]

Teaser stallions that check mares for heat but do not breed can develop problems. If used too often to tease without the opportunity to breed, the horse's libido can decline. These stallions also tend to develop stereotypic behaviors, particularly stall weaving.[6] The ratio of teasing to mating opportunities necessary to avoid problems differs for individuals, but it should be taken into account.

Physical problems must be corrected to improve the stallion's performance, especially those that relate to genital or limb injuries.[6] Other types of musculoskeletal pain, hormonal abnormalities, and obesity can result in diminished interest. Ground (standing) collections instead of mounted ones may be needed if physical problems cannot be helped with time for healing or medication.

Environmental factors must be considered when stallions are reluctant to breed. The physical environment where mating or collection take place should not have slippery floors, low ceilings/roofs, windows that cast light on the floor, reflections, noise, floor patterns, or protrusions on the floor. These can create an aversion to the location. Spaces with a dimension of less than $25\,ft^2$ ($7.6\,m^2$) are potentially dangerous to people in the area.[140] Excessive cold weather can aggravate arthritis of the limbs or spine.

The stallion handler is probably the most important factor in hand-breeding stallions. Handlers can either encourage calm and correct behaviors or, if inexperienced or incompetent, they can cause aggression and make a bad situation worse when working with an already hard to control horse. For this reason, it is critical that the person be familiar with the routine of the breeding operation and experienced in controlling horses.[6] Certain traits have been identified as particularly important to establish a good working rapport with the stallion.[140,141] These include, but are not limited to, being firm but respectful of the horse; recognizing normal stallion sexual behavior; providing positive reinforcement rather than punishment; using clear and consistent cues; and avoiding jerking on the stallion's shank during the initial teasing before washing, approach to the mare or phantom, and through mounting and collection.[28]

Prepping the young stallion reduces the likelihood that he will be a long-term problem.[142] Simulating a natural environment is helpful, but providing access to mares, even if a safe fence separates the horses, is beneficial. Letting the horse approach the mare's head, allowing mounting even without an erection, and giving access to multiple estrous mares can be helpful for him to develop normal, controlled behavioral patterns. For shy breeders being collected for artificial insemination, if is preferable to limit the number of people in the area to minimize distractions.

Several little tricks may also be useful to increase the stallion's interest in breeding.[142] Some techniques involve mare movement by walking the mare forward a few steps and stopping her suddenly or by backing her up alongside a phantom to allow the stallion to reach across and touch her. Chest pressure on the phantom while he is stretching to reach her can stimulate him to mount. Other techniques include using blinkers or a blindfold to focus the stallion's attention, using urine or vaginal wipes from a stallion's favorite mare to rub on other mares, or dribbling a small amount of estrous mare urine on the phantom before he approaches. When frustration levels start to creep up, recruit an experienced stallion handler to help.

Medications can be useful in certain situations. When a previous bad experience has occurred or the stress associated with a new environment is contributory, benzodiazepines can be helpful.[143,144] Medical problems like arthritis may require the use of pain-relieving drugs. Hormonal manipulation is complicated and not within the scope of this discussion.

Aversion to Washing the Penis

A stallion that has been disciplined for sexual arousal may develop an aversion to having anyone come near his penis. The approach of a person will result in aggressive or aversive behaviors and loss of the erection. Such stallions can be rehabilitated using desensitization. The concept is to reward small, progressive moves that approach, first, the flank area, and over time the prepuce and penis. Handlers must remain calm and patient—no yelling, hitting, or other negative corrections may be used. The process begins with no mare present, brushing the stallion's sides and flanks, gradually getting closer to the prepuce, penis, and scrotum with each lesson. Food treats may also be used as positive reinforcement with each step.[145] The key is to progress at a pace slow enough to lessen the likelihood of triggering undesired behavior and to reward the acceptance of touching. When the stallion is comfortable with people touching his genitals, he is collected several times in rapid succession (the reward) without washing. This is positive reinforcement for his allowing people to touch that area. The touching sequence is repeated as before and continues until he tolerates washing of his penis.

Aggressive Breeders

Sex-related aggression occurs during mating. The classic example is the nape bite (*pinching*) by the stallion of the mare's neck. Although this is rarely a problem, there are occasional stallions that are particularly aggressive in their approach.[146] Stallions can show extreme aggression toward a mare or handler, making them particularly dangerous.[25] For some, the aggressive behavior is consistent, but for others, it is unpredictable. Aggression can also be tied to the season, with stallions used outside the normal breeding season more likely to be difficult to handle.[147] In a few unique situations, stallions have been known to charge boldly into the breeding shed and aggressively attack the mounting dummy. They may bite and even kick at it. Then the behavior reverts to that more typical of a stallion preparing to breed a mare.[148] One explanation for such behavior is that it mimics that of a harem stallion chasing away competing stallions prior to mating with the estrous mare. A second possible reason for the charging is that early in the horse's training with a mounting dummy, the stallion may have been allowed to get by with that behavior. By not immediately correcting his inappropriate charging and allowing him to continue and be collected, the stallion's bad behavior was reinforced. It is essential, especially in the earliest lessons for mounting and collecting, that desirable routines be established. If a mount is bad, back the stallion off the phantom mount, wait for erection to recover, and then make a measured, next attempt. Only when the process is done as desired, and safely, will the artificial vagina be placed on the penis and the stallion's ejaculate collected.[28] In either case, the behavior is dangerous to handlers. Retraining does not have to be accompanied by punishment. It begins with a new, novel collection area and dummy mount. With a very experienced handler, the stallion is well exercised and then introduced to the collection area. Methodical, quiet training is essential to undo bad practices. The "one step forward, stop, turn, and go away" approach teaches the stallion that a-step-at-a-time is preferred to charging. In the most extreme cases, it may be necessary to blindfold the stallion to capture his attention by forcing him to listen to the handler.[148] Additionally, it slows the stallion's advances.

Older Stallions

Older stallions are likely to develop physical ailments that make mating unpleasant, with musculoskeletal pain being the most common problem. It is desirable to maintain the usefulness of valuable breeding animals for as long as possible, so precautions need to be taken to prevent negative experiences that ultimately reduce the horse's willingness to breed. Most obvious is to maintain good physical health, including pain management and appropriate levels of physical fitness based on the individual's capabilities and needs. To minimize the energy demand associated with breeding, management aids and therapeutic enhancement of libido or induction of ejaculation are helpful. Managing the stallion's breeding schedule, the size of his book of mares bred during the season, and frequency of and intervals between collections are important.[149]

Another age-related problem in stallions is "crankiness." A confident, well-adjusted stallion may become the equine equivalent of a "grumpy old man" as androgen levels go down. These stallions appear to become less confident and more dependent on aggressive behavior toward herdmates.[150]

Stud-Like Behavior at the Wrong Time

Stallions that are being shown or raced are expected to behave in the presence of other horses with minimal nickering and no signs of sexual arousal. While many horses come close to that expectation, some are extremely difficult to handle. They climb stall walls when a mare is walked by or they whinny excessively. They may be difficult to handle and suddenly blow up while being ridden. Finding the best way to minimize this behavior is difficult. A significant factor necessary to work with these horses is the development of a strong bond and trust between the stallion and his handler, rider, or driver. Some people try to intimidate the horse, but in the long run, this can actually create a more dangerous situation. This is the horse that will suddenly grab the person by the shoulder, arm, or clothing if the opportunity arises. Because stallion behavior is hormonally related, there is interest in using medication to control it. The list of possible drugs includes tranquilizers, progestins, gonadotropin-releasing hormone (GnRH) vaccinations, GnRH antagonists, and GnRH agonists.[151] Progestins reduce the stallion's aggressive behavior, and their use may be linked with a delayed return to breeding soundness. GnRH vaccines result in long-term suppression of libdo, with a very prolonged suppression of sexual behavior in most horses. If vaccines are boostered, suppression may become permanent.[28] GnRH agonists can have very mixed reactions and antagonists work for a few weeks.[151] Important considerations for the use of drugs to suppress stallion behavior are medication rules that might apply at the horse show or race, and the possibility of long-term effects on breeding soundness.

Stallion-like behaviors are common sequelae to the use of anabolic steroids in stallions, mares, and geldings. Their application is associated with attempts to build muscle in halter horses. Resulting behaviors may include stallion-like vocalizations, increased attention to approaching horses, sexual investigation, handling problems, and increased aggression.[152]

Infanticide

In several species, when a new male takes over control of a female group, he may kill the young that are still nursing. This results in the females returning to estrus sooner so that the new male's offspring will arrive in the next birthing season. Stallions are known to aggress toward and even kill the young of other stallions. When a mare aborts a fetus or her foal dies within the first 4 months after birth, the mare is significantly more likely to return to estrus, be bred by the new stallion, and foal his offspring in the following year compared to mares raising a foal to weaning or to barren mares.[58,153] In free-roaming bands, stallions will show aggression toward foals that they did not sire at a rate of once every 15 h. They are more likely to show aggression toward male foals, and they do not aggress at all toward their own offspring.[16,97]

Domestic stallions are also known to kill foals. Owners who have attempted to prevent the attack will frequently have the stallion turn on them first. Predicting the behavior is difficult, but any stallion that has previously shown aggression to foals should be prevented access.

Stallion-Like Behavior in Geldings

It has been shown that 20%–30% of geldings continue to show stallion-like behavior.[43] These horses, commonly called a *false rig*, can show behaviors that include sniffing with flehmen, herding other horses, arching of the neck, aggression, mounting with pelvic thrusts, and penile intromission. While these horses may not be particularly difficult to work with for humans, it is best to keep them away from other horses that receive the unwanted attention.[154] This might mean confining the gelding with one or two other horses that do not stimulate his sexual arousal. Even if geldings and mares are kept in separate groups, this stallion-like behavior is not affected.[155,156] In same sex or mixed groups, the level of injuries and physical contact due to the gelding's aggression is the same.

Common beliefs hold that geldings that behave like stallions have remaining testicular tissue, as could happen with a cryptorchid horse when only the descended testis was removed, or part of the epididymis remains (*proud cut*). The epididymis is closely attached to the testicle, making it unlikely that any would be left behind. Regardless, the epididymis neither produces nor releases hormones, so remaining epididymal tissue cannot be a source of testosterone.[157] Breed incidences suggest that cryptorchidism is overrepresented in Percherons, American Saddle Horses, Quarter Horses, and ponies, and that it is underrepresented in Thoroughbreds, Standardbreds, Morgans, Arabians, and Tennessee Walking Horses.[158] The adrenal gland is also associated with testosterone production, but for most of these stallion-like geldings, testosterone, dihydrotestosterone, and estrogen levels are the same as they are in unaffected geldings.[5,157,159] Hormone assays are helpful in determining whether there is an internal source of testosterone. Geldings generally have a basal serum testosterone concentration of less than 40 pg/mL, while horses with testicular tissue have greater than 100 pg/mL.[157,159,160] Conjugated estrogen levels in geldings are very low to undetectable, but if testicular tissue is present, levels exceed 400 pg/mL.[159]

When the stallion-like behaviors begin in an older gelding, it may be the result of a pituitary tumor. These horses may respond to cyproheptadine to block the release of GnRH.[161]

Problems in Female Sexual Behavior

Aspects of reproductive behavior and foaling usually occur without problems, allowing the horse owner to watch the special bonding between dam and neonate without interference. Unfortunately, there are times when things go awry, and the resulting behaviors are problematic. Abnormal mare behavior is a common reason that bilateral ovariectomies are requested for a misbehaving mare.[162,163] Behavioral improvement is significant in spayed mares compared to those with similar problems but left intact.[164] Improvement occurs in 86% of mares with aggression problems, 81% of those that are disagreeable, 75% of mares that are easily excited, 73% known for kicking and biting, 72% with problems in training, 64% of those frequently urinating, and 64% of those having problems with other horses.[163] Ovariectomies in normally cycling mares with problems attributed to hormones often have different results. In a 4-month period, intact mares demonstrated behavioral estrous behavior for 44 days compared to 152 days for ovariectomized mares, so just an objection to estrous behavior is not likely to be satisfactorily addressed with this surgery.[66]

Silent Heats

Some mares with normal ovarian cycles show little to no outward behavioral signs, a condition called *silent heats*. Other mares show very subtle sexual behavior and are said to have subnormal estrus. The frequency of these conditions is approximately one in every four mares, even when they have unrestricted exposure to a stallion.[29] The problem tends to happen in individual mares and recurs with each cycle. Physiological reasons for anestrus behaviors must be ruled out before this diagnosis can be made. Stress, such as trailering the mare to the stallion and stallion preferences, can suppress estrous behaviors.[53] Ultimately, serial examinations to follow the mare's reproductive activity using transrectal palpation and ultrasound examinations are necessary for a successful pregnancy.

Reproductive Behavior in Nonovulatory Mares

Noncycling mares can show estrous-like behaviors. This happens in about 5% of pregnant mares, typically during days 40–70 of pregnancy in association with secondary follicles that ultimately become secondary corpora lutea.[29,53] Estrous-like behavior has also been correlated to progesterone concentrations that are lower than normally associated with pregnancy.[29,165] Sexual behavior can be shown by seasonally anestrus mares and by 35% of ovariectomized mares.[162] In both types of mares, sexual behaviors are demonstrated when exposed to a stallion. Almost all tolerate mounting, and 70% allow intromission.[65]

Sex-Related Aggression

Mares are more likely to show aggression during mating. If the mare is not in estrus, she may kick or strike the approaching stallion. Such reactions can cause severe injuries such as broken bones, penile hematomas, or testicular bruising. For this reason, artificial insemination and breeding hobbles are popular. In harem groups, dominant mares may aggressively prevent subordinate mares from breeding by biting until the low-ranking mare leaves.[146,166]

Stallion-Like Behavior in Mares

Occasional mares will show stallion-like behavior, most typically that of mounting other mares (Figure 6-5). This usually happens when the mounting mare is in the early luteal phase of her cycle. On that day, her plasma testosterone concentration is higher than for the mare being mounted.[167] Mounting behavior can be directed toward mares that are in heat as well as mares that are not in heat. The stallion-like behavior can be more severe than mounting, however. Granulosa cell tumors are associated with stallion-like behavior (an arched neck, vocalization, and sniffing) in 50% of mares, aggression in 31%, prolonged estrus in 19%, and persistent anestrus in 8%.[5, 50, 68, 69, 164]

Mares can have testosterone levels that approach or even exceed levels found in stallions.[168,169] These mares have very toned musculature and show aggression to people and other horses. In some cases, affected mares had normal external genitalia, normal karyotyping, but abnormal internal genitalia.[169] A genetic component might be involved, at least in some cases. High testosterone levels have been found in three affected sister mares and in a mare and her filly.[168,169] While the behaviors are similar because of the elevated testosterone levels, the causes may be different. Impairments in cholesterol enzymatic transformation or tumors of the adrenal cortex or ovary are possible.[168]

Foal Rejection or Ambivalence

Ambivalence toward a foal by the mother is probably the most common cause of inadequate mare-foal bonding.[170] While 82% of normal mares lick their foals after birth, only 38% of rejecting mares do.[171] Ambivalence to a foal is often observed as a sequel to illness or injury of either the mare or foal, early separation, or excessive human interference during the critical first few hours. Early separation of the foal for treatment can change its odor or appearance, which can then affect its acceptance by the mare.[171] Outright rejection of the foal is a serious problem for its survival.

Rejection occurs in multiple breeds, at various ages, and at variable times after the foal's birth.[172] The majority of rejecting mares are between 4 and 10 years old. Primiparous mares are the ones most likely to completely reject their foal, constituting about 75% of the rejections.[173] This can happen even if the mare shows normal sniffing/licking behavior after giving birth. In some cases, there appears to be a genetic component. The rate of rejection of Arabian foals is 5%, but only 2% in Paint horses and less than 1% in Thoroughbreds.[173–175] Most rejections (70.4%) occur immediately after birth. Another 8.1% occur within the next 12 h, and 10.4% of the foals are abandoned between 12 and 24 h after birth.[173] Sometimes rejection occurs several weeks after foaling.[100,172] Hormone levels of mares that reject foals are different from normal mares, in that prefoaling progesterone levels are significantly lower and estrogen is also somewhat lower in rejecting mares.[173] The postpartum estradiol to progesterone ratio is significantly higher in nonrejecting mares and it increases even more in the first 3 days. This is because the progesterone levels decrease so much more in rejecting mares from a mean of 3.14 ng/dL on day one to 0.49 ng/dL on day three compared to 0.8–0.43 ng/dL in the same period for normal mares.[174]

Rejection can range from mild disinterest to more extreme degrees. With the most common form, the mare rejects the foal's attempts to nurse and progresses to her physically attacking the foal. Primiparous mares, in particular, may block the foal's attempts to nurse. This can be associated with the mare trying to keep the foal in view, so she keeps turning her body to accomplish this.[89] Other mares move away, as if there is a fear component to her rejection. The fearful mare may suddenly turn aggressively on her foal, a situation that requires immediate supervision. The pair should be moved to a large paddock so the mare can avoid the foal if she chooses.[170] Giving the dam time for her hormones to equilibrate postpartum before gradually reintroducing her foal may be sufficient for her to accept the neonate.[92] Most foals are accepted by the dam after a few successful nursing bouts during which the mare is physically restrained. Antianxiety medications such as a benzodiazepine may be indicated.[59]

Painful udders or other sources of pain can cause nursing-only avoidance, with otherwise normal bonding and protectiveness.[170] Infections require medical treatment. If the udder is just swollen, hand-milking or physical restraint until the foal can nurse relieves the pressure and her discomfort. If the mare allows humans to strip out milk, it is likely that novelty of the foal's attempts is causing rejection, and only short-term management will be required.[59] Prefoaling, brief, gentle massage of the mare's teats can acclimate her to touch and minimize rejection of the foal's suckling efforts once it is born.[28]

The most serious form of foal rejection is when the dam savages the foal. She may bite the foal's neck and back, kick it severely, shake, toss or stomp on it, and hold the foal on the ground.[59,170,173] There may be aspects of normal maternal behavior, but affected mares are less likely to lick, nicker, and defend their foals postpartum than normal mares.[176] The behavior can occur soon after birth or several days later. It usually happens when the foal is standing. Restraint and tranquilization with acepromazine to stimulate prolactin may help for a while, but relapses are common. Separation may become the only safe action.[170,175] A genetic origin is suggested because this problem happens primarily in Arabians.[59, 172, 174, 176] Savaging will recur with each new foal, and because there is a possible genetic component, embryo transfer would only pass the behavior on to the next generation.

Foal rejection is serious and requires immediate attention to ensure appropriate bonding and intake of colostrum.[59] Hand-milking the mare is a short-term solution that can help with colostrum administration. This is often inadequate for providing maximum passive immunity or keeping the foal adequately hydrated. Sedation may allow suckling but doing it often is not in the mare's best interest. In some cases, tying her head and hobbling her rear legs will provide an opportunity for suckling. Stocks or a tie stall allow safe nursing by preventing the mare from moving sideways or forward.[59,170] If the mare continues to kick at the foal, it may also be necessary to cover one side of the stocks with a sheet of plywood. A large rectangle is then cut from the plywood or from the solid side of an existing stocks, large enough for the foal to safely puts its head through. It is positioned so the foal can reach the udder. For mares that squat to prevent nursing, a hay bale or bag of shavings can be put under the abdomen to limit the downward motion.[173] When mare restraint is necessary, it tends to be difficult to do so often enough to maintain lactation.[173]

Ultimately, it is necessary to get the mare to bond to the foal. Restraint feeding should be done a few days, at least. If it is needed longer than 1 week, acceptance is unlikely.[59,170] Alprazolam, a benzodiazepine and dopamine blocker, may successfully reduce the anxiety of the mare and facilitate bonding with her foal.[177] The advantage of alprazolam is that by blocking dopamine, prolactin release is stimulated to help milk production.[173] Oxytocin administration is another option used to help promote milk let down. The third option is progesterone, used because postpartum levels are quite low.

Techniques can be used to try to stimulate a mare's maternal behavior. Separating the foal from the mare will trigger anxiety, agitation, and neighing in normal mares. This may be enough to trigger a rejecting mare's interest in her newborn. If not, it indicates a poor prognosis for successful bonding. Maternal protection might be triggered by the presence of a dog, just as would happen in the wild if a carnivore began stalking the baby. Mares also protect their foals from the approach of other horses, so putting the pair in a paddock with other mares or geldings may be sufficient to stimulate maternal behavior.[59]

The outcome for rejected foals is variable. Of 95 rejected foals with a known outcome, 36 (37.9%) were eventually accepted by the mares, 49 (51.6%) required fostering or hand-feeding, and 10 (10.5%) were killed or died.[172] Eventual acceptance depends on keeping the foal near the mare and allowing it to suckle approximately every 3 h, approximating conditions that are normal.

As with all behavior problems, prevention is better than treatment. Foaling mares should be kept in a quiet area, and any type of disturbance should be prevented. Even the ability to see other horses nearby can be disturbing and result in the mare redirecting her aggressive intentions from them to her foal.[59,170]

Excessive Maternal Behavior

Mares may show excessive maternal behavior toward their own foals, other young, or humans. Excessive licking, especially that which continues past the normal first few days, would be an example of excessive behavior directed toward the mare's own foal. In other cases, a mare may adopt another mare's foal. The adopter usually is a pregnant mare that soon will be foaling, and the behavior is probably related to changes going on with the various hormones. These mares abandon the adopted foal as soon as their own foal is born, which can drastically interfere with the stolen foal's ability to bond with its real mother. Occasionally, a mare may adopt a foal, or even a calf, if she has been separated from her own baby or if it died.[5,59]

Excessive maternal behavior is more likely to take the form of excessive protectiveness of the foal. Some aggression is normal, especially in the first few weeks, but when the mare becomes aggressive to humans or other foals in addition to other adult horses, it is excessive. Care must be taken not to set off this response when the foal is between the mare and the potential target of her aggression. Safety is enhanced if the pair are in a paddock rather than in a stall.[170]

REFERENCES

1. Santos MM, Maia LL, Nobre DM, Neto JFO, Garcia TR, Lage MCGR, de Melo MIV, Viana WS, Palhares MS, da Silva Filho JM, Santos RL, Valle GR. Sex ratio of equine offspring is affected by the ages of the mare and stallion. *Theriogenology* 2015;**84**(7):1238–45.
2. McPherson FJ, Chenoweth PJ. Mammalian sexual dimorphism. *Anim Reprod Sci* 2012;**131** (3–4):109–22.
3. Crowell-Davis S, Houpt KA. The ontogeny of flehmen in horses. *Anim Behav* 1985;**33** (3):739–45.
4. Tyler SJ. The behavior and social organization of the New Forest ponies. *Anim Behav Monogr* 1972;**5**(2):87–196.
5. Waring GH. *Horse behavior*. 2nd ed. Norwich, NY: Noyes Publications; 2003. p. 442.
6. Houpt KA, Lein D. The sexual behavior of stallions. *Equine Pract* 1980;**2**(5):810, 12–13, 16, 19–22.
7. McDonnell SM. Raising a stud cold. *The Horse* 2008;**XXV**(3):68–9.
8. McDonnell SM. Young guns. *The Horse* 2009;**XXV**(10):52.
9. Naden J, Amann RP, Squires EL. Testicular growth, hormone concentrations, seminal characteristics and sexual behavior in stallions. *J Reprod Fertil* 1990;**88**(1):167–76.
10. McDonnell SM. When is he mature? *The Horse* 2001;**XVIII**(11):79.
11. Kirkpatrick JF, Turner Jr JW. Comparative reproductive biology of North American feral horses. *Equine Vet Sci* 1986;**6**(5):224–30.
12. Turner Jr JW, Kirkpatrick JF. Hormones and reproduction in feral horses. *Equine Vet Sci* 1986;**6**(5):250–8.
13. McDonnell SM. Reproductive behavior of stallions and mares: comparison of free-running and domestic in-hand breeding. *Anim Reprod Sci* 2000;**60-61**:211–9.
14. Keiper RR. Social structure. *Vet Clin N Am Equine Pract* 1986;**2**(3):465–84.
15. McCort WD. Behavior of feral horses and ponies. *J Anim Sci* 1984;**58**(2):493–9.
16. Berger J. *Wild horses of the Great Basin: social competition and population size*. Chicago: University of Chicago Press; 1986. p. 326.
17. Jezierski T, Jaworski Z, Sobczyńsk M, Kamińska B, Górecka-Bruzda A, Walczak M. Excreta-mediated olfactory communication in Konik stallions: a preliminary study. *J Vet Behav* 2015;**10**(4):353–64.
18. Salter RE, Hudson RJ. Social organization of feral horses in western Canada. *Appl Anim Ethol* 1982;**8**(3):207–23.
19. Lindsay FEF, Burton FL. Observational study of "urine testing" in the horse and donkey stallion. *Equine Vet J* 1983;**15**(4):330–6.
20. Turner Jr JW, Perkins A, Kirkpatrick JF. Elimination marking behavior in feral horses. *Can J Zool* 1981;**59**(8):1561–6.
21. Kimura R. Volatile substances in feces, urine and urine-marked feces of feral horses. *Can J Anim Sci* 2001;**81**(3):411–20.
22. Saslow CA. Understanding the perceptual world of horses. *Appl Anim Behav Sci* 2002;**78** (2–4):209–24.

23. Stahlbaum CC, Houpt KA. The role of the flehmen response in the behavioral repertoire of the stallion. *Physiol Behav* 1989;**45**(6):1207–14.
24. Ransom JI, Cade BS. *Quantifying equid behavior—a research ethogram for free-roaming feral horses*. U.S. Geological Survey; 2009.*https://pubs.usgs.gov/tm/02a09/pdf/TM2A9.pdf*. [Accessed 15 March 2017].
25. McDonnell S. Reproductive behavior of the stallion. *Vet Clin N Am Equine Pract* 1986;**2**(3):535–55.
26. Curry MR, Eady PE, Mills DS. Reflections on mare behavior: social and sexual perspectives. *J Vet Behav* 2007;**2**(5):149–57.
27. Powell DM. Female-female competition or male mate choice? Patterns of courtship and breeding behavior among feral horses (*Equus caballus*) on Assateague Island. *J Ethol* 2008;**26**(1):137–44.
28. Carleton, C. (2018): Personal communication.
29. Asa CS. Sexual behavior of mares. *Vet Clin N Am Equine Pract* 1986;**2**(3):519–34.
30. Heitor F, do Mar Oom M, Vicente L. Social relationships in a herd of Sorraia horses. Part II. Factors affecting affiliative relationships and sexual behaviours. *Behav Process* 2006;**73**(3):231–9.
31. Linklater WL, Cameron EZ, Minot EO, Stafford KJ. Stallion harassment and the mating system of horses. *Anim Behav* 1999;**58**(2):295–306.
32. Feh C. Alliances and reproductive success in Camargue stallions. *Anim Behav* 1999;**57**(3):705–13.
33. McDonnell S. Breeding on cue. *The Horse* 2006;**XXIII**(1):51–2.
34. McDonnell SM. Stallion libido. *The Horse* 2001;**XVIII**(3):29–30.
35. Turner RM, McDonnell SM. Mounting expectations for thoroughbred stallions. *J Am Vet Med Assoc* 2007;**230**(10):1458–60.
36. Dinger JE, Noiles EE. Effect of controlled exercise on libido in 2-yr-old stallions. *J Anim Sci* 1986;**62**:1220–3.
37. McDonnell S. To breed, or not to breed. *The Horse* 2009;**XXVI**(4):41–2.
38. Dowsett KF, Pattie WA. Characteristics and fertility of stallion semen. *J Reprod Fertil Suppl* 1982;**32**:1–8.
39. Parlevliet JM, Kemp B, Colenbrander B. Reproductive characteristics and semen quality in maiden Dutch Warmblood stallions. *J Reprod Fertil* 1994;**101**(1):183–7.
40. Burger D, Dolivo G, Wedekind C. Ejaculate characteristics depend on social environment in the horse (*Equus caballus*). *PLoS One* 2015;**10**(11) https://doi.org/10.1371/journal.pone.0143185.
41. McDonnell SM, Henry M, Bristoif F. Spontaneous erection and masturbation in equids. *J Reprod Fertil Suppl* 1991;**44**:664–5.
42. Wilcox S, Dusza K, Houpt K. The relationship between recumbent rest and masturbation in stallions. *J Equine Vet Sci* 1991;**11**(1):23–6.
43. Line SW, Hart BL, Sanders L. Effect of prepubertal versus postpubertal castration on sexual and aggressive behavior in male horses. *J Am Vet Med Assoc* 1985;**16**(3):249–51.
44. McDonnell S. Stallion-like behaviors. *The Horse* 2006;**XXIII**(12):53–4.
45. Franzen JL, Aurich C, Habersetzer J. Description of a well preserved fetus of the European Eocene Equoid *Eurohippus messelensis*. *PLoS One* 2015;**10**(10) https://doi.org/10.1371/journal. pone.0137985.
46. Monard A-M, Duncan P. Consequences of natal dispersal in female horses. *Anim Behav* 1996;**52**(3):565–79.
47. Aurich C. Reproductive cycles of horses. *Anim Reprod Sci* 2011;**124**(3–4):220–8.
48. Rutberg AT, Greenberg SA. Dominance, aggression frequencies and modes of aggressive competition in feral pony mares. *Anim Behav* 1990;**40**(2):322–31.

49. Matthews RG, Ropiha RT, Butterfield RM. The phenomenon of foal heat in mares. *Aust Vet J* 1967;**43**(12):579–82.
50. Pryor P, Tibary A. Management of estrus in the performance mare. *Clin Tech Equine Pract* 2005;**4**(3):197–209.
51. Beach FA. Sexual attractivity, proceptivity, and receptivity in female mammals. *Horm Behav* Mar, 1976;**7**(1):105–38.
52. Górecka A, Jezierski TA, Słoniewski K. Relationships between sexual behavior, dominant follicle area, uterus ultrasonic image and pregnancy rate in mares of two breeds differing in reproductive efficiency. *Anim Reprod Sci* 2005;**87**(3–4):283–93.
53. Houpt KA, Lein DH. Sexual behavior of the mare. *Equine Pract* 1981;**3**(3):12–8.
54. Asa CS, Goldfoot DA, Ginther OJ. Sociosexual behavior and the ovulatory cycle of pones (*Equus caballus*) observed in harem groups. *Horm Behav* 1979;**13**(1):49–65.
55. Woods GL, Houpt KA. An abnormal facial gesture in an estrous mare. *Appl Anim Behav Sci* 1986;**16**(2):199–202.
56. Hansen KV, Mosley JC. Effects of roundups on behavior and reproduction of feral horses. *J Range Manag* 2000;**5**(3):479–82.
57. Cameron EZ, Setsaas TH, Linklater WL. Social bonds between unrelated females increase reproductive success in feral horses. *Proc Natl Acad Sci* 2009;**106**(33):13850–3.
58. Cameron EZ, Linklater WL, Stafford KJ, Minot EO. Social grouping and maternal behavior in feral horses (*Equus caballus*): the influence of males on maternal protectiveness. *Behav Ecol Sociobiol* 2003;**53**(2):92–101.
59. Crowell-Davis SL, Houpt KA. Maternal behavior. *Vet Clin N Am Equine Pract* 1986;**2**(3):557–71.
60. Campitelli S, Carenzi C, Verga M. Factors which influence parturition in the mare and development of the foal. *Appl Anim Ethol* 1982;**9**(10):7–14.
61. Asa CS, Goldfoot DA, Ginther OJ. Assessment of the sexual behavior of pregnant mares. *Horm Behav* 1983;**17**(4):405–13.
62. Hausberger M, Henry S, Larose C, Richard-Yris M-A. First suckling: a crucial event for mother-young attachment? An experimental study in horses (*Equus caballus*). *J Comp Psychol* 2007;**121**(1):109–12.
63. Wespi B, Sieme H, Wedekind C, Burger D. Exposure to stallion accelerates the onset of mares' cyclicity. *Theriogenology* 2014;**82**(2):189–94.
64. Vanderwall DK. The hormonal basis for reproductive behavior in nonpregnant mares. *Proc Am Assoc Equine Practer* 2017;**63**:117–23.
65. Asa CS, Goldfoot DA, Garcia MC, Ginther OJ. Sexual behavior in ovariectomized and seasonally anovulatory pony mares (*Equus caballus*). *Horm Behav* 1980;**14**(1):46–54.
66. Crabtree JR. Can ovariectomy be justified on grounds of behavior? *Proc Am Assoc Equine Pract* 2017;**63**:124–7.
67. Crowell-Davis SL. Sexual behavior of mares. *Horm Behav* 2007;**2**(1):12–7.
68. Crabtree J. Review of seven cases of granulosa cell tumour of the equine ovary. *Vet Rec* 2011;**169**(10):251–7.
69. Sherlock CE, Lott-Ellis K, Bergren A, Withers JM, Fews D, Mair TS. Granulosa cell tumours in the mare: a review of 52 cases. *Equine Vet Educ* 2016;**28**(2):75–82.
70. Satué K, Gardón JC. A review of the estrous cycle and the neuroendocrine mechanisms in the mare. *J Steroids Hormonal Sci* 2013;**4**(2):115. https://doi.org/10.4172/2157-7536.1000115.
71. Voss JL, Pickett BW. The effect of rectal palpation on the fertility of cyclic mares. *J Reprod Fertil Suppl* 1975;**23**:285–90.
72. Pickerel TM, Crowell-Davis SL, Caudle AB, Estep DQ. Sexual preference of mares (*Equus caballus*) for individual stallions. *Appl Anim Behav Sci* 1993;**38**(1):1–13.

73. McDonnell S. Minimizing foal stress levels in the breeding shed. *The Horse* 2016;**XXXIII**(1):48–9.
74. Rezagholizadeh A, Gharagozlou F, Akbarinejad V, Youssefi R. Left-sided ovulation favors more male foals than right-sided ovulation in thoroughbred mares. *J Equine Vet Sci* 2015;**35**(1):31–5.
75. Malschitzky E, Pimentel AM, Garbade P, Jobim MIM, Gregory RM, Mattos RC. Management strategies aiming to improve horse welfare reduce embryonic death rates in mares. *Reprod Domest Anim* 2015;**50**(4):632–6.
76. Allen WR, Wilsher S, Turnbull C, Stewart F, Ousey J, Rossdale PD, Fowden AL. Influence of maternal size on placental, fetal and postnatal growth in the horse. I. Development *in utero*. *Reproduction* 2002;**123**(3):445–53.
77. Peugnet P, Wimel L, Duchamp G, Sandersen C, Camous S, Guillaume D, Dahirel M, Dubois C, Reigner F, Berthelot V, Chaffaux S, Tarrade A, Serteyn D, Chavatte-Palmer P. Enhanced or reduced fetal growth induced by embryo transfer into smaller or larger breeds alters postnatal growth and metabolism in weaned horses. *J Equine Vet Sci* 2017;**48**:143–53.
78. Peugnet P, Wimel L, Duchamp G, Sandersen C, Camous S, Guillaume D, Dahirel M, Dubois C, Jouneau L, Reigner F, Berthelot V, Chaffaux S, Tarrade A, Serteyn D, Chavatte-Palmer P. Enhanced or reduced fetal growth induced by embryo transfer into smaller or larger breeds alters post-natal growth and metabolism in pre-weaned horses. *PLoS One* 2014;**9**(7) https://doi.org/10.1371/journal.pone.0102044.
79. Walton A, Hammond J. The maternal effects on growth and conformation in Shire horse-Shetland pony crosses. *Proc R Soc Ser B Biol Sci* 1938;**125**(840):311–35.
80. Wessel M. Staging and prediction of parturition in the mare. *Clin Tech Equine Pract* 2005;**4**(3):219–27.
81. Jeffcott LB. Observations on parturition in crossbred pony mares. *Equine Vet J* 1972;**4**(4):209–12.
82. Haywood L, Spike-Pierce DL, Barr B, Mathys D, Mollenkopf D. Gestation length and racing performance in 115 thoroughbred foals with incomplete tarsal ossification. *Equine Vet J* 2018;**50**(1):29–33.
83. Rollins WC, Howell CE. Genetic sources of variation in the gestation length of the horse. *J Anim Sci* 1951;**10**(4):797–806.
84. Rossdale PD, Short RV. The time of foaling of thoroughbred mares. *J Reprod Fertil* 1967;**13**(2):341–3.
85. Wulf M, Erber R, Ille N, Beythien E, Aurich J, Aurich C. Effects of foal sex on some perinatal characteristics in the immediate neonatal period in the horse. *J Vet Behav* 2017;**18**:37–42.
86. Howell CE, Rollins WC. Environmental sources of variation in the gestation length of the horse. *J Anim Sci* 1951;**10**(4):789–96.
87. Robles M, Gautier C, Mendoza L, Peugnet P, Dubois C, Dahirel M, Lejeune J-P, Caudron I, Guenon I, Camous S, Tarrade A, Wimel L, Serteyn D, Bouraima-Lelong H, Chavatte-Palmer P. Maternal nutrition during pregnancy affects testicular and bone development, glucose metabolism and response to overnutrition in weaned horses up to two years. *PLoS One* 2017;**12**(1) https://doi.org/10.1371/journal.pone.0169295.
88. Rossdale PD. Clinical studies on the newborn thoroughbred foal. I. Perinatal behavior. *Br Vet J* 1967;**123**(11):470–81.
89. Rossdale PD. Perinatal behavior in the thoroughbred horse. In: Fox MW, editor. *Abnormal behavior in animals*. Philadelphia: W.B. Saunders Company; 1968. p. 227–37.
90. Aleman M, Weich KM, Madigan JE. Survey of veterinarians using a novel physical compression squeeze procedure in the management of neonatal maladjustment syndrome in foals. *Animals* 2017;**7**(9):1–12.

91. Toth B, Aleman M, Brosnan RJ, Dickinson PJ, Conley AJ, Stanley SD, Nogradi N, Williams CD, Madigan JE. Evaluation of squeeze-induced somnolence in neonatal foals. *Am J Vet Res* 2012;**73**(12):1881–9.
92. Zurek U, Danek J. Maternal behavior of mares and the condition of foals after parturition. *Bull Vet Inst Pulawy* 2011;**55**:451–6.
93. Virga V, Houpt KA. Prevalence of placentophagia in horses. *Equine Vet J* 2001;**33**(2):208–10.
94. Doarn RT, Threlfall WR, Kline R. Umbilical blood flow and the effects of premature severance in the neonatal horse. *Theriogenology* 1987;**28**(6):789–800.
95. Houpt KA. Formation and dissolution of the mare-foal bond. *Appl Anim Behav Sci* 2002; **78**(2–4):319–28.
96. Giannetto C, Bazzano M, Marafioti S, Bertolucci C, Piccione G. Monitoring of total locomotor activity in mares during the prepartum and postpartum period. *J Vet Behav* 2015;**10**(5):427–32.
97. Gray ME, Cameron WZ, Peacock MM, Thain DS, Kirchoff VS. Are low infidelity rates in feral horses due to infanticide? *Behav Ecol Sociobiol* 2012;**66**(4):529–37.
98. Crowell-Davis SL. Developmental behavior. *Vet Clin N Am: Equine Pract* 1986;**2**(3):573–90.
99. Barber JA, Crowell-Davis SL. Maternal behavior of Belgian (*Equus caballus*) mares. *Appl Anim Behav Sci* 1994;**41**(3–4):161–89.
100. Crowell-Davis SL. Nursing behavior and maternal aggression among Welsh ponies (*Equus caballus*). *Appl Anim Behav Sci* 1985;**14**(1):11–25.
101. Wolski TR, Houpt KA, Aronson R. The role of the senses in mare-foal recognition. *Appl Anim Ethol* 1980;**6**(2):121–38.
102. Carson K, Wood-Gush DGM. Behaviour of thoroughbred foals during nursing. *Equine Vet J* 1983;**15**(3):257–62.
103. Crowell-Davis SL. Spatial relations between mares and foals of the Welsh pony (*Equus caballus*). *Anim Behav* 1986;**34**(4):1007–15.
104. Estep DQ, Crowell-Davis SL, Earl-Costello S-A, Beatey SA. Changes in the social behavior of drafthorse (*Equus caballus*) mares coincident with foaling. *Appl Anim Behav Sci* 1993;**35**(3):199–213.
105. York CA, Schulte BA. The relationship of dominance, reproductive state and stress in female horses (*Equus caballus*). *Behav Process* 2014;**107**:15–21.
106. Heitor F, Vicente L. Affiliative relationships among Sorraia mares: influence of age, dominance, kinship and reproductive state. *J Ethol* 2010;**28**(1):133–40.
107. Van Dierendonck MC, Sigurjónsdóttir H, Colenbrander B, Thorhallsdóttir AG. Differences in social behavior between late pregnant, post-partum and barren mares in a herd of Icelandic horses. *Appl Anim Behav Sci* 2004;**89**(3–4):283–97.
108. Crowell-Davis SL, Houpt KA, Burnham JS. Snapping by foals of *Equus caballus*. *Z Tierpsychol* 1985;**69**:42–54.
109. Rho JR, Srygley RB, Choe JC. Behavioral ecology of the Jeju pony (*Equus caballus*): effects of maternal age, maternal dominance hierarchy and foal age on mare aggression. *Ecol Res* 2004;**19**(1):55–63.
110. Stanley CR, Shultz S. Mummy's boys: sex differential maternal-offspring bonds in semi-feral horses. *Behaviour* 2012;**149**(3/4):251–74.
111. Williams M. The effect of artificial rearing on the social behavior of foals. *Equine Vet J* 1974;**6**(1):17–8.
112. McDonnell S. Keeping a stallion with a foaling mare (or mares). *The Horse* 2002;**XIX**(12):43–4.
113. Turner J. Foal care. *Equus* 2008;**364**:53–6.

114. Henry S, Richard-Yris M-A, Tordjman S, Hausberger M. Neonatal handling affects durably bonding and social development. *PLoS One* 2009;**4**(4)https://doi.org/10.1371/journal.ppone.0005216.
115. Christensen JW. Early-life object exposure with a habituated mother reduces fear reactions in foals. *Anim Cogn* 2016;**19**(1):171–9.
116. Henry S, Hemery D, Richard M-A, Hausberger M. Human-mare relationships and behavior of foals toward humans. *Appl Anim Behav Sci* 2005;**93**(3–4):341–62.
117. Henry S, Briefer S, Richard-Yris M-A, Hausberger M. Are 6-month-old foals sensitive to dam's influence? *Dev Psychobiol* 2007;**49**(5):514–21.
118. Apter RC, Householder DD. Weaning and weaning management of foals: a review and some recommendations. *J Equine Vet Sci* 1996;**16**(10):428–35.
119. Boyd LE. Ontogeny of behavior in Przewalski horses. *Appl Anim Behav Sci* 1988;**21**(1–2):41–69.
120. Heitor F, Vicente L. Maternal care and foal social relationships in a herd of Sorraia horses: influence of maternal rank and experience. *Appl Anim Behav Sci* 2008;**113**(1–3):189–205.
121. Houpt KA, Hintz HF. Some effects of maternal deprivation on maintenance behavior, spatial relationships and responses to environmental novelty in foals. *Appl Anim Ethol* 1983;**9**(3–4):221–30.
122. Palmer E, Guillaume D. Some mechanisms involved in the response of mares to photoperiodic stimulation of reproductive activity. *Reprod Domest Anim* 1998;**33**(3–4):205–8.
123. Nagy P, Guillaume D, Daels P. Seasonality in mares. *Anim Reprod Sci* 2000;**60-61**:245–62.
124. Guerin MV, Wang XJ. Environmental temperature has an influence on timing of the first ovulation of seasonal estrus in the mare. *Theriogenology* 1994;**42**(6):1053–60.
125. Kubiak JR, Crawford BH, Squires EL, Wrigley RH, Ward GM. The influence of energy intake and percentage of body fat on the reproductive performance of nonpregnant mares. *Theriogenology* 1987;**28**(5):587–98.
126. Lemazurier E, Toquet MPL, Fortier G, Séralini GE. Sex steroids in serum of prepubertal male and female horses and correlation with bone characteristics. *Steroids* 2002;**67**(5):361–9.
127. Pickett BW, Voss JL, Clay CM. Management of shuttle stallions for maximum reproductive efficiency—Part 2. *J Equine Vet Sci* 1998;**18**(5):280–7.
128. Clay CM, Squires EL, Amann RP, Pickett BW. Influences of season and artificial photoperiod on stallions: testicular size, seminal characteristics and sexual behavior. *J Anim Sci* 1987;**64**(2):517–25.
129. Waheed MM, Ghoneim IM, Abdou MS. Sexual behavior and hormonal profiles in Arab stallions. *J Equine Vet Sci* 2015;**35**(6):499–504.
130. Khalil AM, Nobouro M. Effects of harem size on the testosterone level in Misaki feral horses, In: *Poster at the meeting of the American Veterinary Society of Animal Behavior*; 2000.
131. McDonnell SM, Murray SC. Bachelor and harem stallion behavior and endocrinology. *Biol Reprod Monogr* 1995;**1**:577–90.
132. Holtan DW, Nett TM, Estergreen VL. Plasma progestins in pregnant, postpartum and cycling mares. *J Anim Sci* 1975;**40**(2):251–60.
133. Kirkpatrick JF, Turner JW,J. Seasonal ovarian function in feral mares: seasonal patterns of LH, progestins and estrogens in feral mares. *J Equine Vet Sci* 1983;**3**(4):113–8.
134. Mottershead J. *The mare's estrous cycle*. http://www.equine-reproduction.com/articles/estrous.htm. [Accessed 2 May 2017].
135. Terblanche HM, Maree L. Plasma progesterone levels in the mare during the oestrous cycle and pregnancy. *J S Afr Vet Assoc* 1981;**52**(3):181–5.

136. Sato K, Miyake M, Yoshikawa T, Kambegawa A. Studies on serum oestrogen and progesterone levels during the oestrous cycle and early pregnancy in mares. *Equine Vet J* 1977;**9**(2):57–60.
137. Brann DW, Mahesh VB. Role of corticosteroids in female reproduction. *FASEB J* 1991;**5** (12):2691–8.
138. Burger D, Dolivo G, Marti E, Sieme H, Wedekind C. Female major histocompatibility complex type affects male testosterone levels and sperm number in the horse (*Equus caballus*). *Proc R Soc B* 2015;**284**(1805):20150407.
139. Burger D, Thomas S, Aeplli H, Dreyer M, Fabre G, Marti E, Sieme H, Robinson MR, Wedekind C. Major histocompatibility complex-linked social signaling affects female fertility. *Proc R Soc B* 2017;**284**(1868).
140. McDonnell SM. Starting a novice breeding stallion. *Clin Tech Equine Pract* 2007;**6**(4):232–8.
141. McDonnell SM. Stallion behavior, In: *Proceedings of the Wild West Veterinary Symposium, Reno, Nevada*; 1998. p. 1–12.
142. McDonnell S. Slow-starting stallion. *The Horse* 2003;**XX**(7):59–61.
143. McDonnell SM, Kenney RM, Meckley PE, Garcia MC. Novel environment suppression of stallion sexual behavior and effects of diazepam. *Physiol Behav* 1986;**37**(3):503–5.
144. McDonnell SM, Kenney RM, Meckley PE, Garcia MC. Conditioned suppression of sexual behavior in stallions and reversal with diazepam. *Physiol Behav* 1985;**34**(6):951–6.
145. McDonnell S. Stallion washing aversion. *The Horse* 2005;**XXII**(3):109–10.
146. Beaver BV. Aggressive behavior problems. *Vet Clin N Am Equine Pract* 1986;**2**(3):635–44.
147. Houpt KA. *Domestic animal behavior for veterinarians and animal scientists*. 3rd ed. Ames: Iowa State University Press; 1998. p. 495.
148. McDonnell S. Savage stallion. *The Horse* 2006;**XXIII**(7):65–6.
149. McDonnell SM. Techniques for extending the breeding career of aging and disabled stallions. *Clin Tech Equine Pract* 2005;**4**(3):269–76.
150. McDonnell S. Understanding male aggression. *The Horse* 2003;**XX**(10):58.60.
151. Cooke CD. What can I give to calm this stallion down? *Equine Vet Educ* 2015;**27**(9):496.499.
152. McDonnell S. Anabolic steroid effects. *The Horse* 2007;**XXIV**(11):53–4.
153. Berger J. Induced abortion and social factors in wild horses. *Nature* 1983;**303**(5912):59–61.
154. McDonnell S. Herding the mares. *The Horse* 2006;**XXIII**(11):97–8.
155. Hartmann E, Søndergaard E, Keeling LJ. Keeping horses in groups: a review. *Appl Anim Behav Sci* 2012;**136**(2–4):77–87.
156. Jørgensen GHM, Borsheim L, Mejdell CM, Søndergaard E, Bøe KE. Grouping horses according to gender—effects on aggression, spacing and injuries. *Appl Anim Behav Sci* 2009;**120** (1–2):94–9.
157. Schumacher J. Why do some castrated horses still act like stallions, and what can be done about it? *Compendium: Equine Ed* 2006;**1**(3):142–6.
158. Hayes HM. Epidemiological features of 5009 cases of equine cryptorchidism. *Equine Vet J* 1986;**18**(6):467–71.
159. Cox JE, Redhead PH, Dawson FE. Comparison of the measurement of plasma testosterone and plasma oestrogens for the diagnosis of cryptorchidism in the horse. *Equine Vet J* 1986;**18** (3):179–82.
160. Cox JE. Experiences with a diagnostic test for equine cryptorchidism. *Equine Vet J* 1975;**7** (4):179–83.
161. Houpt, K.A. (1994): Personal communication.
162. Hooper RN, Taylor TS, Varner DD, Blanchard TL. Effects of bilateral ovariectomy via colpotomy in mares: 23 cases (1984–1990). *J Am Vet Med Assoc* 1993;**203**(7):1043–6.

163. Kamm JL, Hendrickson DA. Clients' perspectives on the effects of laparoscopic ovariectomy on equine behavior and medical problems. *J Equine Vet Sci* 2007;**27**(10):435–7.
164. Bliss S. Options for managing estrus related behavior problems in performance mares. *Remuda* 2011;**4**(2):13–6.
165. Tomasgard G, Benjaminsen E. *Plasma progesterone in mares showing oestrus during early pregnancy. Nordisk Veterinaermed* 1975;**27**(11):570–4. Author's translation abstract from https://www.ncbi.nlm.nih.gov/pubmed/1196852. [Accessed 3 May 2017].
166. Houpt KA, Wolski TR. *Domestic animal behavior for veterinarians and animal scientists.* Ames: The Iowa State University Press; 1982. p. 356.
167. Gastal MO, Gastal EL, Beg MA, Ginther OJ. Stallion-like behavior in mares: review of incidence, characteristics, ovarian activity, and role of testosterone. *J Equine Vet Sci* 2007;**27** (9):390–3.
168. Beaver BV, Amoss Jr MS. Aggressive behavior associated with naturally elevated serum testosterone in mares. *Appl Anim Ethol* 1982;**8**(5):425–8.
169. Cammaert S, Coryn M, de Kruif A, Spincemaille J, Delbeke FT, Debackere M. Three full sister mares with a stallion-like behavior and a high blood testosterone concentration: a case report. *Equine Vet Sci* 1993;**13**(4):220–2.
170. Grogan EH, McDonnell SM. Mare and foal bonding and problems. *Clin Tech Equine Pract* 2005;**4**(3):228–37.
171. Zurek U, Danek J. Foal rejection—characteristics and therapy of inadequate maternal behavior in mares. *Ann Anim Sci* 2012;**12**(2):141–9.
172. Houpt K, Lieb S. A survey of foal rejecting mares. *Appl Anim Behav Sci* 1994;**39**(2):188.
173. Houpt, K.A. (2011): Equine maternal behavior and its aberrations. In: Recent advances in companion animal behavior problems, Houpt, K.A., ed., International Veterinary Information Service (www.ivis.org) [downloaded October 27, 2017].
174. Berlin D, Steinman A, Raz T. Post-partum concentrations of serum progesterone, oestradiol and prolactin in Arabian mares demonstrating normal maternal behavior and Arabian mares demonstrating foal rejection behavior. *Vet J* 2018;**232**:40–5.
175. Houpt KA. Behavior in horses, In: *Notes from talk given at the Southwest Veterinary Symposium, Ft. Worth, TX*; 2010.
176. Juarbe-Daíz SV, Houpt KA, Kusunose R. Prevalence and characteristics of foal rejection in Arabian mares. *Equine Vet J* 1998;**30**(5):424–8.
177. Wong DM, Alcott CJ, Davis JL, Hepworth KL, Wulf L, Coetzee JH. Use of alprazolam to facilitate mare-foal bonding in an aggressive postparturient mare. *J Vet Intern Med* 2015;**29**:414–6.

Chapter 7

Equine Maintenance Behaviors

The term *maintenance behavior* includes a group of behaviors with a specific purpose. They are necessary for an animal's overall survival and health, although other purposes may exist as well. The list of maintenance behaviors can be short or complex, depending on what is included.[1,2] Eating, drinking, and eliminating are three groups of behaviors typically listed.

A subset of maintenance behaviors, *comfort behavior*, defines actions that increase physical comfort, in addition to their importance in maintaining overall health. Grooming behavior is the primary group within this category, but others like resting, yawning, and stretching would fit under this definition too. These later behaviors also fit within the broad category of locomotor functions and will be described there. Grooming will be included in this chapter.

INGESTIVE BEHAVIORS

Eating behavior is the best researched of the maintenance behaviors because of its overall importance to a broad range of needs—from survival, to maximum performance. Within this category are a broad range of actions: suckling, grazing, feeding, browsing, and drinking. Suckling foals put their lips around the mare's teat, and by creating negative pressure, cause milk to enter the mouth. When grazing and browsing, the horse isolates preferred plant material with its upper lip and then grasps the vegetation between upper and lower incisors before tearing it loose to chew and swallow. Feeding involves the consumption of concentrates, hay, and grain. For this, the lips and tongue manipulate the food into the mouth for chewing. Drinking involves the intake of liquids. The horse puts its nose at or below the surface of the liquid and draws the liquid in through mostly closed lips by creating negative pressure.

Suckling Behaviors

Attempts to nurse begin shortly after the foal is able to stand, although foals that are prevented from standing will make spontaneous sucking attempts in midair.[3] Once the foal is standing, nursing begins within the first few hours.[3,4] The range generally given is 35 min after birth up to 420 min, with the average

Equine Behavioral Medicine. https://doi.org/10.1016/B978-0-12-812106-1.00007-3

being 111 min.[3,5] At first, attempts are clumsy and can be directed between the mare's forelimbs or abdomen or at an inanimate object. The dam will help the foal by repositioning herself.[3] Foals may approach the mare's udder in three ways. In the first, they orient themselves parallel to the mare and face her hindquarters as they move in to nuzzle the udder. If the mare and foal are slightly separated, the foal may show intention movements by flattening its ears, tossing its head, and moving the nose directly to the udder. For the third type of approach, the foal walks from one side of the mare to the other by passing under the mother's neck first and then sliding its head under her flank. The move has been called *crossing of the bow*.[6] This latter approach may signal the mare to stand still and perhaps stimulate milk letdown.[7]

While nursing, foals typically stand along the mare's side, facing her flank, and position their nose upward toward the udder (Figure 7-1). In this position, it

(a)

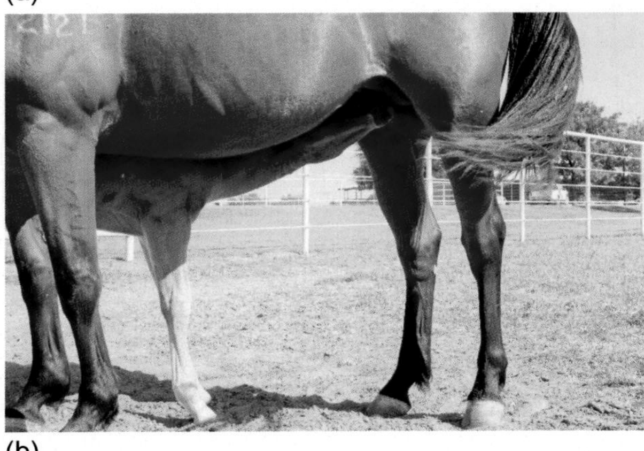

(b)

FIGURE 7-1 (a) Foals usually nurse while facing toward the rear of the mare. (b) This position allows the foal to be able to reach both teats of the udder from the same side.

FIGURE 7-2 Although most nursing occurs with the foal pointed toward the mare's rear, it can also occur when the two are facing in the same direction, as shown here. The foal may also be perpendicular to its dam, and occasionally it will nurse from the back, between the mare's rear legs.

is able to reach both teats from the same location. Occasionally the foal will stand facing in the same direction while nursing (Figure 7-2). Rarely, a foal will nurse with its head between the mare's hind limbs.[3]

As followers, foals have ready access to their milk supply and will initiate nursing several times a day. Each nursing bout has bursts of sucking with interspersed pauses.[3] During the first week, a foal will drink 15% of its body weight in the first 24 h, working up to 25% (about 4 gal or 15 L) by the end of the week.[8] This happens during four to seven nursing bouts an hour, with each lasting just under 2 min. By 4 weeks, the foals nurse three times per hour, and by 6 months of age, the frequency is reduced to once an hour for just over 1 min.[5,7,9–11] Colts nurse slightly longer than fillies, and nursing that occurs in a stall tends to be longer than that occurring in a pasture or paddock.[3] In areas of good grazing, foals will nurse longer than when nutrition is poor, probably due to better milk production by the mare.[12]

For the first few weeks, the mare terminates the nursing bouts 30%–45% of the time by moving away. She can also hinder nursing if she continues to graze. In free-ranging horses, foals terminate nursing bouts about 75% of the time.[3]

Within the first few days, foals will eat small amounts of fresh feces passed by their mothers, usually within 30 s of her defecating.[13,14] Coprophagy continues for several weeks, being the greatest during the first 2 months. Occasionally the feces may originate from the foal itself or other members of the herd, but these are exceptions. The primary reason for coprophagy is probably to introduce microflora into the foal's gastrointestinal system.[13,15,16] It may also provide nutrients to the foal. Alternatively, coprophagy may be triggered by maternal pheromones to signal the presence of specific components needed for gut immunocompetence, nervous system myelination, growth acceleration, and/or sexual maturation.[15] In approximately one-quarter of the events, the foal will urinate before, during, or after eating the feces.[15]

Transitional Eating Behaviors

Transitional eating behaviors occur while a foal is still nursing on its mother but begins its transition to adult foods. Their long legs minimize the amount of grazing that young foals can do. During the first week, foals will nibble grass during the daytime but for no more than 5 min per hour, and at 3 months, the time spent grazing remains low, less than 15 min per hour. As the neck length catches up with limb length by 4 months of age, more time is spent grazing and less time nursing. By 7 months, grazing increases to over 40 min per hour.[2,14] Foals also spend more of that grazing time in the late afternoon at the same time their mothers are grazing.[3,7,14]

Nursing serves two purposes. It not only provides nutrition, it also provides maternal reassurance and security.[7,17] Milk supplies all the nutrition a foal needs for the first 6–8 weeks of life, but it only supplies 30% of the foal's energy requirements at 4 months. By then, suckling behavior continues to meet security needs and occurs most commonly when the youngster appears stressed.[3,17,18]

Ultimately, foals are weaned from their mothers—a stressful time for both. Exactly when it is best to wean a foal is governed by several factors, the most important of which should be to minimize stress. In free-ranging horses, weaning usually takes place in the eighth or ninth month.[12,19] To prepare for weaning, foraging behavior increases dramatically. It changes from nibbling 13% of the time in the early weeks to a more adult-like 62% when ready to be weaned.[20] With access to good quality grazing, the mare may delay weaning until shortly before the birth of her next foal 1 or 2 years later. Under very poor grazing conditions, weaning might occur as early as 3 months of age.

Managed mare-foal separations usually occur when the youngster is 4–6 months old, but this can vary by management style, mare health, or nutritional status. The distance from the dam and tendency to play with other foals are good predictors of sensitivity to social separation and reactions to novel objects.[21] Prior to separating mares and foals, the youngsters should be well started in eating the foods they will have access to postweaning. This generally means they have experienced creep feeding for several weeks.[8,17] Preweaning diets have a postweaning effect on behavior. The weanlings on diets higher in fat and fiber are more inquisitive, more social, and easier to handle; all of which suggests they are less stressed than foals fed diets higher in starch and sugar before weaning.[22]

Abrupt separations are common within the horse industry and they can occur in several ways. Younger foals are more apt to be weaned by abrupt separation and barn separations, especially if mares are rebred every year. Older foals are weaned more gradually.[23] Neonatal weaning occurs within a few days after birth if the mare dies or rejects the foal.[17] In a few situations, it may be done if the mare needs to be shipped a long distance, causing concern about stress or injury to the foal. While the separation stress is not strong for the foal in those early days, it is hard on the mare. The foal will then need to be raised as an orphan.

Older foals can also experience abrupt weaning. For 2-month-old foals this should only be done if the foal is strong and healthy. There are no controlled studies on resulting behaviors other than to show these foals ultimately tend to rank low in the social order as adults.[17] A few horse owners will wait until 7 or 8 months to wean foals.

Abruptly weaned foals show distress behaviors—significant increased amounts of walking, urinations, defecations, and vocalizing compared to before weaning.[24–26] Cortisol levels rise significantly within 24 h of weaning, although they may still be within the normal range for a foal.[27] Fillies have higher overall cortisol responses postweaning, and they lose more weight that do colts.[26] Colts vocalize and defecate more than do fillies.[26] Although cell-mediated immune responses suggest individual weaning is less stressful, behavioral indicators suggest there is less stress if foals are weaned in pairs or small groups instead of individually.[17,25,28] Single foals in a stall are more likely to show aberrant behaviors, including licking or chewing the wall, kicking the wall, pawing, bucking, and rearing.[22,24] They also are more likely to develop stereotypies. Foals subjected to periodic 10-min separations from their mothers only show less walking behavior postweaning, indicating short separations are minimally effective at reducing weaning stress.[29] A two-stage separation to first prevent nursing and then physically separate the mare and foal is also minimally effective at reducing stress.[30] Equine appeasing pheromone has also been tried 30 min before separation and then twice daily for the next 2 days but did not prove useful.[27]

The only technique used for weaning that results in no difference in behavior between the weaned and nonweaned foals is gradual separation. With creep feed available, mares and foals are allowed visual and nose contact but are physically separated by a fence.[17,31] If several mares and foals are pastured together, removing one mare at a time is another gradual weaning process that seems to be less stressful than abrupt separation (Figure 7-3).

Adult Eating Behaviors

Grazing and feeding are the primary ingestive behaviors for domestic horses. Free-ranging horses would replace feeding with browsing instead. Grazing makes up the majority of a horse's eating behavior. The head is down, the teeth clutch several blades of grass, and the horse tears off the grass, chews and swallows it. It then takes a step or two before repeating the process (Figure 7-4). The

FIGURE 7-3 Weaning is less stressful if the foal spends most of its time with other weanlings but has access to its mother for reassurance. One mare was removed at a time until the stocking-legged foal is the last of this group to be weaned. She (a) can see her mother in a small pen in one corner of the pasture, and (b) go to her for maternal comfort but net wire along the mare pen fence prevents nursing.

head moves left to right and back to access the opening made as the opposite forelimb advances. This lowered head position is preferred, even for grain, probably because it allows the horse to see in all directions as an antipredator strategy (see Figure 2-4 in Chapter 2).[32] The direction of travel is typically parallel to the direction of the wind, which for free-ranging horses allows them to maintain sensory vigilance too.[3]

The feeding rate as measured by the number of bites per minute and the amount of forage consumed within a given period of time varies in different situation. Following the intake of grass, a horse uses the cheek teeth to grind

FIGURE 7-4 Horses continue to move as they graze. Each will take a step or two and then nibble grass as its head moves slowly in a side-to-side direction.

the vegetation into smaller pieces at a rate of 1.0–1.7 movements per second.[2,3] Horses eat more if they can see another horse eating, making it a useful strategy for an anorexic animal.[32,33]

In free-ranging horses, some grazing occurs every hour of a 24-hour cycle, and 70%–80% of the total time is spent eating.[3,34,35] The duration of each bout varies by season, time of day, and lactation status. More time is spent grazing and browsing in winter, but as the grass grows again in spring and summer, the amount of time spent eating decreases. In all seasons, grazing occurs preferentially during hours of light. Depending on grass quality and sex of the horse, between 55% and 78% of daylight hours are spent grazing. Peak activity occurs from dawn to midmorning and then again in the late afternoon and evening.[2,3,12,14,36–38] The exception occurs in barrier island ponies on the East Coast of the United States, which show extensive nocturnal foraging patterns.[35,39] In the summer, horses graze longer in the early morning and less in the afternoon, so they can seek shelter when flies are the most active. Heavy fly activity at other times can also change grazing patterns.[12,37] The energy needed for lactation requires mares to graze longer than nonlactating horses.[12,40]

The length of each grazing bout is generally 1–2 min but can last 12 h. Each is followed by an interval between 11.5 min and 3 h before grazing begins again. The length of the interbout interval is not related to the length of the previous meal.[14,37,41,42] Pastured horses show a pattern similar to that of free-ranging horses. They eat 10 to 12 h a day in 30- to 180-min bouts, with a diurnal emphasis.[41–44]

Regardless of the size of the pasture, horse owners maintain a heavy reliance on supplemental feed, with approximately 95% of owners giving concentrates or supplements and 85% feeding hay daily.[45] Almost 26% of pastured horses eat tree bark, and 17% lick or eat dirt or sand. They will also eat bedding, snow, and feces, although the latter is most common in young foals and rare in adults.[3,45–47] Except for the sodium in salt, horses have little nutritional wisdom for minerals.[48] The eating of sand or dirt is not correlated to the specific mineral content needed.[41]

Horses are selective grazers. While there seems to be a fascination with eating on the outside of a fence line (Figure 7-5), there usually is more of an attraction to certain plants. A horse will keep returning to the same feeding site until it is overgrazed. Switching sites occurs if there is nothing left to eat or if another horse, especially a higher-ranking one, is already there.[49] In heavily grazed areas, the dry matter content of the grass is significantly higher than in lesser-eaten grasses. Constant grazing helps prevent the reduction of this content.[14] Availability of food has little effect on the length of a grazing bout, but the amount of fiber does.[37] High-fiber foods reduce grazing time. The amount eaten is dependent on the intragastric nutrients, including caloric content, generated satiety cues, the taste and texture of the feed, and other external cues.[41,44,50,51] It is not related to the amount of food present.

Taste tests confirm that the general nutritional value, but not energy content, can be the primary factor in diet selection.[41,52,53] Preferred grass selection is not uniform across the species. Additionally, grasses that score high for selection can become undesirable if they get too tall.[3,54,55] Certain tastes are likely the equine equivalent of candy. Sweet tastes and odors can influence choice of feeds. Free-ranging ponies will preferentially spend large portions of time

FIGURE 7-5 Horses seem to have a natural tendency to eat grass on the other side of a fence, regardless of the quantity and quality of grass in their own pasture.

eating acorns and bracken ferns.[2,14,52] Persimmons and apples are favorites for some horses. Others will paw up and eat the roots of milk vetch (*Astragulus* sp.) and winter fat (*Erotia lanata*) plants.[41] Grain preferences rank sweet mixed grain as the favorite, followed in order by oats and corn. Barley, wheat, and rye were the lowest preferences.[48,56] Taste preference tests have identified flavors that might make concentrate feeds more palatable. Fenugreek is the most palatable, followed in order by banana, cherry, rosemary, cumin, carrot, peppermint and then oregano.[57] Consumption times for these flavors were also less than for unflavored feed equivalents. Horses have some ability to associate taste with the nutritional level of a food. They will change from a preferred flavor on a low quality food to a less-favored flavor on a more nutritious food.[58] The acceptance of novel foods occurs faster if the new food has a taste and/or odor that is familiar.[52,59] For pelleted feeds, size and relative hardness strongly influence desirability.[41,60]

Horses are good at being able to separate less desirable things out of feeds. Dirt is shaken from the roots of grasses pulled up, as an example. They also do well at separating medication from grains, even when owners go to great measures to hide the taste or make the medication stick to the feed. Olfaction is the primary sense used for avoidance. Horses can learn to avoid low or moderately palatable feeds that result in abdominal pain if it happens within 30 min of consumption.[48] Avoidance does not happen if the food is highly palatable.[41] Strong odors like citrus are used to prevent the horse from eating something undesirable.

Horses avoid grazing near feces if at all possible, resulting in the grass in elimination areas growing taller (called *roughs*) than in other grazed areas of the pasture (called *lawns*) (Figure 7-6).[61,62] When there is high grazing intensity, horses eat in both tall, rough areas and short areas equally.[63,64] The grass itself remains palatable, and horses readily eat grass cut from the elimination areas. They no longer do so if a fecal odor is present.[61]

Of concern for stabled horses is that their housing results in a greatly reduced amount of time spent eating roughage compared to what would be happening in a pasture setting. Digestive problems and stereotypic behavior are associated with this reduction, so several things have been tried to increase time spent eating. The one usually deferred to is feeding lesser amounts multiple times a day, instead of the standard one or two times. Mechanical feeders are not particularly helpful because it is roughage, not grain, that should be fed multiple times. Small openings in hay nets help slow down consumption but not to a highly significant degree. Using multilayered hay nets works better.[65,66] Straw bedding is desirable because horses can nibble it when hay is not available. High grain diets with lower digestibility are associated with slower eating and a reduction in water intake—another available option.[67] In paddocks, continuous access to hay in otherwise bare lots reduces the amount of time a horse spends standing and moving, with a corresponding increase in time spent eating and interacting positively with others in the same paddock.[68,69]

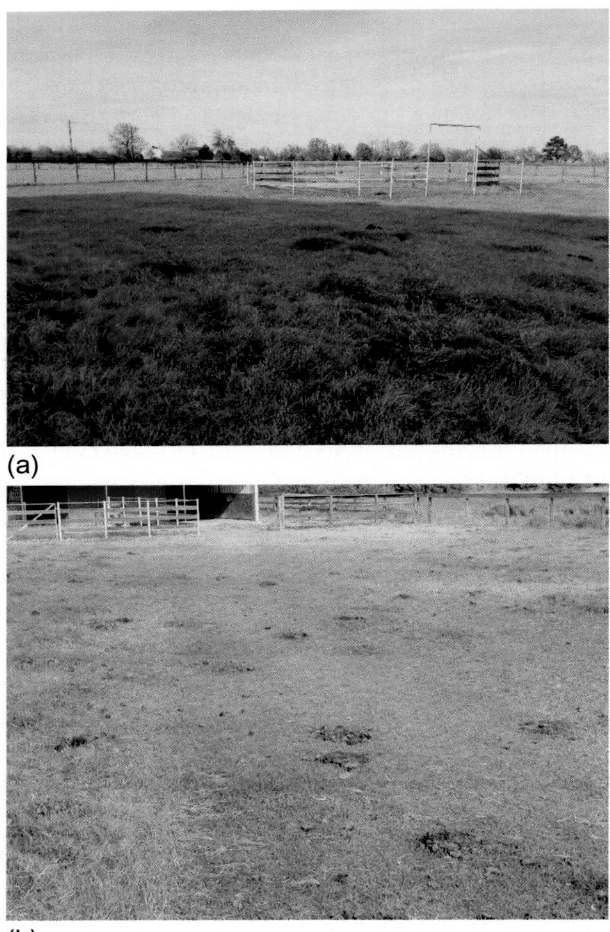

FIGURE 7-6 Elimination areas are specific locations where horses urinate and defecate. (a) Domestic horses usually eliminate in specific latrine areas that continue to get larger because mares and geldings walk to the area, leaving their eliminations to accumulate at the edges. These areas are called roughs since horse do not graze there. (b) Free-roaming horses and those raised in very large pastures do not eliminate in specific areas, but deposit excreta randomly instead, even if put into smaller areas later in life.

The type of diet can play a role in the behavior of the horse, and this is particularly relevant for stalled animals. While the natural diet is high in fiber, stalled horses tend to have diets higher in starch and carbohydrates. These are associated with shorter eating times, significantly higher heart rates when stressed, increased reactivity to handling, and increased vigilance.[70,71] Diets high in fats reduce spontaneous activity levels and reactivity.[72]

There are ways to encourage peaceful mealtimes in pastures and paddocks. To minimize aggression between strange horses being fed as a group, concentrates should be spread throughout numerous troughs that are low to the ground and far apart. Spacing between troughs is the only real consideration if the horses are familiar with each other.[69,73] If buckets are used instead of troughs, they should be spread apart. A couple of extra buckets allow displaced horses to find another place to eat. Bossy, high-ranking horses and shy, low-ranking ones can be removed and fed separately. This minimizes disruptions to the remaining herd members and allows the shy horse to eat in peace. Individual feedbags are another option.

Ponies present a slightly different scenario. Because most can be traced back to ancestors that lived in harsh environments in northern Europe, they tend to be efficient in their use of nutrients. When raised on the rich, nutritious diets typically fed today, ponies gain excessive weight. It is suspected that ponies are genetically predisposed to certain conditions such as insulin resistance, laminitis, and equine metabolic syndrome. Obesity may trigger health problems, but keeping an appropriate body weight is complicated. Owners often feel that food is a good way to pamper the pony, setting it up for additional weight gain. Almost 90% of horse owners, and particularly pony owners, do not recognize overweight animals.[74] Additionally, it is almost expected that show horses carry extra "bloom."

Preventing obesity or encouraging weight loss in ponies is difficult to accomplish by simply reducing the amount of food to which they have access. Grazing muzzles are useful to reduce roughage intake by at least 30% regardless of the type of grass available.[75–77] They can reduce the rate of intake of pelleted feed too.[78] If a grazing muzzle is used, it is important to factor in the amount of time the pony does not wear a muzzle. If given unmuzzled access to food for part of a 24-h period, a pony can make up the deficiency by eating faster, so weight gain continues.[79] Another method that has been tried to treat obesity in ponies involves dispensing small amounts of food from a stationary mechanical feeder. The feeder randomly dispenses food with prolonged intervals of no food. This encourages exercise by having to travel back and forth to the feeder. This can also reduce body fat.[80]

DRINKING BEHAVIOR

While some liquid is taken in by eating succulent plants or snow, most is acquired by drinking water. Horses touch the water with an almost closed mouth and suck the water in before swallowing (Figure 7-7). Occasionally animals will put their nose deeper into the water to drink, but many times when doing so, they do not actually drink until the lips are at the surface again. The interswallow interval is about 2 s and there is an average of 15 swallows per drinking bout.[81]

The frequency of drinking varies. Nursing foals rarely drink, but adults will drink five to seven times during the summer and two to three times in winter if

FIGURE 7-7 Most horses drink by sipping water with their nose barely touching the surface.

water is readily available. Drinking is done most often in the afternoon, especially during hot weather.[3,36] Lactating mares drink more often when it is hot but not for longer periods.[41] If water is near the food, horses will commonly drink immediately after eating.

Free-ranging horses may differ in the frequency of drinking compared to domesticated horses, but seasonal differences are similar. For them, the number of watering sites available, the distance between water holes, grazing and shelter areas, and ambient temperatures play a role in how often they drink. Drinking occurs most often during early and late daylight hours. Free-ranging horses are generally within 3 mi (4.8 km) of water sources in the summer and may be farther away in winter months.[82] If the distance is great, drinking may only occur once daily or even every other day.[2,3,41] All members of a harem group go to the watering hole at one time with lactating mares arriving first and being the last to finish.[2] If space is limited, dominant animals drink first. Then, each horse waits for the rest to finish drinking before moving off again. Groups rarely stay at the site longer than 30 min.[2,3]

Several things affect the amount of water consumed during each drinking bout. Some of these include the animal's age and size, pregnancy status, ambient temperature, water temperature, availability of water, amount of salt consumed, type of diet, and moisture content of available food. Ponies and horses on exclusive roughage diets consume 8–10 gal (31.4–38.4 kg of water per day.[81,83–85] Those on a combination of roughage and grain or kept in a cool resting environment drink about half that amount (4.62 gal = 17.5 kg).[83] Two pastured horses during extremely hot weather have been known to drink approximately 100 gal (approximately 380 kg) in a 24-h period.

ELIMINATIVE BEHAVIORS

Eliminative behaviors are used to get rid of accumulated body waste, and as is true in some other species, urination and defecation can also have a social function—marking. For the harem stallion, the mere action of another horse eliminating may trigger a similar response. As a result, the frequency of urinations and defecations by that individual stallion is significantly higher than for other group members.[61]

Domesticated horses are known to urinate and defecate in specific locations, even walking some distance to get to the specific location. While they are less likely to disrupt a grazing session to defecate, they will do so to urinate.[86] Mares and geldings will walk to the latrine area and stop just as they enter it. The result is that the urine and feces is deposited at or near the edge of the patch, causing it to gradually enlarge in diameter. Stallions will typically walk through the area and stop on the far side, or they may back into it instead. This results in their urine and feces being deposited within the existing area.[62,86] The tendency to use a specific location is why most horses are relatively clean in a stall, depositing feces and urine in one or two specific locations. There are, however, exceptions. Horses that are free-roaming and those raised in very large pastures tend not to create elimination areas, instead urinating and defecating in random locations (Figure 7-6b).[3,87] Careful evaluation has shown that some free-ranging ponies do have specific latrine areas, though.[88]

Stallions, in particular, use urine and feces to mark eliminations of other horses—a trait shared by all equids.[61] Horses are capable of identifying their own feces and are particularly interested in smelling feces from horses previously encountered in aggressive actions.[89] This suggests an important communicative function. Stallions defecate on 65% of the feces they encounter, while mares will urinate on the feces 32% of the time or reject it completely 53% of the time.[2] Harem stallions respond to approximately 90% of mare eliminations during breeding season, but very rarely do so during the anestrus months.[14,90–93] The high concentration of cresols in stallion urine during the breeding season is thought to change the chemical composition of estrous mare feces to resemble that of an anestrous mare instead.[91] Breeding season testosterone levels positively correlate with the duration of a feces sniffing bout. Stallions also defecate more often on feces of the lowest-ranking male.[94]

Scent marking is a trait of many species, but in horses it is associated with three specific situations.[2,3,61,92] Most commonly, the behavior occurs when the stallion eliminates on top of newly deposited urine or feces from another horse. The behavior begins when the horse first smells the area. It then eliminates over the deposit and backs up to resmell the area.[90,92,95] Occasionally the horse will defecate or urinate over the deposit multiple times before moving on. Stallions are most likely to defecate on other stallion's feces and urinate on mare excrement.[2,47] The remaining two situations of scent marking are shown most by territorial and free-roaming equids. Repeated deposition of feces within a relatively small area can result in dung piles (*dung heaps*). Some horses casually eliminate on any communal dung pile it passes even though no other male is present. This tends to occur most often near trails to water sources, particularly if several different groups of horses use the same path. The third situation occurs when a stallion eliminates on a dung pile as part of a ritualized aggressive encounter, which happens in 25% of aggressive encounters between free-roaming stallions. They seem to compete in alternating eliminations as to which horse will be the last to defecate on top of the other horse's feces. Dominant stallions are usually the most frequent and last.[92,94]

Mares rarely smell urine or feces. When they do, they usually respond by either urinating or showing flehmen. Defecation is uncommon. Foals standing near their mother as she urinates will sniff the urine, flehmen, and then urinate over the dam's urine.[14]

Urination

Neonates may first urinate within 3h after birth, but more typically they do so between the third and ninth hour. Colts tend to urinate sooner than fillies, at about 6h compared to 11h.[16] Foals then urinate approximately once an hour for the first 2 weeks of life.[3,14] The frequency gradually increases to once every 3.8h in summer and 4.5h in winter by the time the horse is a yearling.[2,3]

Immediately prior to urinating, 60% of horses will stop eating, suggesting that this behavior is a relatively active process.[2,96] The urination posture differs only slightly by sex (Figure 7-8). In both males and females, the head and neck are lowered slightly, and the tail is raised. The rear legs spread apart and extend behind the body as the front legs move slightly forward. Mares will tilt the pelvis which raises the back somewhat, and their rear legs often do not extend back as far as with males. Geldings and stallions will protrude the penis slightly while urinating. The mare's urination ends with vulvar winking.

Urination sessions are short, about 10s, except for estrous mares.[3] Horses average six urinations per day, although stallions may urinate over 12 times a day.[3,97] Approximately 1.45gal (5.5kg) of urine are eliminated daily in healthy horses.[85] Preferences are shown to urinating in resting areas and other soft-footing locations. As an example, as long as they can assume the urination posture, geldings often do not urinate in horse trailers that only have rubber mat flooring unless the trip is extremely long. There is no reluctance when wood shavings cover the trailer mats. Owners are often upset because a horse seems to wait until it enters its stall to be fed before it urinates, rather than doing so while it is still outside. This relates to favorite locations and to the use of absorbent materials that minimize urine splashing on the horse's legs.

FIGURE 7-8 The elimination postures for horses: (a) mares urinate by spreading their hind legs, lifting their tail, and tilting their pelvis slightly; (b) geldings and stallions spread their hind legs a little farther behind the normal standing position than do mares and lift their tail slightly; and (c) defecation only involves lifting the tail.

Defecation

Meconium is the first thing defecated by a newborn foal, usually within the first 0.75–4.5 h.[16] More of the meconium will be passed in the next few hours.

The defecation posture is similar between males and females. The tail is raised, which is the only posture change used by free-ranging ponies.[14] If the horse is just standing, the pelvis is tilted such that the back rises slightly (Figure 7-8c). Defecation requires less concentration than urination because horses seldom stop grazing, and may even keep moving.[2,3,14] Performance horses usually continue their routines while defecating, although younger animals may stop first. In the pasture and on the range, mares simply walk away when they finish defecating. Stallions will often turn and sniff the feces, and they may paw it as well.[14,96]

Defecation occurs approximately once every 2–2.5 h.[2,3] In mares, this amounts to six or seven defecations per day, but stallions may eliminate at twice that frequency when marking is involved.[3] Normal fluid loss is part of defecation, not just urination. Horses lose approximately 3.7 gal (14 kg) of water through defecation each day.[85]

GROOMING BEHAVIOR

Grooming behavior is part of the comfort behavior subset of the maintenance behaviors. It is complicated because of its many functions. The three primary ones include the most basic of comfort behaviors—scratching areas that itch. A second function is more important—maintenance of skin health. Grooming is used to remove dead skin and hair. It also helps distribute body oils. But grooming is associated with health in other ways too. It removes parasites, most of which are blood-sucking flies and mosquitoes.[98] This not only minimizes blood loss, it also reduces exposure to vectorborne diseases like encephalitis. Lastly, but also important, is the function of social facilitation, particularly allogrooming.

Self-Grooming

Grooming has several forms, but each is limited by body shape and flexibility. The first, self-grooming, involves the use of one part of the body to groom another. The horse uses its teeth, tongue, and/or upper lip. Flexibility limits their use of facial areas to the chest, forelimbs, side of the chest, and lateral hip/pelvic limb areas. Foals spend 60% of their grooming bouts nibbling on their trunk or hind legs, almost twice that of their mothers (Figure 7-9).[3,99]

Forelimbs are also used in self-grooming. The horse will rub its face on the limb, particularly the area around the eyes. This is useful for removing flies and other insect pests. The same goal of insect removal occurs when the horse

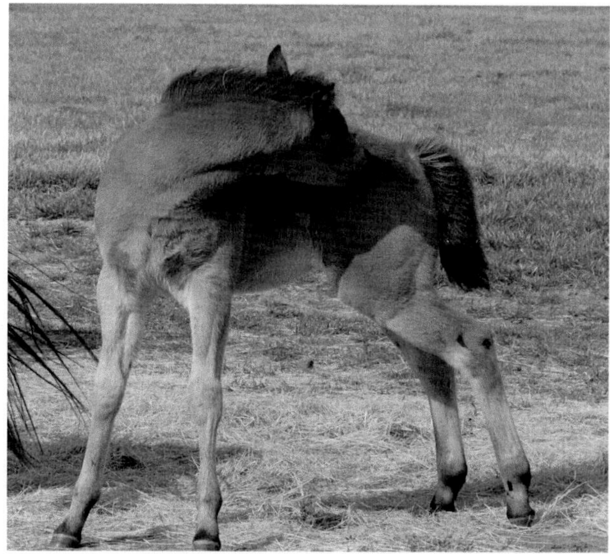

FIGURE 7-9 It is common for foals to use their mouth and teeth to groom parts of their legs or body. They do this more often than do adult horses.

sharply strikes the ground with its hoof after the foot is rapidly raised—a behavior called a *stamp* or *stomp*.[96]

Hind limbs reach forward to remove insects from the ventral abdomen and to scratch the side of the head, neck, and shoulder regions (Figure 7-10). The scratching with a rear foot accounts for 17% of foal self-grooming bouts but only 4% of those for adult mares.[3,99]

Shaking involves the entire body. There is a rapid, rhythmic rotation of the trunk along its longitudinal axis while the feet remain solidly planted on the ground.[96] Shaking helps remove excess dust and dander, so it is common after the horse rolls. Head shaking (along the longitudinal axis), side-to-side tail swishing, and skin twitching are used to get rid of insects and other irritants.

Mares self-groom once or twice an hour, while foals do so much more frequently. By 2 months of age, they are self-grooming about 12 times an hour. This gradually decreases by half at 6 months.[99]

Grooming Using Inanimate Objects

The second type of grooming pattern involves the use of inanimate objects. These provide a way for horses to groom areas that are difficult for them to reach, particularly the topline and rump. Rolling uses the ground to groom and rub the back. There are other functions as well. It rapidly dries wet hair, restoring insulative properties, aids shedding, protects the hair from the sun,

FIGURE 7-10 The hind feet are used to groom parts of the head and neck.

and adds a protective dust layer against insects.[100] Additionally, rolling helps the horse stretch its muscles to maintain flexibility.[101] There are social implications to rolling as well. When one horse rolls, it is common for others to do the same, particularly young foals.[100]

Rolling is common before a lying horse rises, when a horse gets wet, and when horses are near watering sites. If not already lying, the horse will do so. It then rotates its body along the long axis while keeping its legs tucked against its body (Figure 7-11a).[96] The amount of twist from a lateral position varies from 45 to 180 degrees. Stalled horses roll to 90 degrees so their feet are on top, and then return to the lateral starting position. For horses that roll over to the opposite side, they will roll back to the original side before getting up.[102] Before getting up, a horse resting in a pasture will roll 30.4% of the time, and approximately 95% of horses roll at least once a day.[102] If stalled, the frequency of rolling decreases from 95% to approximately 30%, and that usually happens when the horse is turned out into a paddock. This is more likely if the horse is kept in a small stall.[103] Rolling is not a common behavior for foals, but it accounts for 13% of self-grooming bouts in adult horses.[3,99]

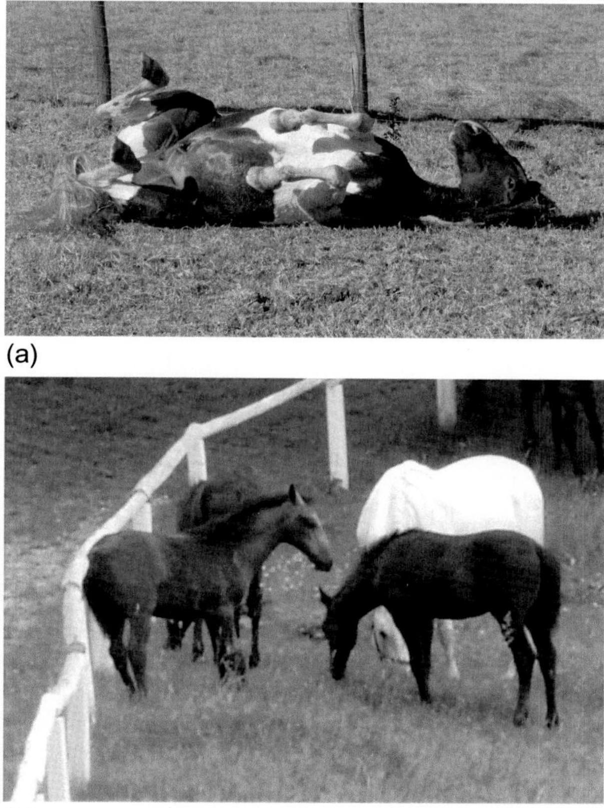

FIGURE 7-11 Because of limited flexibility, horses rub against inanimate objects to groom inaccessible areas and scratch when it itches. (a) Rolling on the ground allows the horse to groom its back and the croup. (b) Rubbing against objects like the fence allows this Lipizzaner foal to scratch its rump.

Of the various natural surfaces on which a horse could roll, loose soil is strongly preferred to grass, compacted dirt, loose sand, or straw.[3,101] This is part of the reason pawing is common before rolling and why specific, favored rolling sites develop that over time become bowl-shaped. In free-roaming bands, members will successively roll at the same site, with the harem stallion being the last to do so. Individual horse odors are left on the dust to be detected by other horses. Stallions, in particular, explore these odors and groups of the stallions will roll in these sites in a specific order, with the highest-ranking horse rolling last.[100] Stallions will also roll in mud and water.[2]

Rubbing against inanimate objects like fences, tree limbs, feeders, and similar structures is another way to scratch a difficult spot (Figure 7-11b). This behavior can be facilitated by attaching broom heads to sturdy objects so that horses can rub various body regions against the stiff bristles.[104] The ventral abdomen can be groomed when the horse sits with forelimbs extended and rubs its belly against the ground by moving back and forth. Mares use rubbing in 42% of self-grooming bouts, but foals do so only 13% of the time.[3,99]

Allogrooming

Allogrooming (mutual grooming) is the third grooming pattern, and it occurs between horses that are close associates ("friends"). The two horses stand close to each other, facing head-to-tail. Tail swishing by one helps keep insects off the face of the other, a passive type of allogrooming. Horses also show an active form called *withers nibbling*, during which they mutually nibble the general area of each other's withers (Figure 7-12). When humans rub this same area on a horse that does not get to physically interact with others, the horse often responds with an extension of its neck and upper lip (Figure 7-13). Introductory sniffing often precedes allogrooming bouts between horses, with nibbling then beginning at the crest of the neck and moving on toward the withers area. Other areas may also receive the nibbling behavior, but less often (Figure 7-14).[2,3]

Foals begin allogrooming behaviors within a few weeks of birth, and the behavior peaks by the third month. Fillies show mutual grooming 1.6 times per hour and direct their efforts equally between fillies and colts. Young males groom less often, only 0.9 times an hour, and direct their bouts almost exclusively toward fillies, especially yearlings if they are present.[105,106] Colts are also less likely to groom their mothers than are filly foals.[106]

FIGURE 7-12 Withers nibbling is a type of allogrooming between horses that are closely bonded. It is associated with the release of brain endorphins and lowering of the heart rate.

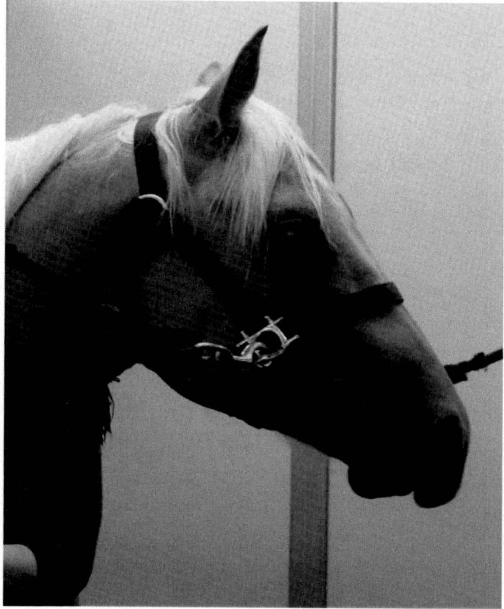

FIGURE 7-13 Mutual grooming can occur with one partner being human. In this case, the withers and neck areas are being rubbed and the stallion responds with several relaxed postures, especially the relaxed upper lip and rounded eyelids.

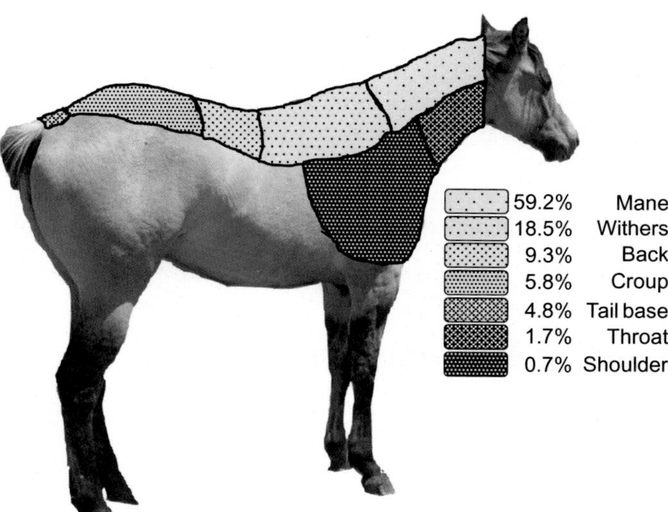

FIGURE 7-14 Several areas receive allogrooming from herdmates, but the dorsal neck and withers areas receive more than three-fourths of the interactions.[2, 3]

The importance of allogrooming can be demonstrated in various ways. First, all horses show the behavior, indicating that there is an evolutionary aspect. If prevented from mutual grooming, chronic stress occurs as evidenced by a rebound effect when the horse gets to interact with another again. The rebound is shown as excessive and prolonged grooming bouts. Thirdly, the behavior is self-rewarding—brain endorphins, the body's reward system, are released.[107] This is also indicated by the heart rate slowing when withers nibbling occurs, regardless of whether the massage comes from a conspecific or a human.[3,108–111]

Allogrooming is part of the bonding process between mares and foals, as well as between close associates and herdmates. It also lessens social tensions. When grooming partners are tightly bonded, individuals are likely to intervene in allogrooming events between their partner and some other horse.[112] Lower-ranking individuals initiate over 60% of grooming bouts, but dominant individuals usually end them.[3] This facilitates nonconfrontational reinforcement of social hierarchies, minimizing the need for aggression within a group. The frequency of the mutual grooming bouts is inversely proportional to the amount of time spent grazing and stage of pregnancy.[113–115]

SHELTER SEEKING

Shelter seeking by pastured and free-ranging horses is usually done for protection from the weather. In good weather, horses seek shelter less than 10% of the time. However, there is a significant increase when winds exceed 4.9 mph or 2.2 m/s, particularly if paired with rain or snow.[116,117] Shelter seeking also increases when the ambient temperatures fall outside the range of 44.8–77.4° F (7.1–25.2°C).[117] When it is snowing and wind speeds exceed 11.0 mph (4.9 m/s), shelter seeking increases to 62%.

Bothersome insects can cause horses to try to find shelter. However, as wind speeds increase, horses are less likely to need protection against the pests and thus are less likely to seek shelter.[118] When they do seek protection against insects, horses prefer shelters that have one open side in contrast to those that with a roof only.[118]

COMMON ORAL BEHAVIOR PROBLEMS

Modern equine diets often fail to take into account normal equine ingestive behavior. They evolved on grass—a high roughage, low calorie diet. What many now get is a low roughage, high calorie one (Table 7-1). This has changed how the horse behaves and misbehaves.[119] With more available time and fairly bland environments, horses find ways to create their own enrichment. Those ways can be problematic.

TABLE 7-1 Comparative Behaviors Between Horses Eating Hay or Pelleted Feed Exclusively[119]

Behavior	Hay Diet Only	Pellet Diet Only
Time spent eating/24h	51%	10%
Foraging wood shavings	1%	12%
Time spent standing	35%	58%
Jaw movements while eating	60/min	70/min
Estimated jaw movements/day	43,000	10,000

Bolting Food

Some horses seem to "inhale" concentrated food because they are eating it so fast. Animals that bolt down food are often ones that experienced food deprivation at some time in their life. They might also be trying to eat as much of a favorite food as possible before being displaced by a higher-ranking individual.

There are physical concerns for horses that bolt down food. The most significant is the possibility of choking, as they try to swallow large amounts of relatively dry material.[78] Inhalation of grain into the trachea can result in coughing or aspiration pneumonia. For ponies with partial tracheal collapse, this becomes even more likely. As significant, insufficient chewing does not allow appropriate mixing with saliva to start the digestive process. As a result, undigested grain may pass completely through the gastrointestinal tract.

The classic treatment for bolting of food is to place several large stones in the feed bin so that the horse has to move them around in order to get to the grain. Another technique is to feed hay first, waiting a little while to feed grain when the horse's appetite has been partially satisfied. Concentrates can also be changed to be less palatable or fed in several small meals instead of one or two large ones. Various types of grazing muzzles have been shown to slow the rate of intake.[78] When food bolting happens in a pasture, it works best to bring the individual horse into a separate area so that there is no social competition.

Coprophagy

Coprophagy is the eating of feces. The behavior is normal in foals for about a 3-week period beginning when they are 2 weeks old. They eat their mother's feces.[120–122] It is thought the behavior helps establish normal intestinal microbial flora in the foals. The behavior is not normal in adult horses. When it does occur, it is more likely to be done by companion horses ridden for pleasure than by event horses.[123] The problem has been linked to diets that are low in protein or inadequate in volume and roughage.[120,121,124,125] Horses fed concentrates have a significantly higher rate of coprophagy too. There may be a relation to cecal acidity. It has been shown experimentally that increasing cecal pH causes horses to spend significantly more time standing and less time in coprophagy.[3,126]

Destructive Chewing

Destructive chewing is a term applied to the behavior of a horse destroying non-food items by chewing on them. The items take a variety of forms, from mangers to blankets, from lead ropes and halters to electric cords. Because of the potential danger, owners must take care that horses do not have access to items that could cause harm or that they do not want destroyed.

Bark Chewing

One of the more common problems reported by owners is the horse that chews the bark from trees.[45] The reasons for this destruction are probably numerous, and include things like deficiencies in roughage or minerals and lack of other environmental activities. Free-ranging horses will eat tree bark too. The behavior is more common in cold, wet weather, suggesting a connection with the need for more roughage.[127] Treatments are directed toward preventing access to trees by fencing them off or wrapping the trunks with something like chicken wire. It is advisable to check the diet to be sure it is adequate.

Wood Chewing

Wood chewing is a common form of destructive chewing that can result in costly damage. One pony can consume 2 lb (1 kg) of wood per day, to cause significant destruction.[120] Reports of the incidence of wood chewing vary considerably, from 0.2% to 34% of horses.[128–132]

This problem is generally considered to be a variation of ingestive behavior rather than a stereotypy because it is common in horses on diets that are high in concentrates but low in roughage, have less than 6.8 kg of roughage, or have less than 10% protein.[120,121,125–127,131,133,134] The behavior is also more common in cold, wet weather and the late winter months, which is another suggestion supporting a relationship to the consumption of roughage.[127,135] Horses fed concentrates have a significantly lower cecal pH and cecal acetate and a higher cecal propionate 4–6 h after eating compared to horses that only eat hay. Although this relationship is unclear, the horses getting grain show a higher rate of wood chewing.[126] So too do horses fed cubed alfalfa compared to those getting the loose form.[136] Wood chewing tends to be diurnal, being more common at night when less food would be moving through the intestines.[131,137] Other relationships to wood chewing include a salt- or mineral-deficient diet.[127]

Nondietary factors are also associated with chewing wood. Horses that get exercise are less likely to do so.[137] More specifically, the type of riding a horse gets is related, with a greater tendency in reining, show jumping, and general pleasure horses compared to those that perform in other events.[123] Wood chewing is also more common in stalled horses that cannot see other horses and in individually stalled weanlings.[24,127,133,138] Foals of dominant mares, those weaned in stalls, and those housed as part of a group are also more likely to chew wood.[131]

Palatability of the wood being chewed must also be considered. While horses chew both soft and hard woods, redwood planks and certain types of trees, particularly willow trees in spring, are favored.[131] Obviously, they will do more damage on fences made from soft wood.[3]

In theory, eating wood could result in the intake of splinters, but if this happens, it is not thought to be common. It has been shown that ponies can eat a diet of 50% sawdust with no apparent ill effects.[120]

Wood chewing is not considered to be a stereotypy, but it often precedes the development of cribbing. While about one-third of young horses show wood chewing, 74% of cribbers had previously been known to do so.[139]

For many years, the treatment of choice has been the use of taste aversion, but this does not address the reason the behavior started. A nasty tasting substance is painted onto the surfaces the horse chews, and the behavior is supposed to stop. If it does, it stops on that surface and moves to another. There are several problems associated with how taste aversion is used. Most of the products that taste bad are compounds that stain clothes, discolor fences, and chemically burn human skin. This tends to make owners shy away from frequent use. Another problem is that the process is not done correctly. Bad-tasting substances like pepper sauce, bitter apple, and even creosote are used. Teaching taste aversion is a two-step process, rather than simply coating the object with a bad taste. Step one involves sensitizing the horse to the bad taste using both olfaction and taste. Several milliliters of any particularly bad tasting substance are drawn into a needleless syringe. The horse is allowed to get a good smell of the substance, and this is followed quickly by the substance being squirted into the horse's mouth. The desired reaction should be what appears as obvious disgust. Then in step two, the substance is coated on whatever object should not be chewed or eaten. The goal is for the horse to couple the odor with a particular bad experience. Taste aversion does not work particularly well in horses because coated objects are usually outdoors and the coating is diluted by sun, wind, rain, and time. If the technique works, it will need to be repeated periodically and the object frequently recoated.

Other treatments for wood chewing include increasing the exercise and the use of bars or grills between stalls so that horses can see or interact with each other (see Figure 5-3 in Chapter 5).[127,133] Adding more hay or increasing grazing time can help, and the use of low-quality hay may be needed if obesity is a factor.[121] Another substitute would be the use of straw bedding as a nonhay roughage source.[133] It is also possible to enrich the environment so the horse has other activities. Objects like large balls or empty water jugs suspended from the ceiling are used as play objects by some horses.[121] A companion may be a sufficient diversion, including companions of other species like goats or chickens.

When hay or pasture time is restricted, other enrichment opportunities can be designed. As an example, operant conditioning is used to teach the horse to push a lever to get a small amount of pelleted food. Once the horse learns that pushing a lever brings food, the number of times it has to push for the food reward is gradually increased, and then randomized. The amount of food dispensed should be small so that the activity will consume a great deal of time.[120]

Eating the Bedding

Approximately 21% of horses eat their bedding.[130] While the amount and frequency varies considerably, the cause generally relates to the animal's diet. Horses that get low amounts of roughage are more likely to eat bedding. Straw tends to be preferred to wood shavings, but either one can be ingested. The behavior is considered one of opportunity.

Pica

Pica is the eating of nonfood objects. Technically wood chewing, bark chewing, and eating of wood shavings would be considered pica, but more often the term is used for objects like plastic, dirt, electric cords, and potentially food items that are not natural for equids. Horses will commonly eat bread, and in Iceland, pastured horses are often fed salted fish as a protein, mineral, and salt supplement.[140] But horses also eat things like hot dogs or chicken meat. In most of these situations, the horse does a lot of oral environmental exploration or readily takes treats offered by people.[140] Pica is not common, accounting for only one animal in a pasture group in most places.[45] This suggests that there are individual reasons for it rather than one serious stress triggering the problem in multiple animals. When pica occurs, it is best to try to determine why, rather than to just punish the behavior.[141]

Dirt eating can be associated with "self-medication" when minerals are lacking in the diet. It might also be associated with a need to calm gastric irritation, particularly if there is kaolin containing clay.[141] Other forms of pica might relate to low forage intake, insufficient grazing time, or the lack of environmental stimuli. Young horses will also explore their environment orally (Figure 7-15). While they usually drop the objects, sometimes dirt and smaller objects are swallowed.

If specific causes cannot be found, it is important to limit access to items that might be targeted, particularly if they are dangerous. In addition to removing objects from the environment, the horse could wear a grazing muzzle.

FIGURE 7-15 Young horses investigate their environments by taking objects like dirt clots (as shown), sticks, ropes, and even plastic bags. They usually toss their head up and down before letting go of the object. The object is rarely eaten by the youngster.

Mane and Tail Eating

A small amount of mane and tail chewing is normal between foals. Young horses on pelleted diets have a significantly higher rate of this behavior, as well as wood chewing, compared to horses on high roughage diets.[3,138] Treatment consists of adding more dietary fiber. In extreme cases, it may be necessary to use a grazing muzzle when the horses are together or to wrap the victim's tail in a tail bag or sock to prevent access.

Obesity

The body condition of free-ranging horses varies by the seasons, and it does so in domestic horses too. As an example, the prevalence of obesity is 27% at the end of winter and 35% during the summer.[142] Obesity is the most common body condition problem, especially for ponies, even pastured ponies. It is also common in horses that show in halter classes: show people call it "bloom." In ponies other than Shetlands, the incidence of obesity is 30%, and in cob-type horses, it is 20%. This compares to a 12.5% prevalence in lightweight horses.[142] The well-known association between obesity, insulin resistance, and laminitis is termed the equine metabolic syndrome (EMS).[143–145]

Managing obesity, particularly in ponies, is an ongoing challenge. While it revolves around limiting caloric intake relative to caloric use, in pastured animals that can be a challenge. Keeping the animal on pasture is desirable for many reasons, so grazing muzzles can be used. These muzzles reduce the intake of pasture dry matter by 77%–83% and water-soluble carbohydrate intake was also significantly lowered.[76] Even if used for only 10 h a day, ponies will still gain weight, just less.[79] Long grass is the most difficult to eat when a grazing muzzle is used.[76] For horses or ponies kept in a paddock or stall, a diet of plain grass hay (with low sugar content if insulin-resistance is a problem) is sufficient to reduce weight in some breeds, such as Standardbreds. Ponies and Andalusians are relatively resistant to the loss of body fat and require daily exercise to improve weight loss.[146] While a hay diet provides a more normal intake of roughage, care must be taken to ensure the horse gets needed protein, minerals, or vitamins.[145]

Psychogenic Polydipsia

Excessive water drinking is common on hot days, but it can be problematic at other times. Approximately 16% of horses show psychogenic polydipsia.[130] Even though the condition can be associated with medical problems such as chronic renal failure, pituitary adenoma, or diabetes insipidus, most cases have a psychogenic cause. The problem usually is noticed by the person who cleans the stall because the bedding is excessively wet.[32,147] If the water source is filled manually, the rapid rate of consumption compared to other horses can initiate the concern. True psychogenic polydipsia is a diagnosis of exclusion, and one where stable management should also be evaluated. Affected horses might be considered to be "bored" or stressed, so feeding practices, turnout time, and interactions with social peers need to be evaluated. Most of the time, adjusting management practices to increase enrichment will reduce the problem. If water restriction is considered, it is critical to rule out medical causes of polydipsia and evaluate reasonable amounts of available water to avoid dehydration.

Refusing to Drink

Foals rarely drink water prior to weaning, but when this happens in an older animal, there usually is a reason. The horses appear gaunt and tucked up in the flank. Some show signs of pain, some have no interest in any water, and some will readily drink from another water source. Things that can result in a horse not drinking include a number of physical problems, such as severe dental pain, temporomandibular joint issues, tongue abnormalities, or swallowing problems. Similar clinical signs are associated with eating too. Environmental issues can cause the horse not to want to approach the water. Extremely dirty water, including that contaminated with the bodies of dead birds or rodents, is repulsive to horses. Phantom voltage can electrify the waterer such that the horse gets a shock every time it tries to drink. A medical evaluation is needed to rule out physical causes, and then problems in the environment are considered.

COMMON PROBLEMS ASSOCIATED WITH GROOMING BEHAVIOR

Splashing in Water

Horses that have access to ponds or lakes may use water as a way to drive off insects. It is also a way for them to remove excess dirt from their skin and hair coat and to cool off when the summer weather turns hot and humid. Once a horse has learned to swim but no longer has access to water, it may find another way to accomplish the same thing. If a water trough is large enough, the horse might climb over the side and lie down in the water. When the tank is smaller, it can put the front feet in (Figure 7-16). This is usually followed by splashing to get the water onto the belly and upper leg, but the splashing muddies the surrounding area, dirties the water, and at least partially empties the tank.

Excessive Grooming and Self-Rubbing

Grooming is an important maintenance behavior for horses. While it is well documented that the lack of social contact is stressful and plays a role in the development of stereotypies, the significance of stress when stablemates are unable to groom each other is not well understood.[107] What role the lack of allogrooming plays in excessive grooming and self-rubbing is not known.

FIGURE 7-16 The mare gets into the water tank when insects are bad and when the weather gets too hot. If given a choice, she will swim in a pond instead. The splashing dirties the water, muddies the area, and partially empties the tank. This makes it necessary to have a second tank available for drinking water.

Hypersensitivity to *Culicoides obsoletus* antigens are fairly common in horses and are a major cause of excessive grooming and self-rubbing. The hair loss tends to occur most often along the mane, tail, and ventral abdomen. While all horses produce allergen-specific antibodies, nonaffected horses are able to mount a better immune response to protect against symptoms.[148] It has also been suggested that the antigens are regionally specific, so moving the horse to a different location may be sufficient to reduce or stop the excessive rubbing.

Self-Biting or Kicking

All horses will show some self-directed grooming behavior that involves biting or kicking, mostly against annoying insects. Approximately 4% of horses show some form of self-biting that is not insect related.[130] The cause is not always obvious. Colicky horses are the classic example of pain-associated self-biting or kicking. Other conditions that cause this are allergies, dermatitis, reproductive tract conditions, and gastrointestinal tracts issues.[149,150] Determining the cause will direct the treatment.

REFERENCES

1. Houpt KA. *Equine maintenance behavior: feeding, drinking, coat care and behavioral thermoregulation.* http://research.vet.upenn.edu/HavemeyerEquineBehaviorLabHomePage/ReferenceLibraryHavemeyerEquineBehaviorLab/HavemeyerWorkshops/HorseBehaviorandWelfare1316June2002/HorseBehaviorandWelfare2/EquineMaintenanceBehaviorFeedingDrinkingCo/tabid/3123/Default.aspx [downloaded June 19, 2017].
2. Waring GH, Wierzbowski S, Hafez ESE. The behaviour of horses. In: Hafez ESE, editor. *The behavior of domestic animals.* 3rd ed. Baltimore: The Williams & Wilkins Company; 1975. p. 330–69.
3. Waring GH. *Horse behavior.* 2nd ed. Norwich, NY: Noyes Publications; 2003. p. 442.
4. Zurek U, Danek J. Maternal behavior of mares and the condition of foals after parturition. *Bull Vet Inst Pulawy* 2011;**55**:451–6.
5. Rossdale PD. Clinical studies on the newborn thoroughbred foal. I. Perinatal behavior. *Br Vet J* 1967;**123**(11):470–81.
6. Joubert E. The social organization and associated behavior in the Hartman zebra (*Equus zebra hartmannae*). *Modoqua* 1972;**1**(6):17–56.
7. Crowell-Davis SL. Developmental behavior. *Vet Clin N Am Equine Pract* 1986;**2**(3):573–90.
8. Lawrence LM, Cymbaluk NF, Freeman DW, Geor RJ, Graham-Thiers PM, Longland AC, Nielsen BD, Sicilianmo PD, Topliff DR, Valdes EV, Van Saun RJ. Unique aspects of Equine nutrition. In: *Nutrient requirements of horses.* 6th revised ed. Washington, DC: The National Academies Press; 2007. p. 235–67.
9. Barber JA, Crowell-Davis SL. Maternal behavior of Belgian (*Equus caballus*) mares. *Appl Anim Behav Sci* 1994;**41**(3–4):161–89.
10. Carson K, Wood-Gush DGM. Behaviour of thoroughbred foals during nursing. *Equine Vet J* 1983;**15**(3):257–62.

11. Crowell-Davis SL. Nursing behavior and maternal aggression among welsh ponies (*Equus caballus*). *Appl Anim Behav Sci* 1985;**14**(1):11–25.
12. Berger J. *Wild horses of the Great Basin: social competition and population size*. Chicago: University of Chicago Press; 1986. p. 326.
13. Francis-Smith K, Wood-Gush DGM. Coprophagia as seen in thoroughbred foals. *Equine Vet J* 1977;**9**(3):155–7.
14. Tyler SJ. The behavior and social organization of the New Forest ponies. *Anim Behav Monogr* 1972;**5**(2):87–196.
15. Crowell-Davis SL, Houpt KA. Coprophagy by foals: effect of age and possible functions. *Equine Vet J* 1985;**17**(1):17–9.
16. Jeffcott LB. Observations on parturition in crossbred pony mares. *Equine Vet J* 1972;**4** (4):209–12.
17. Apter RC, Householder DD. Weaning and weaning management of foals: a review and some recommendations. *J Equine Vet* 1996;**16**(10):428–35.
18. Burns HD, Gibbs PG, Potter GD. Milk-energy production by lactating mares. *J Equine Vet* 1992;**12**(2):118–20.
19. Waran NK, Clarke N, Farnworth M. The effects of weaning on the domestic horse (*Equus caballus*). *Appl Anim Behav Sci* 2008;**110**(1–2):42–57.
20. Boy V, Duncan P. Time-budgets of Camargue horses I. Developmental changes in the time-budgets of foals. *Behaviour* 1979;**71**(3–4):187–202.
21. Hausberger M, Henry S, Larose C, Richard-Yris M-A. First suckling: a crucial event for mother-young attachment? An experimental study in horses (*Equus caballus*). *J Comp Psychol* 2007;**121**(1):109–12.
22. Nicol CJ, Badnell-Waters AJ, Bice R, Kelland A, Wilson AD, Harris PA. The effects of diet and weaning method on the behavior of young horses. *Appl Anim Behav Sci* 2005;**95**(3–4):205–21.
23. Williams C, Randle H. What methods are commonly used during weaning (mare removal) and why? A pilot study. *J Vet Behav* 2016;**15**:89.
24. Heleski CR, Shelle AC, Nielsen BD, Zanella AJ. Influence of housing on weanling horse behavior and subsequent welfare. *Appl Anim Behav Sci* 2002;**78**(2–4):291–302.
25. Houpt KA, Hintz HF, Butler WR. A preliminary study of two methods of weaning foals. *Appl Anim Behav Sci* 1984;**12**(1–2):177–81.
26. Wulf M, Beythien E, Ille N, Aurich J, Aurich C. The stress response of 6-month-old horses to abrupt weaning is influenced by their sex. *J Vet Behav* 2018;**23**:19–24.
27. Berger JM, Spier SJ, Davies R, Gardner IA, Leutenegger CM, Bain M. Behavioral and physiological responses of weaned foals treated with equine appeasing pheromone: a double-blinded, placebo-controlled, randomized trial. *J Vet Behav* 2013;**8**(4):265–77.
28. Malinowski K, Hallquist NA, Helyar L, Sherman AR, Scanes CG. Effect of different separation protocols between mares and foals on plasma cortisol and cell-mediated immune response. *J Equine Vet* 1990;**10**(5):363–8.
29. Moons CPH, Laughlin K, Zanella AJ. Effects of short-term maternal separations on weaning stress in foals. *Appl Anim Behav Sci* 2005;**91**(3–4):321–35.
30. Merkies K, Dubois C, Marshall K, Parois S, Graham L, Haley D. A two-stage method to approach weaning stress in horses using a physical barrier to prevent nursing. *Appl Anim Behav Sci* 2016;**183**:68–76.
31. McCall CA, Potter GD, Kreider JL. Locomotor, vocal and other behavioral responses to varying methods of weaning foals. *Appl Anim Behav Sci* 1985;**14**(1):27–35.
32. Houpt KA. *Domestic animal behavior for veterinarians and animal scientists*. 3rd ed. Ames: Iowa State University Press; 1998. p. 495.

33. Sweeting MP, Houpt CE, Houpt KA. Social facilitation of feeding and time budgets in stabled ponies. *J Anim Sci* 1985;**60**(2):369–74.
34. Kaseda Y. Seasonal changes in time spent grazing and resting of Misaki horses. *Nippon Chikusan Gakkai-ho* 1983;**54**(7):464–9.
35. Keiper RR, Keenan MA. Nocturnal activity patterns of feral ponies. *J Mammal* 1980;**61**(1):116–8.
36. Crowell-Davis SL, Houpt KA, Carnevale J. Feeding and drinking behavior of mares and foals with free access to pasture and water. *J Anim Sci* 1985;**60**(4):883–9.
37. Mayes E, Duncan P. Temporal patterns of feeding behavior in free-ranging horses. *Behavior* 1986;**96**(1):105–29.
38. Salter RE, Hudson RJ. Feeding ecology of feral horses in western Alberta. *J Range Manag* 1979;**32**(2):221–5.
39. Rubenstein DI. Behavioural ecology of island feral horses. *Equine Vet J* 1981;**13**(1):27–34.
40. York CA, Schulte BA. The relationship of dominance, reproductive state and stress in female horses (*Equus caballus*). *Behav Process* 2014;**107**:15–21.
41. Ralston SL. Feeding behavior. *Vet Clin N Am Equine Pract* 1986;**2**(3):609–21.
42. Ralston SL, Van den Broek G, Baile CA. Feed intake patterns and associated blood glucose, free fatty acid and insulin changes in ponies. *J Anim Sci* 1979;**49**(3):838–45.
43. Kern D, Bond J. Eating patterns of ponies fed diets *ad libitum*. *J Anim Sci* 1972;**35**(1):286.
44. Laut JE, Houpt KA, Hintz HF, Houpt TR. The effects of caloric dilution on meal patterns and food intake of ponies. *Physiol Behav* 1985;**35**(4):549–54.
45. van den Berg M, Brown WY, Lee C, Hinch GN. Browse-related behaviors of pastured horses in Australia: a survey. *J Vet Behav* 2015;**10**(1):48–53.
46. Greening L, Shenton V, Wilcockson K, Swanson J. Investigating duration of nocturnal ingestive and sleep behaviors of horses bedded on straw versus shavings. *J Vet Behav* 2013;**8**(2):82–6.
47. Ransom JI, Cade BS. *Quantifying equid behavior—A research ethogram for free-roaming feral horses*. U.S. Geological Survey, https://pubs.usgs.gov/tm/02a09/pdf/TM2A9.pdf; 2009 [downloaded March 15, 2017].
48. Houpt KA. Taste preferences in horses. *Equine Pract* 1983;**5**(4):22–4. and 26.
49. Krüger K, Flauger B. Social feeding decisions in horses (*Equus caballus*). *Behav Process* 2008;**78**(1):76–83.
50. Ralston SL. Controls of feed intake in horses. *J Anim Sci* 1984;**59**(5):1354–61.
51. Ralston SL, Baile CA. Gastrointestinal stimuli in the control of feed intake in ponies. *J Anim Sci* 1982;**55**(2):243–53.
52. van den Berg M, Giagos V, Lee C, Brown WY, Cawdell-Smith AJ, Hinch GN. The influence of odour, taste and nutrients on feeding behavior and food preferences in horses. *Appl Anim Behav Sci* 2016;**184**:41–50.
53. van den Berg M, Giagos V, Lee C, Brown WY, Hinch GN. Acceptance of novel food by horses: the influence of food cues and nutrient composition. *Appl Anim Behav Sci* 2016;**183**:59–67.
54. Archer M. Further studies on palatability of grasses to horses. *J Brit Grassl Soc* 1978;**33**(4):239–43.
55. Archer M. The species preferences of grazing horses. *J Brit Grassl Soc* 1973;**28**(3):123–8.
56. Hawkes J, Hedges M, Daniluk P, Hintz HF, Schryver HF. Feed preferences of ponies. *Equine Vet J* 1985;**17**(1):20–2.
57. Goodwin D, Davidson HPB, Harris P. Selection and acceptance of flavours in concentrate diets for stabled horses. *Appl Anim Behav Sci* 2005;**95**(34):223–32.

58. Cairns MC, Cooper JJ, Davidson HPB, Mills DS. *Ability of horses to associate orosensory characteristics of foods to their post-ingestive consequences in a choice test.* http://research.vet.upenn.edu/HavemeyerEquineBehaviorLabHomePage/ReferenceLibraryHavemeyerEquineBehaviorLab/HavemeyerWorkshops/HorseBehaviorandWelfare1316June2002/HorseBehaviorandWelfare2/AbilityofHorsestoAssociateOrosensoryCharacte/tabid/3114/Default.aspx; 2002[downloaded February 13, 2018].
59. Redgate SE, Cooper JJ, Hall S, Eady P, Harris PA. Dietary experience modifies horses' feeding behavior and selection patterns of three macronutrient rich diets. *J Anim Sci* 2014;**92**(4):1524–30.
60. Hintz HF, Loy RG. Effects of pelleting on the nutritive value of horse rations. *J Anim Sci* 1966;**25**(4):1059–62.
61. Ödberg FO, Francis-Smith K. Studies on the formation of ungrazed eliminative areas in field used by horses. *Appl Anim Ethol* 1977;**3**(1):27–34.
62. Ödberg FO, Francis-Smith K. A study on eliminative and grazing behavior—The use of the field by captive horses. *Equine Vet J* 1976;**8**(4):147–9.
63. Francis-Smith K. Behaviour patterns of horses grazing in paddocks. *Appl Anim Ethol* 1977;**3**(3):292–3.
64. Medica DL, Hanaway MJ, Ralston SL, Sukhdeo MVK. Grazing behavior of horses on pasture: predisposition to strongylid infection? *J Equine Vet* 1996;**16**(10):421–7.
65. Ellis AD, Fell M, Luck K, Gill L, Owen H, Briars H, Barfoot C, Harris P. Effect of forage presentation of feed intake behavior in stabled horses. *Appl Anim Behav Sci* 2015;**165**:88–94.
66. Ellis AD, Redgate S, Zinchenko S, Owen H, Barfoot C, Harris P. The effect of presenting forage in multi-layered haynets and at multiple sites on night time budgets of stabled horses. *Appl Anim Behav Sci* 2015;**171**:108–16.
67. Freire R, Clegg HA, Buckley P, Friend MA, McGreevy PD. The effects of two different amounts of dietary grain on the digestibility of the diet and behavior of intensively managed horses. *Appl Anim Behav Sci* 2009;**117**(1–2):69–73.
68. Benhajali H, Richard-Yris M-A, Ezzaouia M, Charfi F, Hausberger M. Foraging opportunity: a crucial criterion for horse welfare? *Animal* 2009;**3**(9):1308–12.
69. Burla J-B, Ostertag A, Patt A, Bachmann I, Hillmann E. Effects of feeding management and group composition on agonistic behavior of group-housed horses. *Appl Anim Behav Sci* 2016;**176**:32–42.
70. Bulmer L, McBride S, Williams K, Murray J-A. The effects of a high-starch or high-fibre diet on equine reactivity and handling behavior. *Appl Anim Behav Sci* 2015;**165**:95–102.
71. Destrez A, Grimm P, Cézilly F, Julliand V. Changes of the hindgut microbiota due to high-starch diet can be associated with behavioral stress response in horses. *Physiol Behav* 2015;**149**:159–64.
72. Holland JL, Kronfeld DS, Meacham TN. Behavior of horses is affected by soy lecithin and corn oil in the diet. *J Anim Sci* 1996;**74**(6):1252–5.
73. Luz MPF, Maia CM, Pantoja JCF, Neto MC, Filho JNPP. Feeding time and agonistic behavior in horses: influence of distance, proportion, and height of troughs. *J Equine Vet* 2015;**35**(10):843–8.
74. Morrison PK, Harris PA, Maltin CA, Grove-White D, Varfoot CF, Argo CMG. Perceptions of obesity and management practices in a UK population of leisure-horse owners and managers. *J Equine Vet* 2017;**53**:19–29.
75. Glunk EC, Sheaffer CC, Hathaway MR, Martinson KL. Interaction of grazing muzzle use and grass species on forage intake of horses. *J Equine Vet* 2014;**34**(7):930–3.

76. Longland AC, Barfoot C, Harris PA. Effects of grazing muzzles on intakes of dry matter and water-soluble carbohydrates by ponies grazing Spring, Summer, and Autumn swards, as well as Autumn swards of different heights. *J Equine Vet* 2016;**40**:26–33.
77. Longland AC, Barfoot C, Harris PA. The effect of wearing a grazing muzzle vs not wearing a grazing muzzle on pasture dry matter intake by ponies. *J Equine Vet* 2011;**31** (5–6):282–3.
78. Venable EB, Bland S, Braner V, Gulson N, Halpin M. Effect of grazing muzzles on the rate of pelleted feed intake in horses. *J Vet Behav* 2016;**11**:56–9.
79. Longland AC, Barfoot C, Harris PA. Efficacy of wearing grazing muzzles for 10 hours per day on controlling bodyweight in pastured ponies. *J Equine Vet* 2016;**45**:22–7.
80. de Laat MA, Hampson BA, Sillence MN, Pollitt CC. Sustained, low-intensity exercise achieved by a dynamic feeding system decreases body fat in ponies. *J Vet Intern Med* 2016;**30**(5):1726–31.
81. Sweeting MP, Houpt KA. Water consumption and time budgets of stabled pony (*Equus caballus*) geldings. *Appl Anim Behav Sci* 1987;**17**(1–2):1–7.
82. Miller R. Habitat use of feral horses and cattle in Wyoming's Red Desert. *J Range Manag* 1983;**36**(2):195–9.
83. Fonnesbeck PV. Consumption and excretion of water by horses receiving all hay and hay-grain diets. *J Anim Sci* 1968;**27**(5):1350–6.
84. Hinton M. On the watering of horses: a review. *Equine Vet J* 1978;**10**(1):27–31.
85. Tasker JB. Fluid and electrolyte studies in the horse. 3. Intake and output of water, sodium, and potassium in normal horses. *Cornell Vet* 1967;**57**:649–57.
86. Hafez ESE, Williams M, Wierzbowski S. The behaviour of horses. In: Hafez ESE, editor. *The behavior of domestic animals*. 2nd ed. Baltimore: The Williams & Wilkins Company; 1969. p. 391–416.
87. Lamoot I, Callebaut J, Degezelle T, Demeulenaere E, Laquière J, Vandenberghe C, Hoffmann M. Eliminative behavior of free-ranging horses: do they show latrine behaviour or do they defecate where they graze? *Appl Anim Behav Sci* 2004;**86**(1–2):105–21.
88. Edwards PJ, Hollis S. The distribution of excreta on New Forest grassland used by cattle, ponies and deer. *J Appl Ecol* 1982;**19**(3):953–64.
89. Krueger K, Flauger B. Olfactory recognition of individual competitors by means of faeces in horses (*Equus caballus*). *Anim Cogn* 2011;**14**(2):245–57.
90. Keiper RR. Social structure. *Vet Clin N Am Equine Pract* 1986;**2**(3):465–84.
91. Kimura R. Volatile substances in feces, urine and urine-marked feces of feral horses. *Can J Anim Sci* 2001;**81**(3):411–20.
92. McCort WD. Behavior of feral horses and ponies. *J Anim Sci* 1984;**58**(2):493–9.
93. Turner Jr JW, Perkins A, Kirkpatrick JF. Elimination marking behavior in feral horses. *Can J Zool* 1981;**59**(8):1561–6.
94. Jezierski T, Jaworski Z, Sobczyńska M, Kamińska B, Górecka-Bruzda A, Wallczak M. Excreta-mediated olfactory communication in Konik stallions: a preliminary study. *J Vet Behav* 2015;**10**(4):353–64.
95. McDonnell S. Reproductive behavior of the stallion. *Vet Clin N Am Equine Pract* 1986;**2** (3):535–55.
96. McDonnell S. *A practical field guide to horse behavior: the equid ethogram*. Lexington: The Blood-Horse; 2003375.
97. Sambraus HH, Zeitler-Feicht MH. *Urination of horses in free running stables (abstract). Pferdeheilkunde* 2003;**19**(5):521–4. https://eurekamag.com/research/003/998/003998012.php [abstract downloaded November 5, 2015].

98. Duncan P, Vigne N. The effect of group size in horses on the rate of attacks by blood-sucking flies. *Anim Behav* 1979;**27**(2):623–5.
99. Crowell-Davis SL. Self-grooming by mares and foals of the Welsh pony (*Equus caballus*). *Appl Anim Behav Sci* 1987;**17**(3–4):197–208.
100. McDonnell SM. Rolling along. *The Horse* 2001;**XVIII**(5):29–30.
101. Matsui K, Khalil AM, Takeda K. Do horses prefer certain substrates for rolling in grazing pasture? *J Equine Vet* 2009;**29**(7):590–4.
102. Hansen MN, Estvan J, Ladewig J. A note on resting behavior in horses kept on pasture: rolling prior to getting up. *Appl Anim Behav Sci* 2007;**105**(1–3):265–9.
103. Raabymagle P, Ladewig J. Lying behavior in horses in relation to box size. *J Equine Vet* 2006;**26**(1):11–7.
104. Bonner L. An itch to scratch. *Equus* 2003;**305**:49–51.
105. Crowell-Davis SL, Houpt KA, Carini CM. Mutual grooming and nearest-neighbor relationships among foals of *Equus caballus. Appl Anim Behav Sci* 1986;**15**(2):11113–23.
106. Rho JR, Srygley RB, Choe JC. Sex preferences in Jeju pony foals (*Equus caballus*) for mutual grooming and play-fighting behaviors. *Zool Sci* 2007;**24**(8):769–73.
107. van Dierendonck MC, Spruijt BM. Coping in groups of domestic horses—review from a social and neurobiological perspective. *Appl Anim Behav Sci* 2012;**138**(3–4):194–202.
108. Feh C, De Mazières J. Grooming at a preferred site reduces heart rate in horses. *Anim Behav* 1993;**6**(6):1191–4.
109. Kowalik S, Janczarek I, Kędzierski W, Stachurska A, Wilk I. The effect of relaxing massage on heart rate and heart rate variability in purebred Arabian racehorses. *Anim Sci J* 2017;**88**(4):669–77.
110. McBride SD, Hemmings A, Robinson K. A preliminary study on the effect of massage to reduce stress in the horse. *J Equine Vet* 2004;**24**(2):76–81.
111. Normando S, Haverbeke A, Meers L, Ödberg FO, Ibáñez Talegón M, Bono G. Effect of manual imitation of grooming on riding horses' heart rate in different environmental situations. *Vet Res Commun* 2003;**27**(Suppl. 1):615–7.
112. van Dierendonck MC, de Vires H, Schilder MBH, Colenbrander B, orhallsdóttir AG, Sigurjónsdóttir H. Interventions in social behavior in a herd of mares and geldings. *Appl Anim Behav Sci* 2009;**116**(1):67–73.
113. Asa CS, Goldfoot DA, Ginther OJ. Assessment of the sexual behavior of pregnant mares. *Horm Behav* 1983;**17**(4):405–13.
114. Kimura R. Mutual grooming and preferred associate relationships in a band of free-ranging horses. *Appl Anim Behav Sci* 1998;**59**(4):265–76.
115. van Dierendonck MC, Sigurjónsdóttir H, Colenbrander B, Thorhallsdóttir AG. Differences in social behavior between late pregnant, post-partum and barren mares in a herd of Icelandic horses. *Appl Anim Behav Sci* 2004;**89**(3–4):283–97.
116. Heleski CR, Murtazashvili I. Daytime shelter-seeking behavior in domestic horses. *J Vet Behav* 2010;**5**(5):276–82.
117. Snoeks MG, Moons CPH, Ödberg FO, Aviron M, Geers R. Behavior of horses on pasture in relation to weather and shelter—a field study in a temperate climate. *J Vet Behav* 2015;**19**(6):561–8.
118. Hartmann E, Hopkins RJ, Blomgren E, Ventrop M, von Brömssen C, Dahlborn K. Daytime shelter use of individually kept horses during Swedish summer. *J Anim Sci* 2015;**93**(2):802–10.
119. Houpt KA, Elia J. How important is roughage to a horse? In: *Proceedings of the American Veterinary Society of Animal Behavior*; 2003. p. 34–5.

120. Houpt KA. Oral vices of horses. *Equine Pract* 1982;**4**(4). 16, 19, 21 and 25.
121. Marcella KL. Common behavior problems in horses. *Equine Pract* 1988;**10**(6):22–6.
122. McDonnell SM. Young horse habits. *The Horse* 2010;**XXVII**(4):64–5.
123. Leme DP, Parsekian ABH, Kanaan V, Hötzel MJ. Management, health, and abnormal behaviors of horses: a survey in small equestrian centers in Brazil. *J Vet Behav* 2014;**9**(3):114–8.
124. Boyd LE. Time budgets of adult Przewalski horses: effects of sex, reproductive status and enclosure. *Appl Anim Behav Sci* 1988;**21**(1–2):19–39.
125. Schurg WA, Frei DL. Utilization of whole corn plant pellets by horses and rabbits. *J Anim Sci* 1977;**45**(6):1317–21.
126. Willard JG, Willard JC, Wolfram SA, Baker JP. Effect of diet on cecal pH and feeding behavior of horses. *J Anim Sci* 1977;**45**(1):87–93.
127. Simpson BS. Behavior problems in horses: cribbing and wood chewing. *Vet Med* 1998;**93**(11):999–1004.
128. Borstel UKV, Erdmann C, Maier M, Garlipp F. Relationships between owner-reported behavior problems and husbandry; use and management of horses. *J Vet Behav* 2016;**15**:92–3.
129. Dallaire A. Stress and behavior in domestic animals: temperament as a predisposing factor to stereotypies. *Ann N Y Acad Sci* 1993;**697**:269–74.
130. Hockenhull J, Creighton E. The day-to-day management of UK leisure horses and the prevalence of owner-reported stable-related and handling behaviour problems. *Anim Welf* 2015;**24**(1):29–36.
131. Mills DS, Taylor KD, Cooper JJ. Weaving, headshaking, cribbing, and other stereotypies. *Am Assoc Equine Pract Proc* 2005;**51**:221–30.
132. Nicol CJ, Badnell-Waters AJ. Suckling behaviour in domestic foals and the development of abnormal oral behaviour. *Anim Behav* 2005;**70**(1):21–9.
133. McGreevy PD, Cripps PJ, French NP, Green LE, Nicol CJ. Management factors associated with stereotypic and redirected behavior in the thoroughbred horse. *Equine Vet J* 1995;**27**(2):86–91.
134. Normando S, Meers L, Samuels WE, Faustini M, Ödberg FO. Variables affecting the prevalence of behavioural problems in horses. Can riding style and other management factors be significant? *Appl Anim Behav Sci* 2011;**133**(3–4):186–98.
135. Houpt KA. *Behavior in horses. Notes from talk given at the southwest veterinary symposium.* Texas: Ft. Worth; 2010.
136. Haenlein GFW, Holdren RD, Yoon YM. Comparative response of horses and sheep to different physical forms of alfalfa hay. *J Anim Sci* 1966;**25**(3):740–3.
137. Krzak WE, Gonyou HW, Lawrence LM. Wood chewing by stabled horses: diurnal pattern and effects of exercise. *J Anim Sci* 1991;**69**(3):1053–8.
138. Waters AJ, Nicol CJ, French NP. Factors influencing the development of stereotypic and redirected behaviours in young horses: findings of a four year prospective epidemiological study. *Equine Vet J* 2002;**34**(6):572–9.
139. Hothersall B, Nicol C. Role of diet and feeding in normal and stereotypic behaviors in horses. *Vet Clin N Am Equine Pract* 2009;**25**(1):167–81.
140. McDonnell S. Follow-up: carnivorous horses. *The Horse* 2003;**XX**(5):67–9.
141. Houpt KA. A taste for plastic? *Equus* 2016;**471**:66.
142. Giles SL, Rands SA, Nicol CJ, Harris PA. Obesity prevalence and associated risk factors in outdoor living domestic horses and ponies. *PeerJ* 2014;**2**:e299. https://doi.org/10.7717/peerj.299.

143. Carter RA, Treiber KH, Geor RJ, Douglass L, Harris PA. Prediction of incipient pasture-associated laminitis from hyperinsulinaemia, hyperleptinaemia and generalized and localized obesity in a cohort of ponies. *Equine Vet J* 2009;**41**(2):171–8.
144. Geor RJ. Metabolic predispositions to laminitis in horses and ponies: obesity, insulin resistance and metabolic syndromes. *J Equine Vet* 2008;**28**(12):753–9.
145. Geor RJ, Harris P. Dietary management of obesity and insulin resistance: countering risk for laminitis. *Vet Clin N Am Equine Pract* 2009;**25**(1):51–65.
146. Potter SJ, Bamford NJ, Harris PA, Bailey SR. Comparison of weight loss, with or without dietary restriction and exercise, in Standardbreds, Andalusians and mixed breed ponies. *J Equine Vet* 2013;**33**(5):339.
147. Browning AP. Polydipsia and polyuria in two horses caused by psychogenic polydipsia. *Equine Vet Educ* 2000;**2**(4):231–2. 234, 236.
148. Meulenbroeks C, van der Lugt JJ, van der Meide NMA, Willemse T, Rutten VPMG, Zaiss DMW. Allergen-specific cytokine polarization protects Shetland Ponies against *Culicoides obsoletus*-induced insect bite hypersensitivity. *PLoS ONE* 2015;**10**(4)https://doi.org/10.1271/journal.pone.0122090.
149. Marcella KL. Poor socialization can stem from a variety of circumstances. *DVM Newsmag* 2006;2E–5E.
150. McDonnell S. Equine self mutilation. *The Horse* 2000;**XVII**(3):73–8.

Chapter 8

Equine Locomotive Behavior

INFANT GROWTH AND MOVEMENT

Movement of a newborn foal starts within minutes of its birth. First comes the righting to sternal recumbency and then efforts to stand. Getting to its feet and becoming steady is particularly important for a species that does not hide its young. Rapid standing allows the newborn to suckle and gain strength relatively soon. It also allows the foal to follow the mother, avoiding potential predators.

In normal foals, there is a maturation period for locomotor kinetics. Velocity of the walk and trot increases over time as a result of an increased length of stride rather than stride frequency. The vertical force of each foot remains relatively constant, but stance variability decreases.[1] By day nine, a foal is able to travel up to 6.2 mi (10 km) in a 24-h period.[2]

Reflexes involving posture, body orientation, and the locomotor system are present at birth because of the foal's relatively developed neurological state.[3] The extensor thrust reflex is seen as soon as the foal tries to get up, assisting in straightening the limbs and helping it become steady on its feet. Extensor thrust can be evaluated by pressure on the sole of the foot causing limb extension. Other reflexes are associated with balance. The labyrinthine reflex involves righting, as when the foal moves from lying on its side to lying on its sternum. The vestibular reflex is shown when the nose is moved vertically upward, causing the forelimbs to flex and hind limbs extend. Downward movement of the nose has the opposite effect on the legs. The tonic neck reflexes associated with the neck moving up or down while the head is static result in limb responses that are opposite those of the vestibular reflex. The kicking reflex is demonstrated by moving a hand along the hind limb. Lastly, moderate pressure in the midlumbar region of very young foals results in bucking.

During the first week, a new foal will spend 32% of its time in recumbent rest, 96.5% of which is in lateral recumbency. The amount of rest gradually decreases, reaching 6.5% by week 21. As this change occurs, so does the resting posture, which gradually becomes more upright.[4]

The promotion of exercise, even above that which the youngster gets in a paddock, has been shown to have strong positive effects on the musculoskeletal system. Two-year-olds in training have longer and more successful careers in

Equine Behavioral Medicine. https://doi.org/10.1016/B978-0-12-812106-1.00008-5

racing than those started later.[2] Foals born before 325 days of gestation are more likely to have incomplete ossification of carpal and tarsal bones and ultimately do less well on the racetrack and perhaps in other demanding sports.[5]

ADULT MOVEMENT

Horses depend on their ability to move to avoid predators, find nourishment, and obtain water. The species evolved to do well on open, flat terrain where speed could be used for escape. Anything that restricts movement is stressful to untamed animals because it takes away the individual's ability to flee. This is the theory behind the use of hobbles: the horse is limited to going short distances slowly.

Locomotion is more than just the movement of the limbs. It involves interactions of the muscles of the limbs and trunk, movement of the neck and head, respiration, blood flow, neurologic coordination, strength of tendons and ligaments, and involvement of the skeleton, especially the joints, limb bones, and spine. With 60%–65% of the animal's weight carried on the forelimbs, the neck becomes an important counterbalance to the back half of the body. Breathing is synchronized with gait cycles in most animals, and in the horse, this is most evident at speeds faster than the walk. It occurs in a 1:1 ratio—one stride, one breath.[6,7]

It is assumed that because free-ranging horses spend a lot of time moving, that exercise is important. When a horse is first taken out of a stall, there is not a great increase in activity.[8] Observations confirm that horses tend to be lazy. While free-living horses will cover several miles per day, the domestic horse, even in a large pasture, does not. Preference tests can be used to study the degree of a horse's interest in various conditions. The horse is taught to push a gate open to get food, companionship, or exercise. At first, one push is required to open the gate, then two pushes, three pushes, and so on until the horse quits pushing. After 3 weeks of confinement in a stall, horses are willing to push the gate 191 times to get food, 56 times to get the companionship of another horse, but only 37 times to get access to a paddock.[9] Placing hay in various locations around a paddock can be useful in increasing and enriching activity for individual horses. It also reduces the risk of agonistic behaviors between herdmates.[10]

The walk, trot, canter/lope, and gallop are the four natural gaits common to all horses. It should be noted that some people consider the canter/lope to be a variation of the gallop, giving the horse three natural gaits instead. Because there are differences in how the feet touch the ground, the canter/lope can be considered as unique from the gallop. As speed increases from the walk through the other gaits, horses increase both stride frequency and length. However, at the gallop only the length of stride increases.[11] Horses are also known for having a number of special gaits. Not only do these gaits vary by breed, but each breed association tends to use terminology unique to them.

Gaits can be described in a variety of ways but one of the easiest to picture is with the use of footfall patterns (Figure 8-1). A solid line represents the body of

FIGURE 8-1 Footfall patterns are diagrammatic representations of which feet are on the ground at any one time. This figure shows the footfall pattern that corresponds to the picture above it. The horse in this picture is loping, and it is on the right lead both in front and behind.

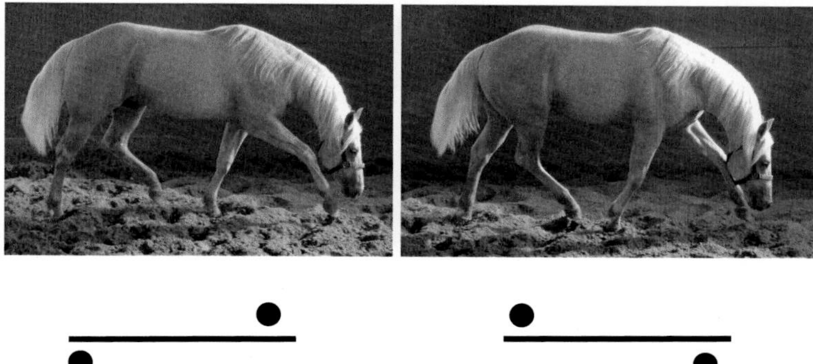

FIGURE 8-2 The trot is a symmetrical gait because the first and second halves of the stride are mirror images.

the horse, moving in a left to right direction. The small circles on either side of the line represent the foot that is on the ground, and the entire sequence represents one complete stride. The advantage of using footfall patterns is their simplicity in understanding the sequence of feet touching the ground and the type of pattern that each gait represents. The disadvantage is that the length of stride and timing between each foot touching are not described. There can be tremendous variation in these last three points based on size of the horse, speed it is going, and breed variations, so discussion of these points becomes complex.

Another way gaits are described is by the symmetry of the feet touching the ground. For a symmetrical gait, the second half of the gait sequence is the mirror image of the first half and the front and hind pairs of feet are evenly spaced in landing (Figure 8-2).[12–14] This mirror image can be *lateral* during which both legs on one side move before the corresponding legs on the other side (as in the pace), or *diagonal* where one leg on each side touches before the opposite legs do (as in the trot). Asymmetrical gaits have footfall patterns that are random in appearance, being repeated with the next complete stride and not within a stride. They are not evenly spaced in hoof landings either.[15]

The beat describes how many times a sound is produced during a stride, and this relates to how many feet touch the ground at the same time. A four-beat gait means a person can hear each foot land separately, while a two-beat gait means that 2 ft. touch the ground at the same time, followed by the other two touching at the same time.

Symmetrical Gaits

Walk

The walk (also called the *flat walk*) is a symmetrical, four-beat, lateral gait, with each foot hitting the ground separately. As diagrammed, it is symmetrical because the second four parts of the stride have the same foot patterns; four-beat because each foot touches the ground alone; and lateral because both feet on one side of the body touch before those on the other side. The order of foot placement on the ground is left hind, left fore, right hind, and right fore. The gait alternates between having 2 or 3 ft. on the ground at any one time.

The speed of the walk varies between 4 and 8 mph or 6.4 and 12.9 km/h.[3,16–18] Front and rear feet spend approximately equal amounts of time on the ground (about 950 ms), giving the walk a 1-2-3-4 cadence of equal spacing.[14,18] As should be expected because of weight distribution differences between the front and rear limbs, the peak vertical force with which the foot hits the ground is greater for the front feet. In front, there is approximately 700 lbs. of force, compared to 600 lbs. of force for rear feet.[18] Horses show head nodding as they walk. Functionally, this helps reduce energy expenditure by redistributing the weight of the head and neck during each stride.[19]

The distance traveled using a walk varies considerably. Free-roaming horses will walk about 11 mi (18 km) a day, particularly if grazing and water areas are far apart. A pastured horse travels about 4 mi (6.5 km) regardless of the season, and a stalled horse, even with time in a paddock, may travel less than 1 mi.[20,21]

The slow walk is a variation used when there is no specific travel goal. The difference from the regular walk is that there is a short period of time when all 4 ft. are on the ground, giving a pattern of 2, 3, 4, 3, 2, 3, 4, 3 ft. down pattern.[12]

Several other four-beat gaits are considered to be variations of the walk because they keep the same footfall patterns but exaggerate certain aspects of the style or rhythm.[14] Only careful observation shows the footfall characteristic of a walk. The Spanish walk is a highly animated, slow-motion walk where the elevated front leg is significantly higher than normal. At the opposite end of the speed spectrum, the running walk is an extremely fast variation that almost looks like the horse is walking in back and trotting in front. Some of the other four-beat gaits will be discussed.

Trot

The trot is a symmetrical, two-beat, diagonal gait where the legs move in unison as a diagonal pair. The trot typically has a short time when no feet are actually on the ground. The pattern sequence would be left rear and right fore, suspension, right rear and left fore, and suspension (also see Figure 8-2).

The speed of a brisk trot is 8–12 mph (12–19 km/h), but racing trotters reach speeds near 30 mph (48 km/h).[3,17,18] The faster speeds are also related to a greater amount of vertical force when that foot hits the ground and a shorter amount of time each hoof is on the ground. At the trot, the front feet exert 1069 lbs. (485 kg) of force and the rear feet exert 1057 lbs. (268 kg) of force. The amount of time each foot is on the ground is slightly over 590 ms.[18]

A variation of the trot commonly seen in the western pleasure classes at Quarter Horse shows is a jog or very slow trot. There is a brief period where all 4 ft. are on the ground before one set of diagonal limbs is lifted.[12]

Named variations of the regular trot include the *passage*, which is a highly animated, exaggerated slow motion trot.[3] The *piaffe* has the trot cadence with little to no forward movement, and the *trote* of Peruvian Paso horses is noted for the "tas tas" sound.

Asymmetry of movement of the hind legs is associated with a rider posting at the trot. There is a greater downward movement of the horse's pelvis when the rider is seated and less upward pelvic movement as the rider rises while posting.[22]

Asymmetrical Gaits

When a pattern seen in the first half of a stride is not repeated as a mirror image during the second half, the gait is said to be asymmetrical. This is most commonly associated with the fastest natural gaits.

Canter/Lope

The canter is a gait that has three beats and is asymmetrical. Because there is no suspension phase, the speed is somewhere between 10 and 15 mph (16–24 km/h).[3,17] The sequence of feet touching the ground as pictured is left rear, left fore and right rear, and right fore. The lope is a slower version of the canter used by western horses. Its maximum speed is 12 mph (19 km/h).[12]

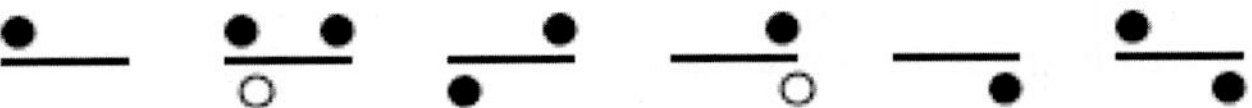

The canter and lope are sometimes considered to be variations of the gallop, giving horses three normal gaits instead of four. Because the footfall patterns differ from the gallop, the canter can be considered a separate gait. Faster variations include the extended canter and hand gallop. Riders have to take care in both cases to maintain the three-beat cadence and not allow the horse to slip into a four-beat rhythm.

Another feature of the canter and its variations is the *lead*. The term refers to which foot of the front and rear pair touches the ground last (and thus farther forward) during a stride.[15] In the preceding diagram, the horse would be on the right lead both in front and behind as indicated by the hollow dots (also see Figure 8-1). Having the same lead foot in front and back gives a much smoother ride and is the safest way for the animal to turn around a curve.

A horse is *cross cantering* or *cross firing* if the front and back have opposite leads, as shown in the following diagram, where there is a left lead in the rear and a right lead in front (also see Figure 8-3). The cross canter is difficult to ride and puts the horse out of balance if turning.

Gallop

The fastest gait a horse can use is the gallop. This is the gait used by racing Thoroughbreds or horses running in pasture. The gallop is a four-beat, asymmetrical gait. Horses typically move into this gait from a trot or canter. The transition from the trot to the gallop occurs at a slower trotting speed when the animal is carrying weigh than when it is running free. It is likely this is a protective mechanism to prevent musculoskeletal self-injury.[23] From a canter, the increase in speed separates the synchronized diagonal landing of a rear and front foot into two separate occurrences, as in the following diagram. In addition, there is a suspension phase during which all feet are off the ground and flexed under the horse's body (also called *gathered suspension*).[15] The sequence of feet touching the ground (in this diagram) is left rear, right rear, left front, right front,

FIGURE 8-3 This horse is crossfiring while galloping. He is on the left lead in front and the right lead behind.

and then a suspended phase after the right front leaves the ground but before the left rear lands again.

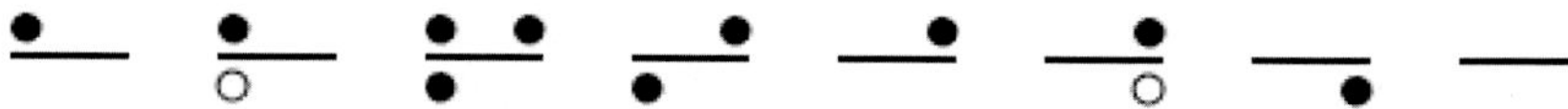

Leads also apply, and the pictured footfall pattern is a right lead both fore and rear as indicated by the hollow circles. During a Thoroughbred race in which horses circle the track in a counterclockwise direction, horses often use the right leads on the straight portions of the track and switch to the inside, left lead as they turn the corners. Doing so provides better balance on the turn. Because the leading foreleg is on the ground up to 1.25 times longer than the hind foot and up to 1.5 times longer in racing horses, changing leads reduces the fatigue on the forelimbs over long distances.[3,15] The feet of a galloping horse land with approximately one-third more force than in a trot, and the lead foot lands with greater force than does its pair.[18]

Speed came as the ancestral horses grew longer legs and single toes while living on dryer, flat land. It originally functioned to escape predators. At a gallop, a horse will travel between 25 and 32 mph (40–51 km/h), about twice as fast as a human can run.[3,17,18,24] Each foot is on the ground for 114–133 ms, and a complete stride takes approximately 450 ms.[18] Running up an incline results in smaller peak vertical forces in the forelimbs and increased ones in the rear. It also takes less time for each stride.[25]

Galloping requires a lot of energy to maintain, so it is not efficient for covering long distances—only a mile or two. Horses can run farther, as Pony Express horses did, but the greater the distance, the slower the speed. Past 20 mi (32 km), the speed approaches 10–11 mph (16–18 km/h).[24]

Racehorses take the gallop to its extreme. The Guinness World Record for speed in a racing Thoroughbred for 0.25 mi (402 m) is 43.97 mph (70.76 km/h) and for 1.5 mi (2414 m) is 37.82 mph (60.86 km/h).[26] Some Quarter Horse sprinters have been clocked at speeds up to 55 mph (88.5 km/h).[17,27] Horses achieve their fastest speeds in the one-quarter to one-third mile (400–600 m) section of the race.[28] Conditioning high-level athletes requires application of proven scientific techniques and sophisticated veterinary monitoring and care. In the month prior to a race, horses that frequently train at high speed or actually race before their main race are more likely to win or place in the race compared to horses that are conditioned with slower cantering.[29]

Many events involve both straight-line and curved pattern running. Thoroughbred races and reining circles have turns with relatively large radiuses, but events like pole-bending, barrel racing, and polo require much tighter curves. Cornering reduces speed, even when the curve is gradual, and the extent of the reduction depends on the coefficient of friction of the hoof-surface interface.[30] When speed determines winners, riders often encourage the horse to go faster by whipping or kicking. Neither increase speeds.[28,31] In barrel racing, whipping and kicking are also correlated with increased tail lashing, kicking out, and difficulty getting the horse to enter the arena.[31]

Heritability of speed is of interest to racing fans and has been highly studied. Figures given for the heritability of any given trait range from 0.0, meaning there is no genetic contribution to the trait, up to 1.0, where inheritance is the only reason for the trait. In general, racing speeds are not very heritable. As an example, racing Standardbreds have a heritability score of 0.29.[32] This means that approximately 29% of the speed may be attributable in some way to the genetics of the individual. It does not mean that 29% of a racing Standardbred's speed is due to his genes and 71% due to the environment.[33] The score is even worse for racing Thoroughbreds. The value attributed to inheritance is less than 0.20, with the number being somewhat higher for short races and lower as the distance increases.[33,34]

Round Pen Training

When training begins, it is common for horses to be worked in round pens and encouraged to gallop around the edge until they gradually slow to a stop. There are many reasons given as to why this type of training is "natural," and even desirable. One commonly cited theory suggests that the person is chasing the horse as would happen between rival stallions or members of the herd. Another is that juvenile play is simulated. The fallacy of either theory is that naturally occurring chases are short in duration, unlike when horses are being worked in a round pen.[35]

More likely, round pen work with naïve horses triggers the horse's predatory survival mode, with the human triggering the flee response just as a predator would. Because of the circular nature of its environment, the horse can only run in a circle and is not able to distance itself from the person. It is not the posture of the approaching human that affects speed of the running horse; rather, it is the speed at which the person moves toward the horse.[35,36] The stated goal for trainers who use round pen training is to accustom the animal to a human and get the horse to cooperate, follow, and recognize the human as the dominant figure. If the horse does not cooperate with the trainer, it is worked until it begins to respond to specific signals.[37] In reality, the horse will gradually become tired and/or learn that it cannot escape the closeness of the human. It may follow the human in the pen, but the lesson is location specific, and the role of dominance is unproven.[37] Naïve horses put into a round pen will show defecation, vocalization, escape attempts, teeth grinding, and a high head position—all being stress behaviors.[38,39]

Special Gaits

Several breeds of horses, and occasionally individuals not purposefully bred, have gaits that differ from the major ones. Special gaits usually replace the trot and are sought after because of their smooth, flowing movement. They were popular in the Middle Ages because they could carry their riders smoothly and rapidly over poor roads.[40] The spread of horses having special gaits corresponded to the spread of the *DMRT3* gene. As the amount of animation increased, anatomical differences in gaited horses also happened, including smaller eyes and jaw widths, proportionally longer front limbs, and thinner lower limb circumferences.[41]

Pace

The pace is a lateral, symmetrical, two-beat gait. It is similar to the trot except that the limbs on the same side move forward at the same time. The sequence would be left front and rear, suspension, right front and rear, suspension. While pacing, the horse has a left-to-right rocking motion as the center of gravity balances against the weigh shifting from side-to-side, somewhat similar to the duck's waddle.

The popularity of harness racing in the United States has resulted in the Standardbred being the fourth most commonly registered horse, according to the American Horse Council. At racing speeds, pacers can move faster than trotters. The record for pacing a mile, set in October 2016, was 1.46 min (33.96 mph, 54.66 km/h).[42] The presence of the *DMRT3* recessive mutation gene is strongly connected to the most successful harness racers in both Standardbred pacers and trotters. Because of this, the gene is found in essentially all the Standardbred racers in the United States. It is speculated that the gene causes an increase in speed due to a longer stride.[43]

At one time, the pace and rack (an ambling gait discussed in more detail shortly) were considered to be the same gait. Both went by the term "rack" even though the footfall patterns had previously been shown to be different.[12] Using a single term probably happened because of the visual illusion of similar foot landings.

Amble

Collectively, all the other special gaited horses are said to *amble* because of their smooth four-beat gaits. In the United States, the gaited breeds are frequently lumped under the heading of plantation horses, which goes back to their favored use in the pre-Civil War South. Recent research has linked the ability to amble to the *DMRT3* gene, which is common in several breeds including the Icelandic Horse, Missouri Fox Trotter, Paso Fino, Peruvian Paso Horse, Rocky Mountain Horse, Kentucky Mountain Saddle Horse, and Tennessee Walker.[43] There remain several gaited breeds that have not been tested for the gene but likely have it also. *DMRT3* does not occur in nonambling horses other than Standardbreds.

The ambling gaits have four beats, and it is the timing of the foot placement that makes each gait so smooth. The leg movement can be diagonal with opposite front and rear legs moving somewhat in unison, or lateral, with legs on the same side most closely related in movement. Different breeds have unique names for their special gaits and it is not the intent to include them all, just some of the more common variations.

Fox Trot

The fox trot is the only diagonal ambling gait. It is similar to the trot except that the front foot lands slightly before the hind foot, eliminating the suspension phase. This changes the two beats of the regular trot to a four-beat gait. The rhythm has a 1–2, 3–4 cadence. One front foot is on the ground at all times, giving the horse a gliding appearance and the illusion that it is walking in front and trotting behind.[16] The fox trot can also be slowed to what is called a fox walk.

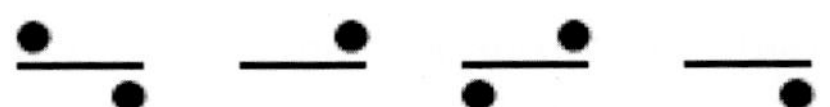

Running Walk

The running walk is typically associated with the Tennessee Walking Horse. Although the footfall pattern of the running walk is the same as for the regular walk, the speed of the gait is much faster. These horses can travel at 10–20 mph (16–32 km/h).[16] At speed, the rear foot reaches 6–18 in (15–46 cm) farther forward than where the front foot on the same side had previously landed. The animation of the front feet results in their spending 10%–20% less time on the ground than do the back feet.[13]

Rack

The rack is a lateral, four-beat gait typically associated with the five-gaited American Saddlebred and the Racking Horse breeds. Each foot hits the ground separately in an equally separated 1-2-3-4 rhythm. The speed is faster than the running walk, very animated, and includes two suspension phases. In this gait the animation reduces the amount of time the front feet are on the ground, compared to the hind feet, by 20%–40%.[13] Visually, the rack looks similar to the pacing gait.

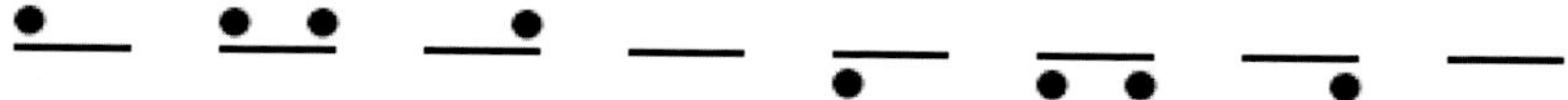

Slow Gait

The slow gait is an animated, four-beat gait that is derived from the pace. In this case, the hind foot hits the ground shortly before the forefoot on the same side.[3] The feet would land as a left rear, left fore, right rear, and then right fore. For the American Saddlebred, emphasis is placed on the animation and not speed. Although similar in appearance, but not speed, the footfall patterns show the slow gait is not the same as a slow rack or pace.[16]

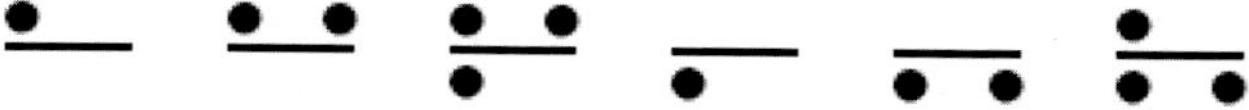

Tölt

Icelandic horses have a very fast, four-beat, lateral, ambling gait named the tölt. The footfall pattern is that of the walk, but horses can obtain speeds as fast as a canter.[16] The prevalence of the *DMRT3* gene in the Icelandic horse population is high and heterozygous horses have the strongest natural ability to perform the gait. In tracing the gene's appearance in Europe, it is now believed the Icelandic horse is the gene's originator.[40]

Other Ambling Gaits

There are a number of ambling gaits that are breed specific. In some cases, they are similar to those already mentioned. Others are unique. The Paso Finos have three four-beat gates that have a 1-2-3-4 steady rhythm and differ only in speed: *paso fino*, *paso corto*, and *paso largo*. They also have a diagonal *trocha* gait. Peruvian Pasos have the *paso llano*, which is similar to a running walk, and the *pasitrote*, a diagonal gait.[16] *Single foot*, a fast rack-like gait with only one foot on the ground at a time, is the unique gait of the Single-Footing Horse. The Florida Cracker Horse has a *coon rack* gait.

Swimming

While swimming is commonly used in the rehabilitation of horses with leg injuries, many horses like to swim on their own if given the opportunity. The movement of the legs is in the trot sequence—diagonal front and rear limbs move forward at the same time.

Jumping

Jumping over obstacles is a sport that certain horses excel at. Serious show jumping classes test the limits of the individual horse's ability. The record height jumped by a horse is 8 ft. 1.25 in (247.02 cm) by Huaso ex-Faithful in 1949.[44]

To make a jump, the horse coordinates much of its musculoskeletal system through a series of phases. The jump begins in the phase of the gallop when only rear limbs are on the ground. The horse uses the back muscles to raise the front higher, while the rear limbs push off with full extension. The forelimbs are flexed under the chest until the jump is cleared and then extended in preparation for landing. The landing occurs on one forelimb, followed by the other.

Jumps can be made over vertical obstacles, as over a wall, or over horizontal ones, such as a ditch. Vertical obstacles are generally easier for the horse because it can gauge the height of the upcoming object. Horses are more dependent on rider cues for horizontal jumps because the horse's binocular vision is not good. Eyesight is an important consideration in another way as well. Dichromatic color vision explains why blue jumps are more likely to be touched than green ones, at least in indoor arenas.[45]

The ability to jump has a relatively low heritability, regardless of whether the horse competes at the highest levels or at an amateur level.[46] A study of the genome of jumpers has detected a quantitative trait locus on chromosome 1, close to a gene encoding a major calcium channel in cardiac muscle, suggesting a possible link.[47] Until more is known, however, long-term success as a jumper is still best predicted by the horse's performance as a young animal.

Backing

Backing more than a few steps is not a natural movement for a horse because it was not needed in the evolutionary environment. Horses can learn the behavior, and when comfortable backing up, they usually use the slow walking or trotting footfall pattern.

Stopping

Horses transition down from various gaits by slowing from faster gaits to slower ones and eventually coming to a stop. When running in a pasture, most can abruptly stop too. With training, a horse is capable of perfecting the abrupt stop from a trot or canter. Highly specialized horses, such as those used in cutting classes, can plant their hind feet deep into the ground not only to stop fast but also to be ready to rapidly turn in the direction the cow or steer is now going. Reining horses also shift their weight to the rear legs when stopping so they can slide long distances (Figure 8-4). The front legs appear to "walk" during the slide.

Circling/Spinning

Turning in very small circles requires one limb to cross over the other. Either the front legs or hind legs can be the center of the turn, with the opposite pair making an outer circle. Horses can also circle with the pivot point being their midsection. As a result, the fore and hind legs make an outer circle about half the distance from the pivot point. Spinning is when the horse pivots at speed, as shown by reining horses.

FIGURE 8-4 While most horses stop by slowing down gradually from faster gaits, cutting horses will rapidly stop so they can follow abrupt turns of cattle. Reiners, like this one, perfect the sliding stop, during which they slide long distances on their hind feet and appear to walk with the front legs.

Pawing

When pawing, a horse will raise a front leg and then drag the toe in a forward-to-backward motion (Figure 8-5). It is a normal behavior used in pastures and on the range to uncover grass after a snowfall. Pawing is common in situations when the animal appears impatient, as when an owner is slow at delivering food to the stall, and in stressful situations, such as social separation. Postworkout pawing is suggestive of pain or an attempt to level the surface to make standing easier.[48] In some horses, the timing and frequency of pawing are indicative of a stereotypy. Excessive pawing for any reason is undesirable because of wear on the feet, shoes, or physical location.

Rearing

Rearing horses stand on their rear feet with the forelimbs elevated. The amount of elevation can be such that the front feet are barely off the ground, or so great that the horse is almost vertical. The behavior evolved for fighting, both in self-defense against predators and against rival stallions. Young horses, particularly stallions, will show rearing as a play behavior used to perfect skills needed as an adult.

In managed horses, rearing occurs primarily in conflict situations. The rider is kicking the horse to go forward while holding the reins tight. The "go but no"

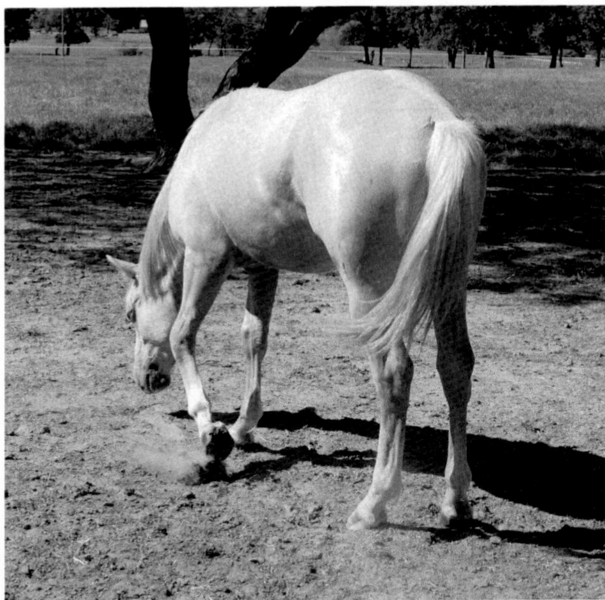

FIGURE 8-5 Pawing is a motion in which the horse repeatedly moves the toe of one front foot backward.

message is frightening, but because the horse cannot move away, it rears. The rider wants to take the horse trail riding alone, but the horse is strongly bonded to a barn buddy and tries to remain close to the other horse. Here the rider wants "go" and the horse wants "no." An overly assertive stallion handler may be quick to jerk on the head restraint as the stallion goes to the breeding shed. The horse responds by rearing, and the handler backs off. Over time, a horse can learn that it will get its way by rearing because each time the frightened handler quits asking. Rearing and even flipping over backward are common in certain bloodlines, suggesting there may be a genetic component.[49] Avoiding situations that trigger the rearing response, particularly with young horses, eliminates the internal self-reward associated with the person backing away. For foals, rather than punishing a fearful response, stroking around the withers can be calming.[50]

There are several variations of rearing used by high-schooled horses, like performing Lipizzaner stallions of the Spanish Riding School. Some variations include deeply flexed hind limbs, some have the forelegs tightly flexed, and still others are associated with a hopping or leaping motion.[3]

Bucking

Bucking behaviors originated when ancient horses tried to get predators off their back. For modern horses, youngsters often show a play version of bucking as they develop motor skills needed in adulthood. The behavior involves an arching of the back and a rapid lowering of the head and neck. The horse then rears, kicks back, or jumps up with all four limbs off the ground.

Laterality (Foot Dominance)

Humans have side preferences typically called *handedness*. We write with our right or left hands. For horses, laterality is expressed as foot preferences. There is much variation in preferences by breed, sex, type of situation, and even amount of training.

Breed variations include Thoroughbreds having a left side preference, while Quarter Horses show little preference.[51–54] However, Quarter Horses do show a left forelimb bias when first stepping on or off an elevated structure and when loading or unloading from a trailer.[53] Between 34.0% and 40.6% of horses show a left leg forward bias when grazing, 9.2%–9.4% have a strong right foreleg bias, and 50.0%–56.8% are ambidextrous. Leg preference increases as horses get older.[52,55,56] Resting horses usually flex one hind limb—12% flex the left limb, 8% have a right side bias, and the remaining 80% show no preference (see Figure 4.3 in Chapter 4).[56] Laterality has not been found for pawing.[56] Horses typically begin a gallop by moving the left leg first, so they are on the right lead. This is strongly expressed in 90% of racehorses, primarily stallions.[57,58] When walking or trotting, 52.5% start movement with the right foreleg and 40% begin with the left. The remaining 7.5% show no consistency.[57] Males are significantly more likely to prefer the left side and females the right side.

A connection between the facial hair whorl and laterality has been described. A horse with a clockwise whorl is more likely to show a right bias than is a horse with counterclockwise whorls (see Figure 2.9 in Chapter 2).[59,60] (Hair whorl direction or position is not correlated with racing performance or show-jumping performance.)[51]

RESTING AND RISING PATTERNS

Resting is an important part of animal survival, yet in prey species letting down one's guard against predators can have deadly consequences. The advantage of group living is that members can alternate guard duty. Some remain alert enough to detect approaching predators, allowing others to relax and even sleep. Even after thousands of years of domestication, it is rare to see over 50% of the horses in a group lying at the same time.[61]

Stretching

Different forms of stretching are common after a horse has been resting, perhaps to prepare the various muscle groups for movement. The front end can move forward to extend the lumbar back and rear limb muscles. In the reverse, the weight can be shifted caudally over the rear legs and the front end lowered as the forelimbs extend forward. Individual legs can be extended, a behavior more common with rear limbs than forelimbs (Figure 8-6). The back can be arched or lowered as well.

Sitting

Horses do not lower their hindquarters into a sitting position as dogs do. Instead, they can only assume a sitting posture after they have been lying down. The horse extends its front legs forward, pushing the chest off the ground. Horses go through the sitting posture briefly while getting up. A few hold the sitting position for a minute or so.

Lying Down and Rising

The process of lying down from a standing positon begins with collecting all four feet under the horse's body. The horse then kneels by flexing the front legs, as the head and neck extend outward (Figure 8-7). As the knees are about to touch the ground, the rear legs flex, allowing the entire underline to touch the ground at the same time.[3]

While lying the horse can assume different positions (Figure 8-8). Sternal recumbency, in the classic definition, would have the horse's ventral midline

FIGURE 8-6 Horses often stretch after they get up and more often stretch rear legs than front ones.

FIGURE 8-7 When lying down, a horse gathers its feet under it and lowers the front half of its body to its knees. Then the rear legs flex to lower the back half of the body as the front half continues to be lowered to the ground.

in contact with the ground. Front limbs would extend forward and rear legs would be parallel to the body and either directed forward or backward. The head and neck would be elevated. Because horses are unable to assume true sternal recumbency, they use a modified version for resting. The front half of the body is in sternal recumbency with the forelimbs flexed slightly to the side, and the back half is more laterally positioned, with the legs flexed close to the body on one side. In lateral recumbency, the horse is lying with one side having complete ground contact, including the head and neck, and all four legs extend

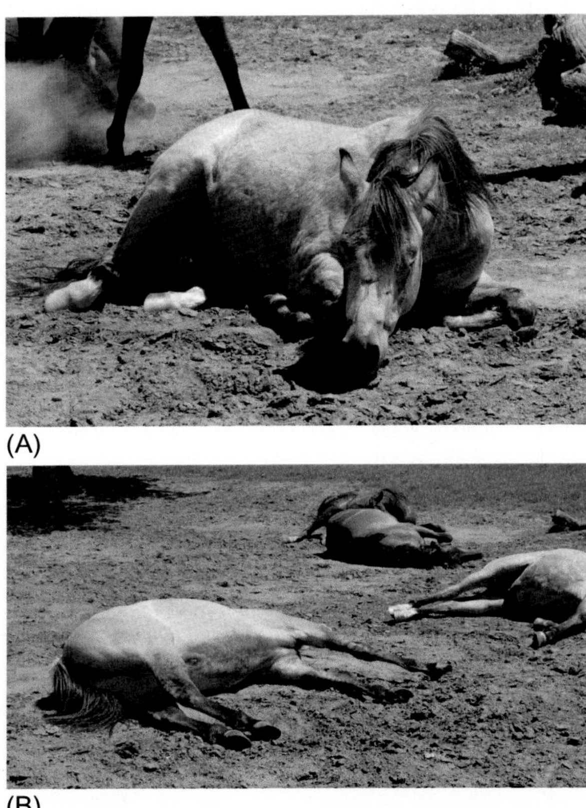

FIGURE 8-8 There are two lying postures used by horses. (A) The horse uses a modified version of sternal recumbency, allowing the ventral midline to have the most ground contact. The legs are folded near the body. (B) Lateral recumbency is characterized by the horse lying on its side.

to the same side of the body. Lateral recumbency is the posture of deep sleep. The shape of the shoulders and withers makes it impossible for horses to lie on their back. They might pause briefly in the position while rolling, but they soon slip to one side or the other.

To rise, a horse shifts its weight caudally from a position of sternal recumbency. One forelimb is extended and then the other (Figure 8-9).[3] As pressure raises the front half of the body, the horse briefly assumes a sitting position. Some horses remain sitting for a while, but most continue to rise. The head and neck move downward to transfer weight forward as the rear half of the body shifts so that the hind legs can be positioned underneath. Extension of the hind legs results in the body rising to a standing position, as the forelimbs balance the animal.

FIGURE 8-9 The process of getting up begins as the horse extends and puts pressure on the front legs. Once the hind legs can be moved underneath the body in a posture that approximates sitting, weight is shifted upwards and the hind legs extended.

Sleeping

Wakefulness to deep sleep is a continuum rather than a stair-step process. For most animals, drowsiness is the first phase moving toward sleep, and it is characterized by inattention that the horse can quickly revert to full attention with an external stimulus. Horses have an *intermediate phase* that occurs even before drowsiness. The two look similar, but arousal from the intermediate phase is even easier.[62] True sleep is unconsciousness, and it can be differentiated into two unique parts. Slow wave sleep (SWS) is the sleep state closest to consciousness. With enough stimuli, the horse will wake up. SWS is also the first and last phase of sleep an animal passes through between wakefulness and deep sleep. The name, SWS, comes from associated electroencephalographic features. Rapid eye movement (REM or paradoxical) sleep is deep sleep from which it is very difficult to awaken an animal. As an example, an electrical stimulus given to a pony will change drowsiness to full wakefulness 100% of the time, SWS to fully awake 80% of the time, but REM to awake just over 60% of the time.[62] REM gets its name because of episodic eye movements that occur in this stage. In addition to the eye movements, a horse may show leg movement, twitching, and blinking.[63] In humans, dreaming occurs during REM sleep. There is no way to prove horses dream, but muscle movements suggest they do. REM is preferentially made up when sleep-deprived people or horses finally get sleep.[63,64]

Electroencephalographic (EEG) recordings are necessary to confirm the actual state of consciousness and differentiate between wakefulness, drowsiness, SWS, and REM sleep (Figure 8-10). Beta waves are present when the animal is awake, and they are characterized by low amplitude peaks around 10–30 μV having a frequency greater than 12 cycles per second (cps) and as high as 40 cps.[62,65] As the horse enters the “intermediate phase” and drowsiness, the EEG wave patterns start to include alpha waves. Alpha waves have a

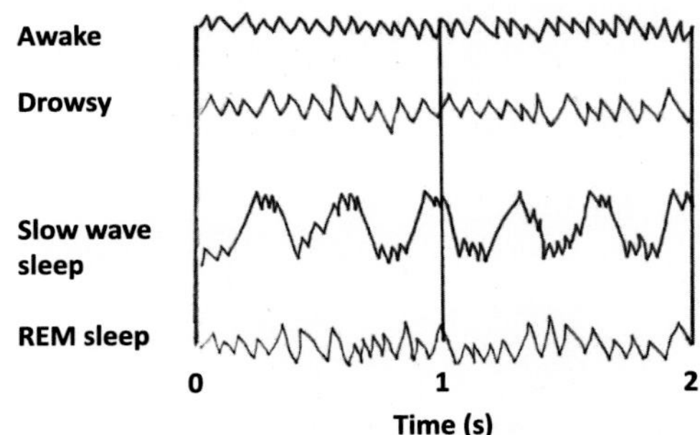

FIGURE 8-10 Diagrammatic representations of electroencephalographic tracings during various stages on consciousness show wake patterns are characterized by beta waves, drowsiness by predominately alpha waves, slow wave sleep by large delta waves with interspersed beta waves, and rapid eye movement (paradoxical) sleep by a combination of beta and theta waves.[62, 65, 66]

medium amplitude and a frequency of 8–12 cps. SWS patterns show a dramatic change, with tall delta waves of 40–90 μV occurring at 1–4 cps.[62,66] Finally, the REM sleep pattern is similar to a wake one, characterized by low voltage with mixed frequencies. It has both beta and theta waves, with the latter being of medium amplitude and slightly slower than alpha waves at 4–8 cps.[62]

Horses evolved to spend 18–19 h a day in full awareness to avoid predators. Drowsiness occurs for another 2 h, and sleep for 3 or 4 h. Of time spent in sleep, 2 to 3 h is for SWS and no more than 1 h for REM sleep.[62,66] Even during that hour, the cycles alternate between SWS and REM sleep.

Sleep sessions are short and scattered throughout the 24-h day. A SWS/REM cycle lasts 10–15 min at a time, with SWS averaging 6.4 min and REM sleep being 4.2 min.[62,67] At night, horses sleep in five to seven bouts, each lasting between 30 and 40 min. Awake segments of at least 45 min are interspersed between each sleep bout.[65,68] Stalled horses sleep mainly at night, although a 2-h nap at noon is common. During this daytime nap, however, REM sleep is rare.[62]

Thanks to the horse's stay apparatus, drowsiness and SWS usually occur while the horse is standing. One hind leg is flexed, the head and neck are lowered, the eyes are partly closed, and the lower lip sags (see Figure 4.3 in Chapter 4).[63,64,67] Drowsiness and SWS can also occur while the horse is lying down.

The normal progression for longer sleep periods begins as drowsiness and progresses to SWS. The horse will then lie down in sternal recumbency and reenter drowsiness and SWS. As sleep progresses into REM, the horse repositions itself into lateral recumbency. Occasionally, it might rest the head and body against a solid object if lateral recumbency is not possible.[62] About 15% of a 24-h period is spent in lying down, with a significantly greater percent of this time occurring between midnight and 4:00 a.m. Lateral recumbency takes up as much as 15% of time for newborn foals and then gradually decreases to about 3% typical for free-ranging adults.[62,69–71] Stalled horses spend considerably more time in both sternal and lateral recumbency than do pastured ones, although the total time spent resting is approximately the same between the two.[62,72] In a stall, a horse is recumbent for a total of approximately 5 h of which about 1 h is in lateral recumbency. In contrast, pastured horses lie down for 3 h, with only a half hour spent in lateral recumbency.[62] This 3-h time is divided into brief cycles of 2–15 min.[72] Up to one-third of pastured horses might not lie down every day.[61]

A horse needs a minimum of 30 min of REM sleep within a 24-h period.[73] If it is deprived of sleep for long periods because it cannot or will not lie down, episodes of partial collapse can occur when muscle tone is lost in REM, causing the standing horse to collapse.[63,66] The falling movement startles the horse awake again.

Environmental factors play a role in the quantity and quality of sleep a horse gets. Most horses will only lie down in familiar surroundings, and if moved to a new location, it usually takes a few days to adjust.[62,65,74] This can be problematic for horses on the show circuit because they lack the rest needed for peak performance. Going to the show facility a few days before competition begins can be useful. A second environmental factor is stall size.[61,75] Horses in large stalls spend more time lying down, especially in sternal recumbency. Low-ranking horses kept in larger paddocks and pastures have longer durations of recumbency and less sleep disturbance.[73] Even then, these horses have different resting patterns than do high-ranking horses. They lie down for shorter periods of time (69 min vs. 91 min for high-ranking horses and 93 min for middle-ranking ones) and have fewer lying bouts (2.5/day vs. 3.2/day).[61] The type of bedding is a third environmental factor that impacts resting. Horses are less apt to lie down on plain ground or bare rubber mats. Significantly more time is spent lying if the area is bedded with either straw, wood shavings, or sand.[61,73]

DAILY ACTIVITY PATTERNS

Graphic versions of the time budget show the major activities of a 24-h day (Figure 8-11). Eating constitutes the major daily activity. When grass or hay is available ad libitum, between 45% and 70% of the day is spent grazing.[62,70,76,78–85] The grazing behavior begins as soon as the first day after birth, but only constitutes a total time of 3.5 min (less than 1.0 min/h) a day nibbling. Time spent grazing increases significantly to 16.3 min/h by the fourth month and 44.4 min/h by a year of age. Weanlings graze significantly more in the late afternoon and early evening.[86] Environmental factors such as weather, insects, availability of food, and ease of obtaining food impact the amount of time spent eating and when it occurs. Even accessibility of hay in hay nets can change the time budget. Small holes that make it harder to get hay prolong time spent eating so that it more closely approximates natural grazing.[87]

When hay or grass is limited, as for stalled horses, eating might take up 15% of the day, and the time spent standing increases to 55% or 60%.[82] Heavily pregnant mares spend more time standing and lying and less time grazing than do postpartum ones.[88]

Locomotion, particularly walking, constitutes 5%–8% of the daily activity.[69,79,84] Horses will occasionally trot and gallop too, but for short periods. Active movement occurs mainly during daylight hours, while nighttime activity is briefer and low key.[89] If grass is readily available, movement is usually just a few steps, a bite of grass, and another few steps. Longer treks occur if water is some distance from the grazing areas. Stalling reduces the amount of active movement for a horse. Foals weaned in a paddock with others of the same age spend more time moving and engaging in other behaviors. Their time budget is similar to that of a free-ranging horse. Stalled weanlings move less and are more likely to show abnormal behaviors like licking, chewing, kicking, and rearing.[90] Adult horses confined exclusively to stalls are much more active than stalled horses also having access to a paddock.[89] Pregnancy also affects

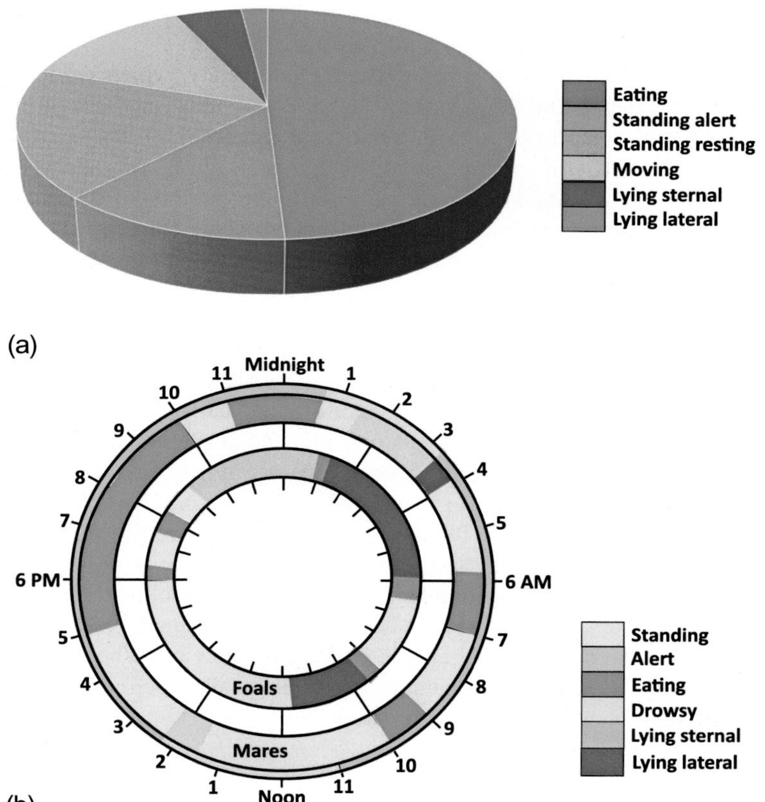

FIGURE 8-11 Representations of different types of time budgets for horses: a) represents the distribution of activities performed during a typical 24-h day; and b) matches the major activities with the time of day each occurs for a mare and for foals up to 60 days. In this case, nursing by foals made up about 20% of the day, but the bouts were too brief to be included as eating behavior, so they are included in with standing.[69, 74, 76–81]

locomotor activity in that late prepartum mares become significantly more active after giving birth.[91]

Standing takes up approximately 25%–36% of a horse's day, of which 60% is done while the horse is alert and 40% while the horse is resting. Foals and yearlings spend more time resting and standing alert than is typical of adult horses. Resting behavior peaks in foals within a few weeks of birth at 31% of the day. It then decreases gradually to 5% at the time of weaning.[62,69,70] In free-ranging horses, standing alert takes up less time than for domestic horses because of their need to travel between water and grazing areas.[70,77] An exception to this can happen in summer months when free-ranging horses stand in shade during the heat of the day.[86] Stalled horses spend more time in nonalert standing if no other horse is visible or if the stall is bedded with shavings rather than straw.[92] It will also spend more time eating hay.[85]

A foal spends 15% of its day in lateral recumbency, which gradually is reduced to 2% by the time of weaning.[69] For adult horses, lying behavior takes up 5%–12% of a 24-h period with most bouts occurring at night. Lying is divided into 80% in sternal recumbency and 20% in lateral posture.[62,70,76,78,80–83,93] Stallions lie laterally for longer periods of time than do mares.[70] Individual preferences in bedding may occur, but most horses do not have a preference.[93] While bedding type does not significantly affect the amount of time a horse spends in sternal recumbency, it may affect the duration in lateral recumbency. Horses bedded with straw will spend three times longer in lateral recumbency than those bedded with shavings.[92,94]

Many things influence how the remaining 10% of time is spent, such as the horse's sex, state of pregnancy, age, presence of vices, daily work schedule, season, weather, location, access to a paddock, diet and availability of food, bothersome insects, and degree of confinement. Free-roaming horses are flexible in their activities and alter them to meet environmental conditions.[95] Working horses, even if worked only a few hours a day, will reduce resting behavior, change their favored position, and show more standing rest at night. If given several days off, nighttime resting shifts toward an increase in sternal recumbency. The amount of lateral recumbency does not change.[96] Crowding is another factor that can have a dramatic impact on behavior. When horses are crowded they will eliminate behaviors like rolling, allogrooming, and lying down, but significantly increase locomotion.[97]

TRANSPORT OF HORSES

Moving horses from one location to another is accomplished with trailers or specialized vans, although flying is common for high-level sport horses and breeding stallions used in both the northern and southern hemispheres. Any type of transport requires that several behavioral considerations be made beforehand so travel is the least stressful and problems prevented. There are many techniques to teach a horse to load into a trailer, just as there are a number of different types of trailers—some have ramps, some are step-ups, some have partitioned stalls, others have no internal partitions. Regardless of the type of trailer or van, the key to successful loading is to teach it early and gradually using positive reinforcement.

Loading

The biggest mistakes people make in loading a horse into a trailer are waiting until they need to take the horse somewhere and being in a hurry to do so. This classic situation leads to negative experiences for the horse, resulting in an association of the trailer with pain and panic. Positive reinforcement, especially with food and social contacts, and lessons taught long before travel needs to be done are critical to establishing a positive experience for the animal.

The first consideration for loading either an experienced horse or neophyte is safety for both horse and handler. Trailer/van design must be considered. The interior space must be wide enough to safely allow both the handler and horse to enter at the same time. Livestock trailers work well, as do side-by-side trailers with center partitions that can be removed or ones where the person can easily enter an adjoining stall. At the very least, there needs to be a safe way for the handler to exit from the front, but many horses still refuse this option, perhaps perceiving it as too space restrictive. An alternative is to use an extra-long line going from the horse, through the trailer stall, out the escape door, and back to the handler. In all cases, the lead rope is intended to line the horse up with the entrance and keep it facing the trailer, not to pull the animal up into the trailer. It is also used to prevent the horse from arcing around the back and sides of the trailer. That becomes counterproductive, so slow, carefully executed lessons are important.

Positive Reinforcement for Loading

Teaching a horse to load can be done with either positive or negative reinforcement. Positive reinforcement produces the most reliable and long-term result. It begins with making the horse comfortable outside the trailer. It is tempting to force the horse to get close, but that urge can backfire.[98] Before starting, the horse should lead readily and be willing to eat treats, grain, or hay. The loading lesson begins by leading the horse toward the trailer entrance and stopping near it for the reward. If there is hesitation during the approach, note the spot of the hesitation. Then put the reinforcer a few feet farther away. As an example, the first approach may go to a distance 20 ft. from the trailer, at which point the horse is to stand calmly and is reinforced for doing so with a treat or small amount of feed in a feed tub. After eating, the animal is led away. The approach to the trailer is repeated, getting gradually closer each time. Eventually the feed tub is put on the ramp or edge of the trailer. The next lessons are rewarded from the tub, which has been advanced a little farther into the trailer each time, eventually onto the manger or makeshift manger. If the horse steps up onto the ramp or trailer and then backs off, do not reward the behavior with the food, but do not punish the horse or try to force it forward. The noise and motion are significantly different from those which occurred with previous steps. Give the horse time to rethink the situation, and it will resume its forward approach. This technique using positive reinforcement is also the best way to retrain horses with trailer loading problems, ultimately taking less time than with negative reinforcement.[99,100]

Once the loading behavior is well established, reduce the amount of food with each repetition to eliminate the long-term need for it. Clicker training and target training are helpful tools in teaching a horse to load because the connection to the reward has previously been learned.[98]

An additional step can be helpful to prepare a horse for trailer loading. Teaching the horse to lead through gates with headers and footers familiarizes it with stationary objects it must pass over and under simultaneously.[38] This technique is particularly useful for reactive animals.

Foals and young horses that readily eat grain can easily learn to load if accompanied by an older horse. The older horse is put into the trailer and then the youngster led up and in. They may hesitate at the entrance, but patience to allow time for them to investigate the surroundings will pay off. Once finished eating, the youngster should exit the trailer before the older horse to prevent panic.

Negative Reinforcement for Loading

Negative reinforcement is not punishment and does not involve hitting the horse. The reward value comes when a negative behavior stops and a correct one occurs. This learning technique teaches the horse that the inside of the trailer is more comfortable and less stressful than outside. Because the technique is intended to "aggravate" the horse, it is particularly important to anticipate attempts to get away and block them. Different tools can be used, one being a thin, stiff pole such as a fishing rod with a plastic bag attached at the tip.[101]

The horse is led toward the trailer. As long as it is headed in the right direction, it is rewarded with treats, rubbing, or kind words—positive reinforcement. If the horse starts to back away, the handler keeps it backing well past the point where the horse wants to stop. Backing is difficult. If the horse tries to turn away from the trailer opening, it is "aggravated" with the flapping plastic on the side it is trying to go. Leading the horse away to reposition would reward the turning behavior and should be avoided. If the horse starts to load but then backs out, the handler flaps the plastic behind the horse until the animal takes a step forward again. When the horse does enter the trailer, the exit should not be blocked by closing the door or using a butt bar.[101] The loading process is repeated several times until the horse is willing to stand comfortably in the trailer and needs to be asked to back out. Only then should the butt bar be secured, the door shut, and the trailer moved.

Unloading

Exiting the trailer calmly is another important lesson a horse must learn, but it is often overlooked. A horse that panics when coming out of a trailer can develop a fear of getting in as well. The technique used to unload a horse will depend on the type of trailer or van. When the horse walks out head first, the handler should turn the horse toward the door and walk slowly, one step at a time toward the opening. The horse should first stop at the exit and wait. The person should step down and left, out of the horse's path, wait a few seconds, and then signal the horse to step down. This slows the horse down so it is less likely to jump out. Immediately after the horse gets out, it should be turned to face the trailer or van entrance so that it learns to turn back instead of jumping out and darting away.

For trailers that require the horse to back out, it is important the horse knows how to back before ever being loaded the first time. The handler can back the horse one step at a time if both horse and person can safely be close. If the handler cannot be next to the horse's head, a long rope is used so the person can stand to the side of the horse's rear while giving a gentle backward pull on the halter. Because changes in footing are unpredictable from the horse's perspective, abrupt changes often trigger bolting. To minimize the unpredictability of the exit, a specific cue can be given just as the horse's hind foot is about to step down—a word, whistle, or halter cue for example. For gentle horses, this cue can be a hand on the rump as the animal backs. Just as the horse is about to step down or step onto the ramp incline, the amount of pressure on the rump is increased.[38] A tug on the tail would be another way to make the signal either to start backing or to step down.[101] After a few repetitions, the horse connects the signal with the immediacy of the abrupt change.

Switching handlers generalizes loading behavior to a trailer and dissociates it from a specific person. Different trailers and different locations increase generalization of trailer loading too. Because lessons fade if not repeated, occasional loading and hauling on short trips are necessary.

Travel

Riding in a trailer or van requires the horse to balance against acceleration, deceleration, sways, and turns. It also must absorb bumps. Transport is stressful, and horses hauled alone show significant levels of pawing, vocalizing, head tossing, and sniffing the ground.[102,103] Physiologically, there are significant stress-related changes in cortisol, plasma total antioxidant status, serum albumin, slowed mitogen-induced proliferation of lymphocytes, packed cell volumes, white blood cell counts, and neutrophil counts.[103–105] These acute phase changes impair the horse's cell-mediated immune response. A horse only needs to be loaded into a trailer to have an increase in cortisol and packed cell volume.[106,107] Even for seasoned veterans, the anticipation of travel increases cortisol as much as an hour before loading.[108] The highest cortisol levels occur toward the end of transport, regardless of whether it is a relatively short trip or up to 8 h.[107] Habituating a horse to trailer loading through repetition does help reduce stress.[109]

Because the horse is working to maintain balance throughout the trip, certain precautions can reduce the likelihood that behavior problems will develop. The interior of the trailer must be adequately sized relative to the horse. It is not reasonable to haul a 17-hand Thoroughbred in a 6-ft. tall trailer, as an example. Full-length partitions between trailer stalls do not allow most full-size horses to spread their feet for balance, especially during cornering. Good ventilation is important, especially during hot weather, and ventilation should provide fresh air, not vehicle exhaust.[98] Acceleration and stopping should be gradual so the horse can adequately brace and not be thrown against the sides. Highway curves must be taken slower than usual so the horse can lean appropriately. If taken too fast, a horse can be thrown against the stall partition and try to regain balance using its feet against the wall (*scrambling*). It only takes one fast curve to trigger scrambling when stall width is inadequate for bracing.

The direction the horse faces while traveling has changed with newer trailer designs. In the past, most trailers were designed so the horse faced the direction of travel. Vans were designed for them to face either forward or back. Slant load trailers have recently become popular for hauling three or more horses and are so named because the horses stand at 35–45 degrees from forward. Given a choice, horses face straight back 28% of the time, face back at an angle 37% of time, and face straight forward only 11% of the time. When not moving, they face forward just over half the time.[110] Horses traveling sideways are less likely to lose their balance or touch the side rails as often. After travel, horses that faced forward have the highest cortisol levels, while those that faced backward moved more but have fewer physiological changes.[111–114] Facing backward seems to be the preferred direction of trailer travel for horses.

Intercontinental transport of sporting horses is becoming quite common, and the biological effects can produce changes in the normal rhythms of life. Jet lag is well recognized in humans and happens in horses too. Associated behavior changes include despondency, loss of appetite, sleep disturbance, fatigue, reduced muscle strength, and reduced performance. The DNA of the "clock" genes is almost identical in both humans and horses.[72] Travel over 500 mi, particularly when the horse is unable to lower its head to the floor, is associated with increased respiratory problems. Also, the white blood cell count is elevated and remains so for at least 24 h after travel.[115] Extra time at the end of travel is needed for the competitive horse to readjust its physiological and sleep states before intense competition begins.

Trailering Problems

Putting a horse in a small, dark "box" to tolerate being slung from one side of that box to the other can be frightening. As with many other things, humans have difficulty experiencing the horse's perspective. While foals traveling with their mothers do not show stress-related changes, adult horses do.[116,117] Even "good" loaders can show anxiety in transit, as evidenced by an elevated heart rate and salivary cortisol.[100,106,118,119] The majority of horses that do not trailer well have been hauled in two-horse, side-by-side trailers (65%).[119] Many of these trailering problems can be treated with management, desensitization, and counterconditioning. Occasionally, anxiolytic medications might be needed in the beginning, and of these, the equine appeasing pheromone was specifically tested on horse loading.[120]

Difficult to Load

Difficulty getting a horse to load into a trailer is associated with 53% of horse trailering problems.[119] It is often the result of rushing, because people wait until they have to load the horse and then underestimate how long it will take. Trailers have several properties associated with fear—strange places, tight spaces, dark holes, unique odors including those from other horses, unsteady flooring, and hollow sounds are frightening.[121] As a curious animal, a horse needs to investigate these new things slowly and cautiously.

Loading horses should be practiced several times before a trip is ever needed. Preloading preparation involves anticipating all things that could go wrong so they can be prevented. Loading another horse first reduces fear for the first-time loader, and a food reward inside will help the initial learning.[122] For foals, loading their mother first makes the lesson go smoother.[123] Another useful technique is to feed the horse's daily grain in the trailer instead of the stall. This reinforces the idea that the trailer is a good place. The halter and lead rope are only used to keep the head pointed in the general direction of the trailer opening, not to pull the horse forward. Time spent getting a horse to load the first time correlates with its age, and probably with the development of trust in humans. Yearlings generally take 368 s, while 2-year-olds take 29.5 s, 3-year-olds take 21.5 s, and those over 3 years take 5.0 s.[106,124] Illuminating the inside of the trailer minimizes the problem of going from a relatively light area into a dark "hole."[106,118] Horse eyes do not adapt well to sudden changes in light intensity so trailer lighting minimizes illumination differences. If mild encouragement is needed, a soft butt rope is handy. With each end held by a different person or one end tied to the trailer and the other held by a person, firm pressure is used to encourage forward motion, not force loading.

Teaching a horse to load that already has a negative connection with a trailer will take longer and is easiest and faster when using desensitization and counterconditioning with positive reinforcement.[99,121,122] Patience is critical because this process will take several days or weeks. Clicker training and targeting may or may not be added depending on the comfort level of the handler in using those techniques.[122]

If the horse is not severely frightened, feeding small amounts of grain closer and closer to the trailer is the start. Then the bucket is placed at the back edge of the trailer and moved gradually forward in successive trials. At some point, the horse will have to step up onto the floor or ramp to reach the reward. If it steps back immediately, give it time to approach again. Punishment should not be used.

When the problem is more severe, a rubber mat can be used to teach the horse to stand on the mat to receive a food reward. Once this is learned, the mat is placed at a distance far behind the trailer where the horse is comfortable approaching. The horse is rewarded for coming up and standing on the mat. In subsequent sessions, the mat is very gradually moved closer to the trailer and eventually into it. The key in this desensitization/counterconditioning training is to NEVER progress to a closer position until the horse is comfortable where it stands for a reward. Also, never punish or speak loudly if it takes what seems like a long time. Mat-standing may need to be repeated several times at a certain position to ensure the horse is relaxed before moving the mat slightly closer to the trailer.

If the trailer has a large opening and is roomy enough inside so the horse can turn around, it can be placed in a paddock with the door open. It is important to secure it to prevent movement, and that it is safe for the horse to enter. Any dividers should be removed. With the problem horse alone in the paddock, the horse's food and water are gradually moved closer and closer to the trailer. Eventually, they are moved inside the trailer and then gradually moved to the front. If this is the only source of food and water, it is more likely to be successful.[123,125]

Scrambling in the Trailer

Horses learn to fight trailers because of careless driving, and once learned, the problem is very difficult to stop. Of horses with trailering problems, 51.5% show the problems while traveling, and of those, 52.8% act up when the trailer starts to move, 47.2% do not take curves well, and 24.5% have trouble with braking.[119] Horses that show problems in the standing trailer (73.6%) may be responding to the internal environment, such as extreme heat. They might also be anticipating when movement will begin or trying to relax after the trip has ended. Problems during movement are associated with horses that have difficulty balancing, are afraid of losing their balance, or are reacting to erratic driving.[121] Scrambling horses usually lean into the trailer wall or divider and move their feet against the opposite wall. Depending on the type of trailer partition, the scrambling may occur when going in one direction around a corner, or it may be generalized so that the horse leans and scrambles regardless of fast/slow or straight/curved. The facial expression of such an animal is that of panic. Changing the trailer or trailer's configuration is usually needed. A livestock trailer, slant-load trailer, or side-by-side with the full-length partition removed or swung over gives adequate room for the horse to spread its feet for balance.[119,121,123]

Difficult to Unload

Before a horse is ever loaded into a trailer, it must know how to back. This is critical in vans where they back into stalls and in trailers where the horse has to back out. The first few times of unloading set the pace for the future. If there is another horse in the trailer, unload the extra horse last so the lesson horse does not panic. Unloading was cited as a problem for 12.6% of horses that had trailering problems, with rushing out being the biggest issue.[119] These horses almost seem panicky. This behavior can result in injuries to the horse and anyone who accidentally gets in the way.

Lessons to stop the rushing out involve one person standing to the side of the trailer exit in a way that they can put their hand on the horse's rump. Before using the technique on a horse in the trailer, however, it should be practiced while backing elsewhere so the horse can learn what hand pressure means.[126] The horse is asked to back. As it starts back, light pressure is maintained on the rump and firmer pressure used when the horse is to back slower or stop. Then as the trailered horse is about to step down or onto the ramp, the increased pressure signals that an awkward step is coming.[127]

The horse will learn that firm pressure means to not move back and lighter pressure means move slowly. The butt chain/bar is released while there is firm pressure and the horse encouraged to stand quietly for a while afterward. Then the "back" command given and pressure lightened. The hand pressure technique teaches the horse to trust the person to guide it out of the trailer. It is important to not pull on the head: direction comes from behind.

COMMON PROBLEMS ASSOCIATED WITH MOVEMENT

Increased locomotor activity is associated with anticipatory behavior.[128] The increase in heart rate correlates to increases in the frequency and duration of standing, moving, arousal, and investigation. It should be no surprise then that movement has associated behavior problems. Incidence data is quite variable depending on what is included and how the information is gathered. As an example, one study found that 22% of horses showed a problem behavior while being ridden, while another suggested the incidence is as high as 84%.[129] Problems while being ridden fall into four broad categories.[130] The first, which includes the 84% of horses, involves the horse inappropriately responding to cues. Examples include not slowing when asked, jogging when asked to walk, shying, moving off when the rider attempts to mount, and pulling or leaning into the bit. Discomfort behaviors such as resisting being turned, bucking, tripping, not taking a specific lead, or refusing to move forward are associated with 61% of horses. For horses that jump, 36% will run out, stop, or rush over jumps. Extreme conflict behaviors are shown by 22% of horses and include bolting, rearing, and bucking.

Problems can result from confusion due to inconsistent cues. They can also be pain-related, as suggested by the ears back, lateral head shaking, kicking out, bolting, or refusing to move forward.[131,132] Most of the time, the cause is poor communication between rider and horse. It is with the human, not the horse, but it is the horse that suffers mishandling and abuse.

Bolting

Bolting has two definitions: running away without control, and rapidly eating food. Bolting food is discussed elsewhere. Bolting as it relates to movement is a potentially dangerous reaction to a frightening event due to association with innate, sympathetic flight or fight responses. It can also be related to pain. Twenty-nine percent of mares with vaginitis and related inflammatory conditions bolt as an associated behavior.[131]

Horses can learn to bolt at times of their choosing. Here, the primary concern is the potential danger for a rider. Bolting results in out-of-control running by a horse that is not paying attention to any rider cues and is not responsive to tightening reins. It might run into objects like fences, lose footing on slippery surfaces like asphalt or ice, or be hit by cars. Riders can be injured if the horse encounters dangerous conditions or if it runs under low hanging tree limbs or into openings of a barn or shed.

Bolting is common in young horses being started under saddle. They may also bolt if taken from the relative safety of the training pen for their first trail ride without an older trail-wise horse from which to take cues. Some older horses bolt or shy unpredictably if they are particularly observant of subtle movements of plastic flags or other objects. Horses can also learn that bolting will result in their being taken back to the barn, and these horses are particularly dangerous for inexperienced riders.

Treatment for bolting depends on the cause. Rider response is important. The recommendation is to get the horse circling. That may require working down one rein to get a hand near the bit or even grabbing the bridle's cheek strap. Once circling, the horse should then be made to continue circling long after the horse tries to stop. A second technique can be used if the area is conducive. The horse is allowed to run after bolting, but when it starts slowing down, the rider keeps it running at speed much longer. Regardless of whether the horse is circled or made to continue to run, the rider is focusing on making the horse do extra work—something horses do not like to do.

Underlying problems should be identified and corrected. Since rider-horse communication may be confusing to a green horse or to a seasoned horse with a green rider, the rider needs to work on sending clear messages and lowering immediate expectations. If the problem appears suddenly without an obvious cause, a medical workup is indicated. Pain can be subtle but must be high on the differential list. For uncooperative mares, vaginitis, cervicitis, or endometritis should be added to the list of differentials, already including musculoskeletal and dental problems.[131]

Bucking

Bucking is a behavior where front legs and back legs alternately leave the ground or all 4 ft. are simultaneously off, with the rump being the most significantly elevated portion. It is a common play behavior, especially for individually stalled weanlings.[90] Bucking was originally used by ancient horses to get predators off their back, and it is still used today to get rid of something, although rarely a predator. While some horses are selectively bred to buck as rodeo stock, the behavior is undesirable in most horses. When first saddled, young stock may try to buck, but careful work can prevent the association between carrying a saddle and bucking.

The abrupt onset of bucking in older horses suggests a problem. In some cases, the behavior is associated with events that appear to frustrate the horse. As an example, a green rider might kick the horse to initiate forward movement but at the same time sending a "whoa" message by holding the reins tight. Other cases may indicate pain. Sharp teeth, painful ovulations, or a sore back are just a few differentials to consider. History taking should determine if the bucking is only associated with riding, when the problem started, frequency, progression, and related events. Video might be helpful.

Treatment for bucking varies with the cause. If the problem started with a new rider or when the horse was young, the horse may have learned that bucking will make the rider get off. That horse should be taught that bucking results in more work, and work is hard. Forcing the horse forward as it starts to buck, and then drawing it into a tight circle that continues well past the time the horse tries to stop is helpful. The older horse that abruptly begins to buck suggests a diagnostic workup is needed for an appropriate treatment plan to be formulated.

Rearing

Horses that rear follow much the same pattern as those that buck. The behavior is common in play and in individually stalled weanlings.[90] It is also common in horses fed high-energy food but not getting much exercise. Rearing can be an initiation to play, especially for orphans, and they may direct it toward humans if not well socialized to horses.[133]

The evolution of rearing goes back to the ancestral horse trying to escape predators. For this reason, the behavior often has a fear or confusion component associated with attempted escape. Starting young horses with older, experienced ones and gradually working farther from herdmates reduces the fear of being alone while building trust in their rider.

As with bucking, pain can trigger rearing, so when an older horse starts rearing, a diagnostic evaluation is appropriate. Regardless of the original cause, learning can increase the significance of the problem. A confused horse might rear and the rider gets off. The animal quickly learns that even small attempts to rear will cause the rider to dismount.

If rearing occurs in a potentially dangerous situation, it is important not to reinforce the behavior. Move, tie, or distract the horse until work can be resumed safely. Rearing as play solicitation may start while a horse is being led. Instead of stopping and jerking on the lead rope, walk "with a purpose." Then when the horse rears, step out of the way, leaving slack in the lead rope, and as soon as the forefeet touch the ground start walking again.[133] Use a rope long enough to ensure no tugging will occur on the halter and the handler is able to remain at a safe distance. Pulling and verbal attention reward the play intention.

The riding horse that rears into a nearly vertical position is particularly dangerous because it increases the likelihood it will lose its footing and fall. Retraining that horse should only be done by experts, and even then, the long-term outlook for it becoming a safe riding horse is not good. Less extreme rearing can be treated similarly to what is done for bucking. As soon as the horse begins to rear, it is kicked into forward motion and then circled for a prolonged time. Since horses do not want to work, they quickly figure out that rearing means more work instead of getting rid of the rider.

Shying

Horses that shy are reacting to movement happening too fast for them to focus on the object. The instinctive reaction is to get away and then assess the situation from a safe distance, and domestication has changed this instinct. In addition, the combination of a high-energy diet and minimal exercise increases the likelihood of overreactions to environmental stimuli. Diets higher in fat, even with controlled caloric densities, reduce startle response intensity and lower resting cortisol levels.[134]

Walking Off

A common complaint from riders is their horse begins walking off as the rider mounts. This is a training problem the rider did not correct. If the horse is positioned to face a wall or fence before the rider puts a foot in the stirrup, its forward motion is blocked. Pulling back on the reins and saying "whoa" when movement starts, or sharply turning the head toward the mounting rider are other techniques used to stop the problem.

Kicking

Kicking stall walls can injure the horse, but it is a problem that happens often, particularly in Arabian horse barns and slightly less so in Standardbred stables.[135] Individually stalled weanlings also show this behavior.[90] Treatment for stall kicking is somewhat dependent on the cause. Kick chains have been used. These are short pieces of chain attached to a band that is fastened around the rear fetlocks. When the horse kicks out, the free end of the chain flies up, striking the horse's leg. The positive punishment technique is advantageous because it occurs every time the horse kicks. Alternatively, the horse could be kept in a in a paddock with in-and-out shelter. Stalls can be modified with a "shelf" extending inward from the wall far enough to prevent the kicking leg from hitting the wall.

Kicking is not limited to problems in a stall. As with rearing and bucking, this behavior can be a form of refusal due to confusion or pain. Orthopedic conditions such as "kissing spines" are associated with horses that kick out while being ridden, particularly at faster speeds. In mares with vaginitis and related inflammatory conditions, 57% kick as an associated behavior.[131]

OTHER PROBLEMS

Horses show other behaviors that do not fit well into larger categories but do relate to movement. While individual horses can come up with their own versions of new issues, the two described next are common enough to be included.

Refusal to Enter a Stall

A horse that stays in one particular stall begins to refuse to reenter that stall but will readily enter any other stall in the barn. There is no problem getting a second horse to enter the initial stall. This situation suggests that the first horse has had an unpleasant experience within the stall and has come to expect that experience will happen again if it reenters. The second horse has not encountered the negative situation so readily enters. In such cases, detective work is necessary to identify what is causing the problem. Effects of stray (phantom) voltage have been studied extensively in dairy cattle.[136] This is not true for horses and barns where they stay. Consideration of the possibility that stray voltage may have electrified mangers or waterers should not be overlooked. Other considerations would include particularly nauseating scents, such as a dead rodent in the manger or water bucket.

Escaping From a Stall or Paddock

The relative boredom of a stall environment and the solitary existence in a paddock provide an ideal scenario for the curious horse to explore its surroundings. Gate and stall door latches are perfect substrates for manipulation. Through trial and error, a horse may accidentally unlatch the mechanism and escape its confines. The first success leads to the second and that to the third. Owners may even encourage the behavior, only later realizing the monster they have fostered. Preventing escape requires mechanisms, like locks, that cannot be reached or easily opened. Also, vigilance is needed to be sure gates and stall doors are secured.

REFERENCES

1. Gorissen BMC, Wolschrijn CF, Serra Bragana FM, Geerts AAJ, Leenders WOJL, Back W, van Weeren PR. The development of locomotor kinetics in the foal and the effect of osteochondrosis. *Equine Vet J* 2017;**49**(4):467–74.
2. Rogers CW, Bolwell CF, Tanner JC, van Weeren PR. Early exercise in the horse. *J Vet Behav* 2012;**7**(6):375–9.
3. Waring GH. *Horse behavior*. 2nd ed. Norwich, NY: Noyes Publications; 2003. p. 442.

4. Crowell-Davis SL. Daytime rest behavior of the Welsh pony (*Equus caballus*) mare and foal. *Appl Anim Behav Sci* 1994;**40**(3–4):197–210.
5. Haywood L, Spike-Pierce DL, Barr B, Mathys D, Mollenkopf D. Gestation length and racing performance in 115 Thoroughbred foals with incomplete tarsal ossification. *Equine Vet J* 2018;**50**(1):29–33.
6. Bramble DM, Carrier DR. Running and breathing in mammals. *Science* 1983;**219** (4582):251–6.
7. Young IS, Alexander RMN, Woakes AJ, Butler PJ, Anderson L. The synchronization of ventilation and locomotion in horses (*Equus caballus*). *J Exp Biol* 1992;**166**(1):19–31.
8. Lesimple C, Fureix C, LeScolan N, Richard-Yris M-A, Hausberger M. Housing conditions and breed are associated with emotionality and cognitive abilities in riding school horses. *Appl Anim Behav Sci* 2011;**129**(2–4):92–9.
9. Houpt KA, Lee J, Floyd T. Operant and two-choice preference applied to equine welfare. In: *Proceedings of the AVSAB annual symposium of animal behavior research*; 2002. p. 47.
10. Jørgensen GHM, Liestøl SH-O, Bøe KE. Effects of enrichment items on activity and social interactions in domestic horses (*Equus caballus*). *Appl Anim Behav Sci* 2011;**129** (2–4):100–10.
11. Heglund NC, Taylor CR, McMahon TA. Scaling stride frequency and gait to animal size: mice to horses. *Science* 1974;**186**(4169):1112–3.
12. Gambaryan PP. *How mammals run*. New York: John Wiley & Sons; 1974367.
13. Hildebrand M. Analysis of tetrapod gaits: general considerations and symmetrical gaits. In: Herman RM, Grillner S, Sterin PS, Stuant DG, editors. *Neural control of locomotion*. New York: Plenum; 1976. p. 203–36.
14. Hildebrand M. Symmetrical gaits of horses. *Science* 1965;**150**(3697):701–8.
15. Hildebrand M. Analysis of asymmetrical gaits. *J Mammal* 1977;**58**(2):131–56.
16. Ambling gait. *Wikipedia*. https://en.wikipedia.org/wiki/Ambling_gait. Accessed 24 August 2017.
17. Horse: *Equus ferus caballus*. Speed of Animals http://www.speedofanimals.com/animals/horse [downloaded August 21, 2017].
18. Ratzlaff MH, Hyde ML, Grant BD, Balch O, Wilson PD. Measurement of vertical forces and temporal components of the strides of horses using instrumented shoes. *J Equine Vet Sci* 1990;**10**(1):23–5. 28–29, 32–35.
19. Loscher DM, Meyer F, Kracht K, Nyakatura JA. Timing of head movements is consistent with energy minimization in walking ungulates. *Proc R Soc B* 2016;**283**.
20. Hampson BA, Morton JM, Mills PC, Trotter MG, Lamb DW, Pollitt CC. Monitoring distances travelled by horses using GPS tracking collars. *Aust Vet J* 2010;**88**(5):176–81.
21. Lewis C, Nadeau J, Hoagland T, Darre M. Effect of season on travel patterns and hoof growth of domestic horses. *J Equine Vet Sci* 2014;**34**(7):918–22.
22. Persson Sjödin E, Hernlund E, Egenvall A, Rhodin M. Influence of the riders position on the movement symmetry of sound horses in straight line trot. *J Vet Behav* 2016;**15**:81–2.
23. Farley CT, Taylor CR. A mechanical trigger for the trot-gallop transition in horses. *Science* 1991;**253**(5017):306–8.
24. Hildebrand M. Motions of the running cheetah and horse. *J Mammal* 1959;**40**(4):481–95.
25. Parsons KJ, Pfau T, Wilson AM. High-speed gallop locomotion in the thoroughbred racehorse. I. The effect of incline on stride parameters. *J Exp Biol* 2008;**211**(6):935–44.
26. Fastest speed for a race horse. *Guinness World Records*. http://www.guinnessworldrecords.com/world-records/fastest-speed-for-a-race-horse. Accessed 23 August 2017.

27. American Quarter Horse. *Wikipedia*. https://en.wikipedia.org/wiki/American_Quarter_Horse. Accessed 23 August 2017.
28. Evans D, McGreevy P. An investigation of racing performance and whip use by jockeys in Thoroughbred races. *PLoS One* 2011;**6**(1). https://doi.org/10.1371/journal.pone.0015622.
29. Verheyen KLP, Price JS, Wood JLN. Exercise during training is associated with racing performance in Thoroughbreds. *Vet J* 2009;**181**(1):43–7.
30. Tan H, Wilson AM. Grip and limb force limits to turning performance in competition horses. *Proc R Soc B* 2011;**278**:2105–11.
31. Waite K, Heleski C, Ewing M. Quantifying aggressive riding behavior of youth barrel racers and conflict behaviors of their horses. *J Vet Behav* 2018;**24**:36–41.
32. Tolley EA, Notter DR, Marlowe TJ. Heritability and repeatability of speed for 2- and 3-year-old Standardbred racehorses. *J Anim Sci* 1983;**56**(6):1294–305.
33. Performance Genetics: heritability of performance, http://performancegenetics.com/heritability-of-thoroughbreds-performance/ [downloaded 4/7/14].
34. Monteiro da Gama MP, Borquis RRA, de Araújo Neto FR, de Oliveira HN, Fernandes GM, Siveira da Mota MD. Genetic parameters for racing performance of thoroughbred horses using Bayesian linear and Thurstonian models. *J Equine Vet Sci* 2016;**42**:39–43.
35. Henshall C, McGreevy PD. The role of ethology in round pen horse training—a review. *Appl Anim Behav Sci* 2014;**155**:1–11.
36. Birke L, Hockenhull J, Creighton E, Pinno L, Mee J, Mills D. Horses' responses to variation in human approach. *Appl Anim Behav Sci* 2011;**134**(1–2):56–63.
37. Krueger K. Behaviour of horses in the "round pen technique" *Appl Anim Behav Sci* 2007;**104** (1–2):162–70.
38. Beaver BV, Höglund DL. *Efficient livestock handling: the practical application of animal welfare and behavioral science*. New York: Academic Press; 2016213.
39. Wilk I, Janczarek I. Relationship between behavior and cardiac response to round pen training. *J Vet Behav* 2015;**10**(3):231–6.
40. Wutke S, Andersson L, Benecke N, Sandoval-Castellanos E, Gonzalez J, Hallsson JH, Lõugas L, Magnell O, Morales-Muniz A, Orlando L, Pálsdóttir AH, Reissmann M, Muñoz-Rodriguez MB, Ruttkay M, Trinks A, Hofreiter M, Ludwig A. The origin of ambling horses. *Curr Biol* 2016;**26**(15):R697–9.
41. Staiger EA, Bellone RR, Sutter NB, Brooks SA. Morphological variation in gaited horse breeds. *J Equine Vet Sci* 2016;**43**:55–65.
42. Giwner D. *Red Mile: Always B Miki paces the fastest mile in Standardbred History Daily Racing Form.com*. http://www.drf.com/news/red-mile-always-b-miki-paces-fastest-mile-standardbred-history; 2016. Accessed 31 August 2017.
43. Andersson LS, Larhammar M, Memic F, Wootz H, Schwochow D, Rubin C-J, Patra K, Arnason T, Wellbring L, Hjälm G, Imsland F, Petersen JL, McCue ME, Mickelson JR, Cothran G, Ahituv N, Roepstorff L, Mikko S, Vallstedt A, Lindgren G, Andersson L, Kullander K. Mutations in *DMRT3* affect locomotion in horses and spinal circuit function in mice. *Nature* 2012;**488**(7413):642–6.
44. Highest jump by a horse. *Guinness World Records*. http://www.guinnessworldrecords.com/world-records/highest-jump-by-a-horse. Accessed June 2016.
45. Spaas J, Helesen WF, Adriaenssens M, Broeckx S, Duchateau L, Spaas JH. Correlation between dichromatic colour vision and jumping performance in horses. *Vet J* 2014;**202**(1):166–71.
46. Ricard A, Danvy S, Legarra A. Computation of deregressed proofs for genomic selection when own phenotypes exist with an application in French show-jumping horses. *J Anim Sci* 2013;**91** (3):1076–85.

47. Brard S, Ricard A. Genome-wide association study for jumping performances in French sport horses. *Anim Genet* 2015;**46**(1):78–81.
48. Butler CL, Houpt KA. Pawing by Standardbred racehorses: frequency and patterns. *J Equine Sci* 2014;**25**(3):57–9.
49. McDonnell S. Rearing and flipping. *The Horse* 2006;**XXIII**(10):79–80.
50. McDonnell S. Playful rearing. *The Horse* 2000;**XVII**(12):24–5.
51. Deesing MJ, Grandin T. Behavior genetics of the horse (*Equus caballus*). In: Grandin T, Deesing MJ, editors. *Genetics and the behavior of domestic animals.* 2nd ed. Waltham, MA: Academic Press; 2014. p. 237–90.
52. McGreevy PD, Thomson PC. Differences in motor laterality between breeds of performance horse. *Appl Anim Behav Sci* 2006;**99**(1–2):183–90.
53. Siniscalchi M, Padalino B, Lusito R, Quaranta A. Is the left forelimb preference indicative of a stressful situation in horses? *Behav Process* 2014;**107**:61–7.
54. Whishaw IQ. Absence of population asymmetry in the American Quarter Horse (*Equus ferus caballus*) performing skilled left and right manoeuvres in reining competition. *Laterality* 2015;**20**(5):604–17.
55. Kuhnke S, König von Borstel U. A comparison of methods to determine equine laterality in Thoroughbreds. *J Vet Behav* 2016;**15**:85.
56. McGreevy PD, Rogers LJ. Motor and sensory laterality in thoroughbred horses. *Appl Anim Behav Sci* 2005;**92**(4):337–52.
57. Murphy J, Sutherland A, Arkins S. Idiosyncratic motor laterality in the horse. *Appl Anim Behav Sci* 2005;**91**(3–4):297–310.
58. Williams DE, Norris BJ. Laterality in stride pattern preferences in racehorses. *Anim Behav* 2007;**74**(4):941–50.
59. Murphy J, Arkins S. Facial hair whorls (trichoglyphs) and the incidence of motor laterality in the horse. *Behav Process* 2008;**79**(1):7–12.
60. Shivley C, Grandin T, Deesing M. Behavioral laterality and facial hair whorls in horses. *Equine Vet Sci* 2016;**44**:62–6.
61. Baumgartner M, Zeitler-Feicht MH, Wöhr A-C, Wöhling H, Erhard MH. Lying behavior of group-housed horses in different designed areas with rubber mats, shavings and sand bedding. *Pferdeheikunde* 2015;**31**(3):211–20.
62. Dallaire A. Rest behavior. *Vet Clin N Am Equine Pract* 1986;**2**(3):591–607.
63. Aleman M, Williams DC, Holliday T. Sleep and sleep disorders in horses. *Am Assoc Equine Pract Proc* 2008;**54**:180–5.
64. Belling Jr TH. Sleep patterns in the horse. *Equine Pract* 1990;**12**(8):22–6.
65. Houpt KA. The characteristics of equine sleep. *Equine Pract* 1980;**2**(4):8–17.
66. Williams DC, Aleman M, Holliday TA, Fletcher DJ, Tharp B, Kass PH, Steffey EP, LeCouteur RA. Qualitative and quantitative characteristics of the electroencephalogram in normal horses during spontaneous drowsiness and sleep. *J Vet Intern Med* 2008;**22**(3):630–8.
67. Dallaire A, Ruckebusch Y. Sleep patterns in the pony with observations on partial perceptual deprivation. *Physiol Behav* 1974;**12**(5):789–96.
68. DuBois C, Zakrajsek E, Haley DB, Merkies K. Validation of triaxial accelerometers to measure the lying behavior of adult domestic horses. *Animal* 2015;**9**(1):110–4.
69. Boy V, Duncan P. Time-budgets of Camargue horses. I. Developmental changes in the time-budgets of foals. *Behaviour* 1979;**71**(3–4):187–202.
70. Duncan P. Time-budgets of Camargue horses II. Time-budgets of adult horses and weaned sub-adults. *Behaviour* 1980;**72**(1–2):26–49.

71. Keiper RR, Keenan MA. Nocturnal activity patterns of feral ponies. *J Mammal* 1980;**61**(1):116–8.
72. Piccione G, Giannetto C. State of the art on daily rhythms of physiology and behavior in horses. *Biol Rhythm Res* 2011;**42**(1):67–88.
73. Burla J-B, Rufener C, Bachmann I, Gygax L, Patt A, Hillmann E. Space allowance of the littered area affects lying behavior in group-housed horses. *Front Vet Sci* 2017;**4**:1–12 [Article 23].
74. Beaver BV. A day in the life of a horse. *Vet Med Small Anim Clin* 1983;**78**(2):227–8.
75. Raabymagle P, Ladewig J. Lying behavior in horses in relation to box size. *J Equine Vet Sci* 2006;**26**(1):11–7.
76. Arnold GW. Comparison of the time budgets and circadian patterns of maintenance activities in sheep, cattle and horses grouped together. *Appl Anim Behav Sci* 1984;**13**(1–2):19–30.
77. Boyd L, Bandi N. Reintroduction of takhi, *Equus ferus przewalskii*, to Hustai National Park, Mongolia: time budget and synchrony of activity pre- and post-release. *Appl Anim Behav Sci* 2002;**78**(2–4):87–102.
78. Boyd LE, Carbonaro DA, Houpt KA. The 24-hour time budget of Przewalski horses. *Appl Anim Behav Sci* 1988;**21**(1–2):5–17.
79. Duncan P. Time-budgets of Camargue horses III. Environmental influences. *Behaviour* 1985;**92**(1–2):188–208.
80. Francis-Smith K. Behaviour patterns of horses grazing in paddocks. *Appl Anim Ethol* 1977;**3**(3):292–3.
81. Sweeting MP, Houpt KA. Water consumption and time budgets of stabled pony (*Equus caballus*) geldings. *Appl Anim Behav Sci* 1987;**17**(1–2):1–7.
82. Kiley-Worthington M. *The behaviour of horses: in relation to management and training*. London: J.A. Allen; 1987. p. 265.
83. Mayes E, Duncan P. Temporal patterns of feeding behavior in free-ranging horses. *Behavior* 1986;**96**(1):105–29.
84. Ransom JI, Cade BS. *Quantifying equid behavior—a research ethogram for free-roaming feral horses*. U.S. Geological Survey; 2009.*https://pubs.usgs.gov/tm/02a09/pdf/TM2A9.pdf*. Accessed 15 March 2017.
85. Sweeting MP, Houpt CE, Houpt KA. Social facilitation of feeding and time budgets in stabled ponies. *J Anim Sci* 1985;**60**(2):369–74.
86. Tyler SJ. The behavior and social organization of the New Forest ponies. *Anim Behav Monogr* 1972;**5**(2):87–196.
87. Ellis AD, Redgate S, Zinchenko S, Owen H, Barfoot C, Harris P. The effect of presenting forage in multi-layered haynets and at multiple sites on night time budgets of stabled horses. *Appl Anim Behav Sci* 2015;**171**:108–16.
88. Houpt KA, O'Connell MF, Houpt TA, Carbonaro DA. Night-time behavior of stabled and pastured peri-parturient ponies. *Appl Anim Behav Sci* 1986;**15**(2):103–11.
89. Piccione G, Costa A, Giannetto C, Caola G. Daily rhythms of activity in horses housed in different stabling conditions. *Biol Rhythm Res* 2008;**39**(1):79–84.
90. Heleski CR, Shelle AC, Nielsen BD, Zanella AJ. Influence of housing on weanling horse behavior and subsequent welfare. *Appl Anim Behav Sci* 2002;**78**(2–3):291–302.
91. Giannetto C, Bazzano M, Marafioti S, Bertolucci C, Piccione G. Monitoring of total locomotor activity in mares during the prepartum and postpartum period. *J Vet Behav* 2015;**10**(5):427–32.
92. Greening L, Shenton V, Wilcockson K, Swanson J. Investigating duration of nocturnal ingestive and sleep behaviors of horses bedded on straw versus shavings. *J Vet Behav* 2013;**8**(2):82–6.

93. Hunter L, Houpt KA. Bedding material preferences of ponies. *J Anim Sci* 1989;**67**(8):1986–91.
94. Pedersen GR, Søndergaard E, Ladewig J. The influence of bedding on the time horses spend recumbent. *J Equine Vet Sci* 2004;**24**(4):153–8.
95. Flannigan G, Stookey JM. Day-time time budgets of pregnant mares housed in tie stalls: a comparison of draft versus light mares. *Appl Anim Behav Sci* 2002;**78**(2–4):125–43.
96. Jones T, Griffin K, Hall C, Stevenson A. Effects of ridden exercise on night time resting behavior of individually housed horses. *J Vet Behav* 2016;**15**:83–4.
97. Benhajali H, Richard-Yris M-A, Leroux M, Ezzaoula M, Charfi F, Hausberger M. A note on the time budget and social behaviour of densely housed horses: a case study in Arab breeding mares. *Appl Anim Behav Sci* 2008;**112**(1–2):196–200.
98. Crowell-Davis SL. Transportation-related behavior problems in horses. *Compendium Equine* 2007;**2**(5):274–8.
99. Hendriksen P, Elmgreen K, Ladewig J. Trailer-loading of horses: is there a difference between positive and negative reinforcement concerning effectiveness and stress-related signs? *J Vet Behav* 2011;**6**(5):261–6.
100. Shanahan S. Trailer loading stress in horses: behavioral and physiological effects of nonaversive training (TTEAM). *J Appl Anim Welf Sci* 2003;**6**(4):263–74.
101. Anderson GF. Trailer loading: a common sense approach, In: *40th Annual AAEP Convention Proceedings*; 1994. p. 149.
102. Kay R, Hall C. The use of a mirror reduces isolation stress in horses being transported by trailer. *Appl Anim Behav Sci* 2009;**116**(2–4):237–43.
103. Knowles TG, Brown SN, Pope SJ, Nicol CJ, Warriss PD, Weeks CA. The response of untamed (unbroken) ponies to conditions of road transport. *Anim Welf* 2010;**19**(1):1–15.
104. Padalino B, Raidal SL, Carter N, Celi P, Muscatello G, Jeffcott L, de Silva K. Immunological, clinical, haematological and oxidative responses to long distance transportation in horses. *Res Vet Sci* 2017;**115**:78–87.
105. Rizzo M, Arfuso F, Giannetto C, Giudice E, Longo F, Di Pietro S, Piccione G. Cortisol levels and leukocyte population values in transported and exercised horses after acupuncture needle stimulation. *J Vet Behav* 2017;**18**:56–61.
106. Padalino B, Maggiolino A, Boccaccio M, Tateo A. Effects of different positions during transport on physiological and behavioral changes of horses. *J Vet Behav* 2012;**7**(3):135–41.
107. Schmidt A, Möstl E, Wehnert C, Aurich J, Müller J, Aurich C. Cortisol release and heart rate variability in horses during road transport. *Horm Behav* 2010;**5**(2):209–15.
108. Schmidt A, Biau S, Möstl E, Becker-Birck M, Morillon B, Aurich J, Faure J-M, Aurich C. Changes in cortisol release and heart rate variability in sport horses during long-distance road transport. *Domest Anim Endocrinol* 2010;**38**(3):179–89.
109. Yngvesson J, de Boussard E, Larsson M, Lundberg A. Loading horses (*Equus caballus*) onto trailers—behaviour of horses and horse owners during loading and habituating. *Appl Anim Behav Sci* 2016;**184**:59–65.
110. Rogers A. Onward and backward. *Equus* 1995;**212**:14–5.
111. Cregier SE. Reducing equine hauling stress: a review. *J Equine Vet Sci* 1982;**2**(6):186–98.
112. Padalino B. Effects of the different transport phases on equine health status, behavior, and welfare: a review. *J Vet Behav* 2015;**10**(3):272–82.
113. Waran NK. The behaviour of horses during and after transport by road. *Equine Vet Educ* 1993;**5**(3):129–32.
114. Waran NK, Robertson V, Cuddeford D, Kokoszko A, Marlin DJ. Effects of transporting horses facing either forwards or backwards on their behavior and heart rate. *Vet Rec* 1996;**139**(1):7–11.

115. Stull C. Box stalls versus slant loads. *Equus* 2014;**440**(87):89.
116. Tischner M, Niezgoda J. Transportation of mares and foals as a stress-causing factor. *Equine Vet J* 1998;**30**(S27):55.
117. Tischner Jr M, Niezgoda J. Effect of transport on the intensity of stress reactions in mares and foals. *J Reprod Fertil Suppl* 2000;**56**:725–30.
118. Cross N, van Doorn F, Versnel C, Cawdell-Smith J, Phillips C. Effects of lighting conditions on the welfare of horses being loaded for transportation. *J Vet Behav* 2008;**3**(1):20–4.
119. Lee J, Houpt K, Doherty O. A survey of trailering problems in horses. *J Equine Vet Sci* 2001;**21**(5):235–8.
120. Cozzi A, Lafont Lecuelle C, Monneret P, Articlaux F, Bougrat L, Bienboire Frosini C, Pageat P. The impact of maternal equine appeasing pheromone on cardiac parameters during a cognitive test in saddle horses after transport. *J Vet Behav* 2013;**8**(2).
121. Houpt KA. Stable vices and trailer problems. *Vet Clin N Am Equine Pract* 1986;**2**(3):623–33.
122. Ferguson DL, Rosales-Ruiz J. Loading the problem loader: the effects of target training and shaping on trailer-loading behavior of horses. *J Appl Behav Anal* 2001;**34**(4):409–23.
123. Houpt KA. Misbehavior of horses: trailer problems. *Equine Pract* 1982;**4**(2):12–416.
124. Waran NK, Cuddeford D. Effects of loading and transport on the heart rate and behavior of horses. *Appl Anim Behav Sci* 1995;**3**(2):71–81.
125. McDonnell S. Trailer resistance. *The Horse* 2003;**XX**(3):61–2.
126. McDonnell SM. Rushing out. *The Horse* 2000;**XVII**(11):28.
127. Beaver B. Too quick an exit. *Equus* 2009;**381**:76–7.
128. Peters SM, Bleijenberg EH, van Dierendonck MC, van der Harst JE, Spruijt BM. Characterization of anticipatory behavior in domesticated horses (*Equus caballus*). *Appl Anim Behav Sci* 2012;**138**(1–2):60–9.
129. Hockenhull J, Creighton E. Management practices associated with owner-reported stable-related and handling behavior problems in UK leisure horses. *Appl Anim Behav Sci* 2014;**155**:49–55.
130. Hockenhull J, Creighton E. Equipment and training risk factors associated with ridden behaviour problems in UK leisure horses. *Appl Anim Behav Sci* 2012;**137**(1–2):36–42.
131. Christoffersen M, Lehn-Jensen H, Bøgh IB. Referred vaginal pain: cause of hypersensitivity and performance problems in mares? A clinical case study. *J Equine Vet Sci* 2007;**27**(1):32–6.
132. Hall C, Barlow R. The relationship between approach behavior and jump clearance in show-jumping. *J Vet Behav* 2016;**15**:91.
133. McDonnell S. Spontaneous rearing and food aggression. *The Horse* 2016;**XXXIII**(6):61–2.
134. Redondo AJ, Carranza J, Trigo P. Fat diet reduces stress and intensity of startle reaction in horses. *Appl Anim Behav Sci* 2009;**118**(1–2):69–75.
135. Luescher UA, McKeown DB, Dean H. A cross-sectional study on compulsive behavior (stable vices) in horses. *Equine Vet J* 1998;**30**(S27):14–8.
136. Reinemann DJ. *Overview of stray voltage in animal housing*. The Merck Veterinary Manual, http://www.merckvetmanual.com/mvm/management_and_nutrition/stray_voltage_in_animal_housing/overview_of_stray_voltage_in_animal_housing.html; 2014. Accessed 17 October 2016.

Chapter 9

Equine Behavioral Medicine

Veterinary medicine has been an integral part of horse health over the years. The profession has experienced significant advances in medicine and surgery, allowing the equine athlete to perform at levels unimagined even a few years ago. At the same time, the horse's lifestyle has changed. Urbanization has resulted in more horses spending much of their lives confined to box stalls. Their owners may be fulfilling a dream of having a horse, but too often the person has little knowledge about handling such a large animal. At the other extreme is the top-level performance horse, an expensive investment that requires the highest quality veterinary treatment. Accompanying these changes is an increasing tendency for humans to anthropomorphize. Horses and humans are not that similar, and the role of dominance in human-animal relationships is overemphasized.[1] Finally, but also significant, is that a horse is a very large animal that evolved as a prey species. It avoids things it does not like and can move quickly. As a result, a horse has the potential to injure those around it. The challenge for working with any horse is how to handle it safely, and that is why understanding behavior is important.

The most significant behavioral trait of the horse is how it evolved to avoid predators. This is the only domestic species that uses flight as its primary survival tactic.[2] Keen senses are needed for quick awareness of potential danger and rapid responses to avoid what it detects. Basically, the horse views its environment at two extremes—things to fear and flee if necessary and things that are not fearful and thus safe to ignore.[2] If a frightening object turns out to be harmless, the horse needs to be able to quickly desensitize to it. This is necessary so the horse does not spend its entire day running from everything like birds flying up and blowing tumbleweeds. Desensitization is dependent on having good long-term memory. A frightening experience can result in a life-long negative association. This can be a very specific association. As an example, harsh treatment by a person who always wears a black western hat can result in fear of the specific person or any person wearing a black cowboy hat. Fear can also be generalized broadly, as with the fluttering plastic triangles along the edge of a show ring being generalized to any flapping plastic, such as a garbage can liner moving on a windy day.

Equine Behavioral Medicine. https://doi.org/10.1016/B978-0-12-812106-1.00009-7

THE VETERINARIAN'S ROLE

In this age of "Dr. Google," horse people have a number of options for information sources. Surveys suggest slightly over 90% go to veterinarians for health advice, with farriers second at 77%.[3,4] Other studies suggest the 90% estimate is high and closer to 70%.[5] Approximately three-fourths of equine operations report using veterinary services each year.[5] The most common sources of behavioral information remain books and magazines.[3] The internet is becoming more important as evidenced by 45% of those involved with horse competitions willing to spend up to an hour on an internet learning session. Fifty-three percent are willing to participate in up to three such sessions.[4]

The general public expects the veterinarian to be the expert in all-things-animal, from the animal's physical health, to its behavioral health and welfare. In the dog and cat world, anyone can claim to be a "trainer" or "animal psychologist," being capable of "fixing" a behavior problem. In the horse world, real solutions for problem behaviors tend to be ignored for a long time while the owner tries multiple things on their own. When those "solutions" do not work, the horse is sent to a "trainer" for correction. Unless veterinarians specifically ask about behavior problems, owners do not think to talk to the practitioner. The problem with "trainers" is that no type of certification exists and many use harsh punishments in an attempt to fix a problem. This is not only poor welfare for the horse; punishment usually creates bigger problems in the end. It is important for the veterinary profession to offer appropriate advice to prevent problem behaviors and have expertise to correctly diagnose and treat problems that do develop. Understanding behavior has an even bigger benefit—it increases the practitioner's safety.

Working Safer

Forcing a horse to do something is fraught with danger, making self-protection one of the major reasons to understand horse behavior. Incidences of work-related injuries and lost work days is the highest for equine practitioners when compared to other professions.[6–14] Over 60% of practitioners work on a difficult horse at least weekly, and 95% do so at least monthly.[15] Techniques used to manage these problem animals are often the very ones that escalate the horse's fear and avoidance. Those used least are the ones that would make the horse safer to handle (Table 9-1). It has been suggested the reason for this is because few practitioners really understand learning theory or how to apply it.[15] While it is impossible to eliminate all injuries, a knowledge of horse behavior and the techniques of learning can reduce the overall incidence. It is important that current practitioners and veterinary students understand equine behavior for longer and safer careers.[7,10,11,13]

Many equine patients are injured, have had a previous negative experience, or are frightened by what is happening to them. With the exception of

TABLE 9-1 Techniques Equine Practitioners Use When Working With Difficult Horses[15]

Restraint Technique	% of Practitioners Using
Sedation	99%
Nose twitch	74%
Neck twitch	69%
Chifney bit	57%
Bridle	49%
Holding up forelimb	47%
Food distraction	40%
Positive reinforcement	20%
Desensitization/overshadowing	8%
Negative reinforcement	7%

Note that the three techniques used the least are the ones that would make horses safer by reducing their reactivity. The top five increase anxiety and aversive learning.

emergency procedures, taking the time to teach the horse how to respond in a calm manner is worth the effort for future encounters. Preventing fear reactions is an important concept because, once frightened, it will take the horse 20–30 min to calm down again.[16] Too often the veterinarian is in a hurry and perhaps even anxious because of previous difficulty with the same horse. Anxiety increases the pace of the approach and the likelihood that maximum restraint is used right away.[17] The veterinarian unconsciously holds his/her breath—something the horse notices.[18] All this elevates the horse's level of fear.[18] While medication can increase the level of safety, there is a corresponding negative association of the needle penetrating the skin. Even though venipuncture is less painful than other restraint techniques, the experience is still negative. The horse is twitched, medicated, or both, and a procedure done. The resulting farm call does not go smoothly, and each subsequent visit is worse than the one before. The frightening lesson of the twitch or needle is very well learned.[7,19,20]

Starting naïve horses correctly does wonders for long-term handling, particularly if the horse is young, infrequently handled, or a young stallion.[21–23] Anticipating potentially fearful steps in any treatment protocol is critical, so they can be introduced slowly. Approaching the horse slowly, letting it investigate equipment, and using gentle but not highly restrictive handling helps. Mild, purposeful distractions can also be helpful. These can come in many forms, such as picking up the left front foot while someone is working on the left hind leg or tapping the horse's forehead when someone is working on a leg.

Several types of assessments provide information about how a horse will respond. The first is to differentiate the anxious or fearful horse from one that is calm. Young horses with "anxious" or "aggressive" temperaments are positively correlated with being difficult to work with during veterinary examinations.[22] If owners and handlers are realistic, a brief behavioral history will provide this information. Physical signs suggesting tension in a horse include tightening facial muscles; tension in the lips, especially the upper lip; lifting the upper eyelids; exposed white sclera; and ears held slightly back. The head and neck are elevated, and the higher the elevation, the greater the fear. Body muscles are tense, and the horse moves around, unable to stand still. The tail can be elevated or clamped tightly against the perineum.[24] Signs of aggression are particularly bothersome. Postures might include teeth bared, ears back against the neck, and the head tossing up and down.[17] An aggressive change can also be a clue of illness: sick animals are not always lethargic or unresponsive to environmental stimuli.[25]

A second assessment is to scan the environment for what could elevate the level of fear and in which horse that is likely. Horses described as "flighty," "highly reactive," or "neurotic" show fearfulness, reactivity, and excitability, making them sensitive to aversion.[7,26] Rapid movement, loud noise, and strange objects are problematic. The first is characteristic of predators, and something the horse's vision does not adjust to very well. Sensitivity to loud noise is greater for horses than for humans. The gentle touch associated with mutual grooming is an underestimated reward.[27] The use of soft sounds, slow movement, and gentle touches is calming. At least, it may help lower the animal's heart rate.[28] Using rewards is often beneficial, particularly for horses described as "extroverts," agreeable," or "curious."[7,26,29] Food treats are associated with increasing a horse's attention toward the person giving them, while rubbing is not.[30] Confinement and tightly restricted movement are fear-provoking for an animal that uses flight for survival. Horses are usually more cooperative if all people stand on the same side of the horse and lead ropes are not excessively restrictive. Putting a horse in stocks or a trailer for the first time is easiest if the process is done incrementally on nonslip surfaces with rewards given for each forward step.

A third safety assessment evaluates the owner, facilities, and amount of training the horse has had. Some people are unwilling or unable to properly control the horse, usually out of ignorance. Owners typically are truthful about their ability to handle the animal, its previous training, and its behavior.[17] Knowing these things allows appropriate handling measures to be in place before the animal is even approached.

Concepts of natural behavior and learning are helpful in reducing tension. Using them will initially take a little longer, but the result is a significantly more pleasant experience for the horse, owner, and practitioner. The first behavioral application is how to approach the horse. Walking straight up to it in a confident, striding manner, just like a predator would, is problematic.[18] It is less stressful for the horse if it is approached from the side, not directly or hastily. The horse should be allowed to smell an outstretched hand before being touched by it. Because they are inquisitive, horses will cautiously but voluntarily approach and smell things that would otherwise frighten them if shoved in their face.[16] When approaching a frightened horse, it is important to not reinforce the wrong thing. Approach very slowly, stopping when the horse holds still, not when it is moving or fidgeting. The reward (approach stopping) is for the desired, quiet behavior, not for flightiness. An additional reward can be added by taking a step or two backward. This "advance and retreat" concept is useful in a number of frightful situations.[18] When next to the horse's side, spending 10 min or so rubbing the animal and observing how it reacts can be a valuable introduction to the veterinarian.[17]

Needle Shy Horses

Needle shy horses are a problem to work with safely, but they can be taught to accept injections using several techniques. Previous experience connected specific signals with impending pain—signals like the sight of the syringe and needle or the tapping on the neck immediately preceding insertion of the needle. The first technique can be used for horses that are not extremely reactive. The person holding the horse should stand on the same side as the veterinarian and place a cupped hand over the horse's eye so the horse cannot see the needle and syringe approaching. The veterinarian puts continuous finger pressure on the specific location to be injected for approximately 10 s. The finger is removed, and the needle is immediately pushed through the skin—no tapping, just the numbing of pressure.

A technique for the reactive horse is desensitization. There are several versions of this "advance and retreat" technique that reward still and relaxed behavior, not anxiety and motion. The first step is to be able to stand next to a nonreactive horse. If the horse starts reacting to the veterinarian's approach, an assistant holds the horse in a roomy box stall with its rear in a far corner. The veterinarian begins a very slow approach toward the stall door (the advancement). If the horse holds still and remains calm, it is rewarded with the veterinarian stopping and taking a step back (the retreat). The assistant should also give a small food treat. If the horse moves and becomes anxious, the assistant repositions the horse while the veterinarian remains stationary. When the horse is back in place, the doctor backs away. This is followed by another approach toward the stall, but the veterinarian stops just before the position where the previous negative reaction happened. Remaining calm is reinforced with stop, a backward step, and food treat. Each subsequent approach is a little closer than the previous one as long as the horse remains calm. It may be necessary to repeat the same distance several times if continuing forward is difficult, making sure the horse understands that quiet is rewarded. Very gradually, the horse is approached. Its quiet demeanor is rewarded, motion is not. Eventually the veterinarian will be close enough to give a shot (but does not do so). This is the desired location and the first goal in desensitization.

The second step in desensitization is for the horse to tolerate being touched on the neck with an empty, flat hand. The reward is withdrawal of the hand and a food treat. This is followed by touch with four fingertips, then three, two, and finally one. Next is having a blunt object like a pen laid flat on the neck for a treat, followed by the point of the pen, a capped needle on a syringe, and finally the injection.[31]

If the horse does not become anxious until it sees the syringe, the desensitization begins there. The veterinarian or an assistant holds the lead rope and stands to the side of the horse. The veterinarian is softly talking to the client while scratching the horse's withers area until the animal relaxes. At this point, the empty hand moves slowly up toward the neck. As the horse begins to tense, the hand motion stops, not moving until the horse relaxes, at which point, the hand is moved back down the neck. The relaxation is rewarded with the withdrawal of the hand. This is repeated until the horse remains relaxed for upward motion. Next, the process is repeated with a pen laid flat on the neck. Successive lessons are with the point of the pen, and then the syringe and capped needle held in the hand as it moves up the neck as the process is repeated. Lastly comes the injection. When the needle is withdrawn, the neck is rubbed. The desensitization hand-syringe movements will need to be repeated if there are multiple injections and on subsequent visits, but each will take less time and fewer trials. Counterconditioning with a favorite food treat (with or without the use of a clicker) can be added to desensitization each time the horse relaxes.

Desensitization of the needle-shy horse can also be combined with the skin twitching and movement as distractions, a combination called *overshadowing*.[7,24] As with the previous technique, this one also begins with scratching the withers area and talking softly to anyone around until the horse calms down. The horse is then asked to take one step back and stop; then take one step forward and stop. Movement is a distraction that will change the horse's focus away from the upcoming procedure. The backward/forward is repeated until the horse easily responds each time it is asked. The veterinarian now puts a hand on the horse's neck. The horse responds by tensing, so the veterinarian maintains the neck touch while asking the horse to move back one step and then forward one step. Initially, this movement may be resisted, but the backward/forward process is repeated until the horse responds easily again. Then the veterinarian grasps a small amount of neck skin and repeats the backward/forward process until the horse relaxes. This progresses by grasping a large amount of skin in an appropriate skin twitch. The next step is to lightly touch the neck with the syringe and capped needle while applying the skin twitch, holding contact until the horse relaxes. Finally, the process is repeated with an uncapped needle and the injection is given. Desensitization and overshadowing should be repeated with subsequent injections, but each session will be shorter. Eventually, the horse will stand calmly for injections.

Giving Oral Medication

Worming pastes and oral medication can become difficult to give, especially when the product tastes bad. The horse may associate the sight of the dose syringe with the negative experience, making administration increasingly difficult. Desensitization and counterconditioning is applied similarly as done for a needle-shy horse. Begin with a hand up the neck and across the cheek. A food treat is placed in the sliding hand so the horse learns that the hand coming forward brings good things. Once comfortable with the hand coming forward with a treat, an empty syringe similar to that used for medication is carried forward, and the horse receives a food reward for accepting the syringe being held on its cheek. The next goals are for the syringe to be held alongside the lips, and finally between the lips.[24] It is desirable to have the owner repeat the process occasionally using a syringe filled with a highly palatable substance such as sweetened applesauce, molasses, honey, or yogurt. This turns an aversive treatment into one that is anticipated instead. The same moist treats can be used to suspend crushed tablet medications as well.

Giving Eye Medication

It is easier to train a horse to accept an eye exam or medication than it is to work with one that has already developed anxiety issues. Shaping the horse to accept these things is done the same way used for a problem animal; it just takes less time. The steps are similar to those for giving oral medication by getting the horse comfortable with the hand touching the cheek. Gradually the hand is advanced closer to the eye in small increments with a "retreat" and reward for calm responses. Ultimately, successful manipulation of the eyelids and administration of an unmedicated ophthalmic ointment are rewarded. Because working around the eye requires the use of both hands, a clicker is particularly useful to give more time for the food reward.[28]

Horses that resist an eye examination or medication are desensitized with the same steps, but signs of relaxation are the determinant for the reward. This process may begin by touching the neck and working in small, rewardable steps toward the cheek and eventually to the eye. Each advance-relax sequence is followed by the hand being moved slightly away and a food treat.

Touching the Ears

Ear twitching for restraint is associated with pain, not with the same sedative effects of lip twitching. It appears to work because of pain-associated distraction. Twitching the ear is not desirable, but if it has to be done, it should not be for long periods and should be carefully done to try to disconnect the ear twitch from pain memory. One technique begins by working a hand up the neck with the advance and retreat approach. Eventually the ear is reached. When the horse is comfortable with an ear touch, the ear is encased in the hand and the person's elbow is rested against the horse's neck. The ear is rotated to point down and away from the body while being squeezed with the fingers. It is not twisted. The process is reversed when releasing the ear by first stopping the squeeze and rubbing the ear while it is still being held. This is followed by a retreat/advance motion going down the neck. The idea is to begin and end with the most pleasant aspects of the procedure.

Once ear shyness develops, desensitization is important so each touch does not require a major production. The process begins with touching the neck and rewarding the relaxed response by backing down a little. Food treats can also be used as part of the reward. Each step advances farther up the neck. Eventually, the ear is touched. The process is similar to that for working around the horse's eyes or mouth.

OVERVIEW OF BEHAVIOR PROBLEMS

Behavior problems have never been reported in free-ranging horses, suggesting bad behavior is a product of management. Reports of the incidence of abnormal behavior in horses vary from 1% to 82%, depending on how data are collected, whether all problems are included or just stereotypies, and for what purpose the horse is used.[32–38] An incidence of 10%–15% is probably a reasonable estimate.

Behavior problems associated with genetic influences, or even epigenetic influences, may not be preventable. However, most are a combination of genetics and environment. In general, preventing behavior problems is easier than treating them, because an associated internal reward does not develop. Prevention means veterinarians must be aware of associated causes likely to lead to an unwanted behavior so they are able to educate their clients. Stereotypies are coping mechanisms for bad environments or management practices.[39,40] Regardless of the cause, horse owners want the behavior to stop. Unfortunately, physically stopping the behavior without addressing the root cause negatively impacts the horse's welfare, so treatment strategies will need to include knowledge of the principles of learning and appropriate environmental management. Some cases can also benefit from drug therapy.

Factors Promoting Abnormal Behavior

Studies have shown strong correlations with certain factors and development of abnormal behaviors. Long-term confinement or genetic factors are obvious, but some are not. Understanding the spectrum of relationships can be helpful for advising clients, to minimize the likelihood their horses will become affected.

Prolonged Stall Rest

Veterinary patients may need long periods of stall rest. When these horses generally feel fine but have an orthopedic or skin issue that requires the confinement, they can develop behavior problems. The horse may begin chewing wood if free-choice hay is not available, as an example.[41] Several recommendations can be made when a horse needs prolonged stall rest.[42] It should begin with a reduction in the amount of energy in the feed given. This is a big issue with racehorses, because trainers want the horse ready to go as soon as possible so it remains on full feed. The surplus energy makes the horse difficult to manage. Recommend instead that tiny amounts of grain or sweet feed be sprinkled on the hay as an occasional treat. It is recommended that a bin or feeder be under the hay to catch loose feed falling through. Another recommendation is to have hay available at all times or at least to feed the horse several times a day to simulate a more natural feeding pattern. This helps minimize the overenthusiastic greeting when food is given. A grazing muzzle can be used to increase the time spent eating if hay must be limited. Small piles of hay put around the edges of the stall promote the "eat a bite, take a step" natural behavior of grazing. Straw bedding can substitute for 24-h access to hay and is associated with a lower risk of stereotypy development.[43,44]

If the horse is mobile but movement needs to be restricted, there are additional suggestions.[42] While the home stall is being cleaned, the horse can be moved to another stall. Then it can remain in the new stall until that stall needs cleaning or remain only while the home stall is cleaned. Hand grazing for 10-15 min, repeated a couple of times each day, gives the horse a change of view. It is important not to forget the social needs of a stalled horse. Visual access to horses in neighboring stalls, a stall mirror, or nearby companion goat, donkey, or pony is particularly important for stallions, since they are often isolated anyway.

High-value horses are usually kept in very restrictive stabling and have a higher than normal frequency of locomotor stereotypies. Horses kept in stalls over 13 h a day are at a greater risk of developing handling issues than are horses stabled for only a few hours.[45]

Feeding Practices

Diets high in concentrates but low in roughage are associated with stereotypy development. Free access to a hay or a hay-only diet is preferable for stalled horses.[32] It also reduces food anticipation.[32] This recommendation is more aligned with the natural 24-h grazing pattern and with the horse's lack of a gallbladder to store bile between meals. Feeding twice a day is common practice, although more frequent meals are encouraged. Care is needed for such a recommendation, however. Four times a day feeding increases the risk for developing a stereotypy by a factor of 2.2.[32] This is blamed on anticipation of being fed. Beyond four daily feedings, the increased risk goes away, perhaps because it approaches constant eating.[32]

Type of Use

How horses are used contributes to the type of abnormal behavior they develop, although why the differences exist is unknown.[34,44,46] Excessive licking is more common in horses used for general riding and reining, while foals and rodeo horses are associated with weaving. Endurance horses show less weaving, cribbing, and wood chewing than do dressage or eventing horses.

The definitive study was done on French Saddlebred geldings that were all living in the same conditions, fed the same diet, and worked in a specific discipline for 1 h per day. Horses used for dressage and high-school performance were significantly more likely to show cribbing and head nodding. Those that were used in eventing or jumping were more likely to show licking and/or biting.[47]

Age of the Horse

Stereotypic behaviors like cribbing, stall walking, and weaving commonly begin within or shortly after the first year of life, sometimes when the foal is only a few weeks old.[34,38,48] Foals that develop problems prior to weaning are more likely to have had early nursing terminations and to have shown more pushing on the udder than normal foals. Postweaning oral behavior problems occur in foals that spend more time suckling and twice as much time teat nuzzling than do nonaffected foals.[48] An association with problems has also been shown in foals born to a dominant mare, weaned abruptly, stalled after weaning, stalled alone, or fed concentrates or other energy-dense hay replacements after weaning.[36,38,49]

Other Factors

The development of problem behaviors is influenced by the time of year, with autumn being most significant.[44] This may relate to the change in availability or nutritional content of grass. Daily access to a paddock or pasture reduces the risk of the development of behavior problems, as does the ability to see social peers and being turned out with a stable group of seven or more horses.[32,43–45] Handling problems are significantly reduced for horses that remain at the same location for at least 5 years.[45] Sex differences exist and are somewhat related to age. Mares show more problems overall; however, before 2 years of age, both geldings and stallions are more likely to have problems.[50]

Approach to Existing Behavior Problems

The role of the veterinarian is critical in dealing with behavior problems correctly. Reaching a diagnosis, both medical and behavioral, depends on an accurate history, physical examination, diagnostic workup, and differential diagnoses. When a client asks for help with a behavior problem, getting the appropriate history and understanding the details of how the problem is expressed in the specific horse are necessary. As with medical problems, the therapeutic protocol is dependent on an accurate diagnosis, and unlike what the owner has already tried, there is a rationale for a specific therapy.

Undesirable behavior can be normal behaviors that the client does not understand.[51] Problems can be attempts to avoid pain or a fear-inducing stimulus, they can be learned, or they may have a genetic or physiologic component. The goal of the history and diagnostic workup is to determine which. To be successful, treatment protocols require information about how easy the horse is to work with for the client, details about the horse's environment and management, and what the client has already tried. If medication is needed, it will be important to discuss how expensive it will be.

Working Toward a Diagnosis

Just as with medical cases, the establishment of a diagnosis for a behavior problem is dependent on a good history. The presenting complaint may be too general to appreciate the specifics, or it may include subjective interpretations. "He is frightened" is an interpretation. Instead, a description of what the ears, head, tail, and limbs were doing is needed. Accessibility of cell phone video makes understanding what is happening easier, but more information is needed.

Behavioral History

The history of a behavior problem is searching for an understanding of "what," "where," "when," and "how".[52,53] The "why" can be speculated, and usually is by the owner, but it is something we can never know for sure. Getting answers to the "what," "where," "when," and "how" questions will often lead to other related questions, so behavioral histories are not quite as simple as four answers.

Free-form history taking is one way to get the important information. Clients describe the problem, although they will require guidance by questions interspersed for clarification or more detail. Veterinarians may prefer to use a specific history form that owners can fill out before the appointment. Some of these forms are intended only to gather general information, ensuring some level of consistency between cases. Retrospective evaluations using that information are useful to identify correlations between potential causes, problems, and outcomes. Another type of form focuses on a specific problem, such as mares that reject their foals, aggression, or trailering problems.[54] A third option is a combination of both free-form and specific questions. This allows for the consistency of general information gathered between cases, and narrows the focus of the problem so it can be explored in greater depth.[54–56] Additional information will need to be obtained during the examination to clarify various points, regardless of the type of form used.

Any history begins with the signalment of the horse: age, sex, breed, and color. These can give clues about what might be contributing to the problem, but alone only serve to introduce the horse to the veterinarian.

Behavioral history-taking includes general questions, regardless of the type of problem.[54–56] These are intended to get information about the environment, such as housing, exercise, diet, early history, training, and other horses, animals and people that interact with the problem animal.

The chief complaint (the "what") can be straightforward or may need clarification. Complaints like rearing or walking in circles in the stall are fairly specific as to the action. The statement "he is frustrated when another horse gets ridden" tells nothing about what the horse is actually doing. This type of statement needs elaboration and refinement to understand the actual complaint. Once the chief complaint is determined, questions can be focused on providing a good mental picture of the specific behaviors. This is where video can be particularly useful.

The "where" question helps differentiate poor learning from environmental influences. Fence walking may only happen when the horse is in a paddock where it can watch one particular horse being ridden away from the barn. A horse that is difficult to load in a trailer may load fine at home, but is difficult when away from home. It may also load fine in the owner's trailer but not in strange trailers. This horse has not generalized the concept of loading to anywhere, any time, and any trailer. If the problem also occurs at home and with a familiar trailer, the horse was never taught to load. It may have been forced, but it was not taught. If the horse "bites" the manger, does it also "bite" the wooden fence? Such questions help determine the rigidity of the pattern and whether it might have progressed to an obsessive-compulsive disorder. The responses also give clues about approaches that might be used to eliminate or minimize the problem during treatment.

"When" determines the frequency and timing of the problem behavior. When did the problem start, and when does it appear? The aggressive episode may be related to the presentation of the feed at the stall. The left-to-right movements are almost continuous while the horse is in the stall but not in the paddock. Occasional expressions of a behavior are usually associated with a particular trigger, and accurately determining that trigger is important for treatment. Behaviors that have existed a long time or are almost constant suggest they will be difficult to treat.

The "how" questions refine descriptions associated with "what" and get environmentally relevant information. Cribbing horses normally grasp a post or board before tensing their neck muscles. Occasionally, a horse is presented that tightens their neck muscles and shows all other signs of cribbing but does not touch an object with their mouth—"air cribbing." A horse frequently opens its latched stall door or a pasture gate to escape. Knowing how it manipulates the latch is important to determining how to prevent it from happening.

Medical Workup

Physical examinations are part of a general health workup and should be done for behavioral workups too. Some medical problems mimic behavior problems and vice versa. In addition, medical issues can coexist with behavioral ones, complicating drug therapy.

Throughout a medical evaluation, it is important not to escalate anxiety or fear in the horse. The physical examination begins with a slow, calm, quiet approach and a "getting to know you" routine. Instead of beginning with temperature taking, the examination should start at the head and neck. This gives the horse more time to feel comfortable. For areas of resistance, the "advance and retreat" technique of approaching the area is helpful.

Drawing blood can be done even on fractious horses if the veterinarian or technician can get to the neck area using the calm approach. Slow movements are particularly important. The neck is stroked with a firm, smooth motion going in the direction of the hair. When the horse is comfortable with that, the jugular vein is held off and the stroking repeated until the horse is again comfortable. The needle is inserted as the person continues softly talking to the animal. At the completion of the blood draw, a repeat of stroking is done. For behavior problems, blood work is valuable to rule out medical causes and to assess the health status of the horse should medications be indicated.

Assessing Treatment Options

In this "a pill for everything" world, it is important to understand that medications are only one tool for dealing with abnormal behavior, and then only for a small percentage of problems. By the time an owner asks for help, they have already tried up to five different things.[57] Some attempts will have been appropriate for the problem, and others not useful. They may or may not have been applied correctly. There is no need to repeat a treatment that was applied correctly but did not work. If applied incorrectly, it might work if appropriate instructions can be given. In general terms, there are five treatment options used in various combinations for managing the problem horse.

Client Education

Preventing the behavior without working on the cause can be particularly harmful to the horse's welfare.[51] Most clients just want the problem to stop, and they do not have an appreciation for how harmful only stopping the behavior can be. This is where client education plays a big role. Not only must the client understand what the problem actually is, they must understand long-term implications for the horse's welfare, what changes will need to be implemented, and the correct way to implement the treatment protocol. There are cases where the behavior is normal but not appreciated as such. Then client education becomes a lesson about normal horse behavior and techniques that either allow the owner to adapt, find ways that minimize the behavior's occurrence, or change timing of the behavior's occurrence to when the owner is not around. As an example, a nervous horse that licks the owner's hand at a horse show may stop the licking if it is being walked or the owner backs away slightly.

Treat Medical Problems

It is important to differentiate medical problems from behavioral ones. Medical problems will need to be treated appropriately so that associated behaviors can be resolved. It is also important to determine if the physical problem or its treatment will interfere with treatments under consideration for behavior problems.

Environmental Modification

Preventing the expression of a problem behavior without addressing the environment is rarely successful. It can actually be harmful, particularly with stereotypies, because the underlying stress continues to build. When the behavior is allowed again, *rebound* occurs as an excessive expression of the problem. Environmental manipulation is an important tool for behavioral therapy that includes both management and enrichment.

Environmental management changes the horse's regular environment in such a way as to prevent the unwanted behavior from happening. In the more difficult cases, it is used to reduce the expression or frequency of the problem instead. Some horses paw while they eat. Management for this problem might include lowering of the feeder to ground level to make the pawing more difficult. For the horse that rubs its teeth on the metal door, management would involve moving the horse to a paddock or coating the metal with thick rubber to change the texture and sound. In another example, the horse that wiggles the stall door latch to escape can be thwarted if the latch is secured with a bull snap or lowered so the horse cannot reach it.

Enrichment gives the horse something else to do. Just adding something different to a stall is not enrichment unless the horse actually interacts with the object. As an example, ropes or toys suspended at the horse's eye level have no effect on general activities or abnormal oral behaviors.[58] While the horse might interact with each object to some degree, the lack of a significant effect on the problem behavior is likely because the object neither addresses the underlying cause nor is highly entertaining to the animal. Access to paddocks or pasture and to social peers are significant forms of enrichment associated with reduction or prevention of behavior problems. Even the use of an aluminum mirror can be helpful if social contact is minimal. Trail riding introduces a variety of new objects for sensory stimulation. Treat dispensers that make the horse work for its reward enrich the stalled horse's environment. Such devices usually work by having the horse swing or roll a movable part and are either available commercially or are homemade. Apples floating in a tub of water will have the same effect. Horses in a paddock have other options including large toys made specifically for them, like big balls and balls with handles. Homemade options would include things like old tractor tires or the 5-gal heavy plastic water bottles. Because novelty is important, rotation of available objects is important.

Behavior Modification

Behavior modification uses the principles of learning to change behavior. Horses constantly learn, not just when a person is trying to teach them something. If they try a behavior and get some type of reward, even an internal reward of accomplishment, the frequency of expressing that behavior increases. For the horse that grows impatient for its food to arrive and begins pawing, it is rewarded when the food appears. The owner is not considering the consequences and continues to reward a behavior they do not want. The horse learns that pawing brings food. Elimination or reduction of the behavior is dependent on learning too. The reward (food) goes away if pawing begins and is given only for acceptable standing behavior. Techniques that apply the concepts of learning are particularly important for understanding the development of and treatment techniques for behavior problems.

Drug Therapy

In today's busy world, people want rapid solutions to problems. For medical issues they want a quick fix in pill or shot form. Occasionally a behavior problem can be reduced or eliminated with medication, like hypothyroid aggression in dogs, but this is the exception. Most behaviorists feel that drug therapy is indicated in about 20% of behavior problems, and its use is to increase the success of other therapies. Alone, drug therapy is seldom successful.[59] Practitioners are used to the medical-treatment paradigm, not behavior modification, so prescriptions feel comfortable. It is tempting to prescribe a drug for a problem without regard to the specific diagnosis, and sometimes on the basis of a layperson's recommendation. These are temptations veterinarians need to resist. Drug therapy for equine behavior problems involves human medications used in an extralabel manner. This increases the importance of anticipating possible contraindications or side effects. Additionally, even generic brands can be particularly pricey for a horse.

Drug therapy for horse behavior problems is not well researched yet. In part, this is because of cost, and in part it is due to the relative lack of veterinary involvement in equine behavioral medicine. Gradually, knowledge will improve and better information will become available.

EQUINE PSYCHOPHARMACOLOGY

Behavior is a function of the brain, and medications used in behavioral medicine act on the primary neurotransmitters. Because brain nuclei in different species can differ in their associated neurotransmitters, understanding what drugs might work across the spectrum of animals treated in veterinary medicine is complicated. There are, however, general functions associated with each of the seven major neurotransmitters.

Neurotransmitters

Throughout the body, chemicals that act as neurotransmitters affect the flow of messages between neurons. Seven of these have been identified as the major neurotransmitters, and they are associated with behavioral functions. While each is associated with specific functions, they also interact in overlapping fashion, complicating the effects of drug therapies (Figure 9-1). This interaction is relative, so if one increases, there is a corresponding reduction in the effect of the other. Another complicating factor is that a drug affecting a neurotransmitter in the brain is also affecting that same neurotransmitter systemically, which can result in significant, unwanted side effects.

Of the seven major neurotransmitters, 5-hydroxytryptamine (*serotonin*) is the one most associated with behavior changes. In general, this monoamine regulates mood, cognitive function, memory, and appetite. It is also associated with behavior problems, including some types of aggression, anxiety, and obsessive-compulsive disorders. Serotonin is a derivative of tryptophan, which is poorly competitive against other amino acids in the intestine.

Dopamine is involved in motivation and drive. Movement is its best-known function because of its relationship to Parkinson's disease in humans. Dopamine also modulates mood, attention, pleasure, and reward. In combination with serotonin, dopamine is involved in aggression. Tyrosine is the parent compound.

A third neurotransmitter that is considered in behavioral drug therapy is *norepinephrine*. It is associated with the energetic components of behavior such as

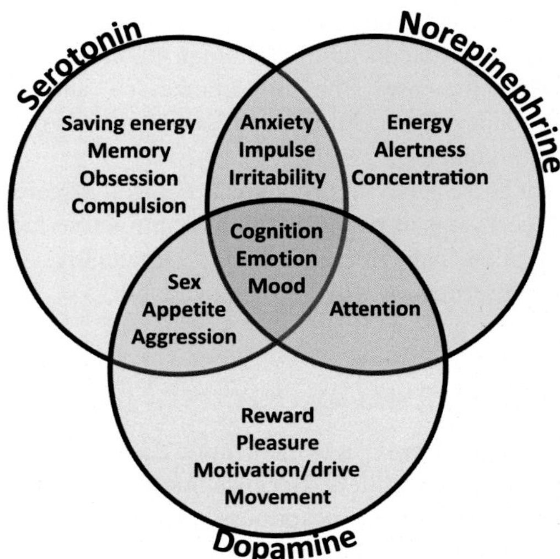

FIGURE 9-1 An example of the overlapping functions of neurotransmitters shows how drug therapy affecting a single neurotransmitter can have multiple effects.

alertness, attention, arousal, and concentration. Behavior problems involving stress responses and anxiety are facilitated. Norepinephrine is derived from tyrosine by way of dopamine.

Gamma-aminobutyric acid (*GABA*) functions to moderate excitability of neurons throughout the nervous system, especially relative to anxiety and movement. It is ironic that GABA, a major inhibitory neurotransmitter, is synthesized from glutamate, the major excitatory neurotransmitter.[60]

The fifth member of this group is *glutamate*, a primary excitatory neurotransmitter. It is involved with the *N*-methyl-D-aspartate (NMDA) receptors that recently have been receiving attention for drug therapy. Glutamate is a derivative of glutamic acid and is the parent compound for GABA.

Acetylcholine is a cholinergic compound whose neurotransmitter functions are related to the autonomic nervous system. It is a derivative of acetic acid and choline, and relates to functions like attention, arousal, memory, and motivation.

Lastly, *histamine*, a derivative of histidine, acts on the H_1, H_2, and H_3 receptors in the brain. Its function is associated with wakefulness, appetite, pain recognition, and cognition. It also decreases release of acetylcholine, norepinephrine, and serotonin neurotransmitters, blunting their effect. Numerous peripheral histamine receptors have other functions.

Neurotransmitters can be affected in several ways at either presynaptic or postsynaptic sites (Table 9-2). From synthesis to release and reception, interfering with the normal progression can affect behavior. The following nine steps are subject to interference in neurotransmitter actions, making them targets for drug therapy too. Precursors must be transported into the neuron, synthesized into the neurotransmitter, stored in specific vesicles, released into the interneuron space, taken back into the original neuron, metabolized to be recycled again, and work competitively or synergistically with other intercellular neurotransmitters in the transmitting neuron.[61] Postsynaptic neurons respond to appropriate neurotransmitters to send the message forward. The sensitivity of the response is regulated by intraneuronal chemical components.[61] In addition, there are various receptor subtypes for each neurotransmitter (serotonin has at least 15), which have different binding profiles.[61–63]

Psychopharmacologic Medications

The ideal drug would effectively eliminate the behavior problem, work on all patients, have no side effects, have a rapid onset of action, have a wide margin of safety, not impair normal motor or mental functions, have an intermediate half-life, and have a defined therapeutic blood level.[64,65] Unfortunately, the ideal drug does not exist. Worse yet, much of the information about how many of the drugs work is not available for horses. Complicating drug therapy even more is the placebo effect. Behavioral studies showed a placebo effect between 30% for "great improvement" and as high as 70% for "some improvement."[66,67]

TABLE 9-2 Ways Pre- and Postsynaptic Neurotransmitter Activities Might Be Altered

Site	Action	Possible Alterations
Presynaptic		
	Precursor chemical transported to the neuron	Insufficient chemical; altered blood-brain barrier
	Precursor chemical synthesized into the neurotransmitter	Genetic diseases
	Neurotransmitter stored in specific vesicles	Leakage of the vesicles causes insufficient neurotransmitter
	Vesicles release the neurotransmitter into the interneuron space	Drugs can increase the amount of neurotransmitter released into the space
	Neurotransmitter taken back into the original neuron	Drug therapy often targets this to inhibit reuptake keeping it triggering receptors
	Catabolic enzymes metabolize the neurotransmitter	Enzyme deficiencies; drugs may block metabolism
	Interneuron relationships between neurotransmitters can affect their actions	Neurons can have multiple neurotransmitters
Postsynaptic		
	Activation of the neuron response	Actions depend on which neurotransmitter subtype receptors are activated; drugs can increase/decrease activation potential
	Sensitivity of the response	Regulation by various other intraneuronal chemical components; selective blocking of subtypes

In addition, blood levels of a drug do not necessarily correlate with levels in the central nervous system.

Medications are classified within broad categories based on chemical type of the active ingredient or by how it is used. The following discussion is one of many categorization systems. Specific medications, their doses, and general uses are listed in Table 9-3. As a point of reference, most drugs currently used in behavioral therapy undergo hepatic metabolism and ultimately can affect other medications that interact with cytochrome P450 enzymes.[68,69] For this reason, the horse's blood chemistry should be evaluated to minimize the

TABLE 9-3 Medications Currently Used for Equine Behavior Problems[60, 102, 104, 105]

Classification	Drug	Dose	Uses
Sedative-hypnotic (benzodiazepine)	Diazepam	10–30 mg/horse PO or IV q8h; 0.02 mg/kg IV with food	Short-term anxiolytic; appetite stimulant; rejection of foal due to fear; dissociation of event from outcome
Sedative-hypnotic (benzodiazepine)	Alprazolam	0.035 mg/kg PO q8h	Short-term anxiolytic; rejection of foal due to fear; dissociation of event from outcome
Tricyclic antidepressant	Clomipramine	500 mg PO q24h, gradually increase to 900 mg PO q24h	Long-term anxiolytic; antidepressive; stereotypies; obsessive-compulsive therapy
Tricyclic antidepressant	Imipramine	2 mg/kg IV q24h; 250–750 mg PO q24h divided	Induce ejaculation; narcolepsy
Selected serotonin reuptake inhibitor	Fluoxetine	0.5–1.0 g/horse PO q24h; 0.25–5.0 mg/kg PO q24h	Long-term anxiolytic; antidepressive; stereotypies; obsessive-compulsive therapy
Selected serotonin reuptake inhibitor	Paroxetine	0.25–5.0 mg/kg PO q24h	Long-term anxiolytic; antidepressive; stereotypies; obsessive-compulsive therapy
Opiate antagonists	Naloxone	0.01–0.04 mg/kg IV	Diagnosis of stereotypies not responding to traditional therapies
Antihistamine	Cyproheptadine	0.3–0.6 mg/kg PO q12h	Headshaking; reduce sexual

Continued

TABLE 9-3 Medications Currently Used for Equine Behavior Problems[60, 102, 104, 105]—cont'd

Classification	Drug	Dose	Uses
			behavior in geldings
NMDA receptor antagonist	Dextromethorphan	1.0 mg/kg IV or PO	Cribbing and other stereotypies
Sodium channel blocker	Carbamazepine	0.3–0.4 mg/kg PO	Headshaking

potential for side effects due to unseen liver problems. Most drugs are also used as extralabel therapy, so owner consent is appropriate.

A number of over-the-counter products cycle through the horse world, primarily claiming to calm excited horses. These include several different herbal preparations and high-potency vitamins or minerals. Rarely is there science to support the claims or to point out potential side effects. The few that have been studied, at least superficially, are mentioned.

Antipsychotic Drugs

As a group, antipsychotic drugs block dopamine receptors, particularly subtype D_2.[50,69] Side effects are common and include seizures, sedation, and motor restlessness.[70] They are used in human obsessive-compulsive patients who are not responding to conventional therapies.

Phenothiazines

In veterinary medicine, the phenothiazine derivatives have antipsychotic and hypnotic effects. The group is not particularly effective in behavioral therapy except as a hypnotic, which is dose dependent. Phenothiazines work by reducing the initiation of motor activity in the brain's basal ganglia.[71]

Acepromazine is the most common type of phenothiazine used. Recently, fluphenazine has become popular in performance horses, making difficult animals easier to handle. Its effects last several days. Fluphenazine is illegal to use in most sanctioned events, but rapid clearance made it difficult to detect until recently.

This class of drugs, and particularly fluphenazine, has some serious side effects. They include increased noise sensitivity, idiosyncratic excitement, sweating, tachypnea, tachycardia, hypotension, extrapyramidal effects, and rarely seizures.[72–74]

Antianxiety (Anxiolytic) Drugs

Anxiety is difficult to quantify in animals because it is internally generated. Fear is related to a specific external event. In the extreme, panic attacks have a rapid onset and rapid crescendo of symptoms from sympathetic overactivity.[75] All levels of anxiety can last from 30 min to several hours. As a result, the duration of action for medications should be considered. Anxiolytic drugs can be divided into sedative-hypnotic and sedative-autonomic groups based on their general actions, but only the first is used for behavior problems in horses.

Sedative-Hypnotic Drugs

The sedative-hypnotic group of anxiolytic drugs produces effects progressing from mild sedation to sleep or hypnosis as the dose increases. They are also associated with muscle relaxation, anticonvulsant action, and the development of tolerance and dependence. Common examples of sedative-hypnotics used in veterinary medicine include the barbiturates and benzodiazepines, with the benzodiazepines being used in behavioral medicine.

The benzodiazepines, particularly alprazolam and diazepam, have been popular for several years as anticonvulsant and preanesthetic agents, but the development of more specific treatments and the abuse potential for humans have reduced their use. The mechanism of action is to augment GABA as an inhibitory neurotransmitter.[70,76] Low doses alleviate anxiety, agitation, and fear by action on receptors in the limbic system. Moderate doses are more anxiolytic and facilitate social interactions, and high doses are associated with confusion.[60,77] Diazepam is useful to increase food intake in adult and young horses.[78] Alprazolam can be useful to facilitate mare-foal bonding for mares that are aggressive toward their foal as it attempts to nurse.[38] It is also useful if given immediately after fearful events such as a barn fire to prevent memory consolidation.

Side effects of benzodiazepines include sedation, cortical depression, ataxia, increased appetite, and paradoxical excitement.[69,77] As fear is removed, aggression may be disinhibited. They may also interfere with antegrade learning and memory.[69]

Antidepressant Drugs

Three subgroups of antidepressant drugs are currently used in horses. While called "antidepressants," most are used for their anxiolytic properties. As a class, these drugs primarily affect neurotransmitters related to glutamate.[79]

Tricyclic Antidepressants

The tricyclic antidepressant drugs (TCAs) are structurally related to phenothiazine and are classified as mixed serotonin and norepinephrine reuptake inhibitors.[69,70,80] TCAs attach to and inhibit presynaptic transporter proteins, thus preventing most neurotransmitter reuptake back into the presynaptic neuron.[81–83] Because serotonin and norepinephrine build up gradually within the interneuron space, it will take 3–4 weeks before there are measurable clinical results. Owners should be committed to a minimum of 6 weeks before deciding if the therapy is useful.[75] Longer term, the increased presence of the neurotransmitter in the interneuron space, particularly serotonin, results in remodeling of the receptors, enhancing transmission efficiency, and ultimately stimulating elements essential to cellular learning and memory.[81] When ending TCA therapy, gradual withdrawal is appropriate.[60,84] Owners might not notice the gradual behavior change that originally occurs until the drug concentration is reduced. If the problem worsens as the drug is reduced, increasing the dose again is easier and more rapidly effective. Because TCAs also block histamine receptors, they have some antipruritic properties.[85] When owners report calming within the first week or so, it is most likely the antihistamine effect.

In equine medicine, clomipramine is the TCA used most for the long-term treatment of generalized anxiety, depression, stereotypies, obsessive-compulsive disorders, and neuropathic pain. Imipramine is used to induce ejaculation and for the treatment of narcolepsy. Amitriptyline is rapidly metabolized, and so is used less often.[60]

Side effects of tricyclics in horses are not well studied but are expected to relate to their anticholinergic properties and blockage of acetylcholine, dopamine, and norepinephrine.[86] Undesirable problems could include mild sedation, especially during the first week, mydriasis, arrhythmia, tachycardia, dry mouth, constipation with the potential for colic, hypotension, urinary retention, muscle fasciculations, hyperresponsiveness to sound, and hemolysis.[70,74,75,82,83,86–89] Nausea, sexual dysfunction, seizures, potentiation of concurrent thyroid conditions, and agranulocytosis have also been reported in humans and other species.[69,70,85,90] Oral forms of TCAs have a bitter taste and will need to be masked in capsules or highly palatable food.

Selected Serotonin Reuptake Inhibitors

Selective serotonin reuptake inhibitors (SSRIs) are often compared to the tricyclics, and in veterinary medicine, both have the same indications for use—long-term relief of anxiety, depression, stereotypies, and obsessive-compulsive disorders. The difference is in their affinity for neurotransmitters, with the SSRIs being quite specific for serotonin. SSRIs work at the presynaptic sites to inhibit reuptake of serotonin back into the presynaptic neuron. As with TCAs, it takes 3–4 weeks to achieve significant levels of serotonin in the interneuron space to produce a behavior change. Similarly, SSRIs will eventually remodel the post-synaptic neuron, and if stopped, it should be done gradually.[81] While there are a number of compounds within the SSRI group, fluoxetine and paroxetine are the two most commonly used in horses.

Side effects of' SSRIs in humans relate to their small norepinephrine component and include nausea, muscle rigidity, anxiety, sexual dysfunction, insomnia, anorexia, diarrhea, nervousness, and headaches.[83,86,89,91] Because there are a large number of serotonin receptors associated with the gastrointestinal tract, the concern for constipation, with the potential for colic, must be considered in horses.[92] Side effects are usually avoided by starting with a half dose the first week or two before using the recommended dose. Alternatively, dosing can be started low for the first month and gradually increased as needed. Interactions with other drugs that work on serotonin neurotransmitters or inhibit CYP2D6 of the P450 enzyme group can lead to the "serotonin syndrome" from a excessive buildup of this neurotransmitter. The syndrome is characterized by agitation or depression, hyperthermia, and altered neuromuscular activity including myoclonus, hyperreflexia, tremors, ataxia, and seizures.[60]

Sympathomimetic Stimulants

Sympathomimetic stimulants are classified as antidepressant drugs even though they are not considered to be particularly good ones. This group consists of amphetamine-related chemicals that work in part by being taken up into the vesicles of the presynaptic nerve terminals replacing the stored norepinephrine.[93] The psychostimulant effects of this class of drugs are mediated at the dopamine synapse.[94,95]

This class of drugs is used to treat attention deficit hyperactivity disorder (ADHD) in people. In veterinary behavioral therapy, sympathomimetic stimulants are used for the treatment of hyperkinesis (ADHD) in dogs.[71,96] Methylphenidate is the drug of first choice for this condition in both children and animals. While methylphenidate is rarely, if ever, prescribed by equine practitioners, it is used illegally in racehorses "to get their mind right."

Beta-Blocking Drugs

Beta-adrenergic receptor blocking drugs first came into vogue as antipsychotic medications for schizophrenia. In humans, beta-blockers such as propranolol have been used to treat situational anxieties because they tend to reduce somatic manifestations like tremors and sweaty palms.[69,70,74] This led to their use in dogs for situational fears. If used therapeutically, desensitization improves treatment success. While beta-blockers have been used in anxious horses, they have been largely replaced by newer and better anxiolytics.

The mechanisms relative to behavior use depend primarily on actions as an antagonist and partial agonist at the serotonin$_{1A}$ receptors.[69] Also, the potential side effects such as excessive sedation, hypotension, bradycardia, and heart failure can be serious.[70]

Opiate Antagonists

In veterinary medicine, opiate antagonists are used primarily to reverse the effects of opiate sedatives. High doses of amphetamine-like drugs can produce stereotypical behavior.[97] Based on this, it has been proposed that stereotypies are self-rewarding behaviors, and their action may be associated with the release of naturally occurring opioids.[98] Opiate antagonists should then block the internal reward, so they are used diagnostically on stereotypic behaviors, particularly for those problems not responding well to SSRIs or TCAs. These drugs have been studied the most in cribbers, where they work well.[99] Unfortunately, as a treatment, opiate antagonists are poor choices because their duration of action is very short.

Antihistamines

Antihistamines are sometimes classified as hypnotic drugs, but their sedative effect on the central nervous system is useful for mild apprehension. Antihistamines are primarily used to counteract histamine release in anaphylactic reactions; however, cyproheptadine has a unique use. Like most antihistamines, cyproheptadine is an H_1-receptor antagonist, which affects cortical activation and arousal mechanisms.[100] It also is a serotonin antagonist.[101] It is probably the combination of actions that makes this a useful therapy in headshaking.[102]

N-*Methyl-*D-*Aspartate Receptor Antagonists*

NMDA receptor antagonists alter glutamatergic neurotransmission, which is thought to be associated with stereotypic and obsessive-compulsive disorders.[103] Because glutamate is the excitatory neurotransmitter, binding some of the receptor sites from activation might reduce a stereotypic expression. This class of drugs has only recently started to be used in behavioral medicine so its effectiveness and side effects are still to be determined for horses. There is evidence that NMDA drugs complement fluoxetine to improve therapeutic results.[103]

Dextromethorphan

In connection with the onset of stereotypies, dextromethorphan, an NMDA receptor antagonist, has been studied on cribbing horses and found to sometimes be helpful.[104] The likelihood of large-scale use, however, is limited because the drug is short acting, expensive, and limited in availability because of its recreational use in humans.

Other Drugs

Several other individual drugs have been used for behavior problems in horses. Some are relatively effective, and some are not.

Carbamazepine

Carbamazepine is a sodium channel blocker that might also work as a serotonin releasing agent and reuptake inhibitor.[106–109] While used primarily in humans to treat epilepsy, this drug has been used alone and in combination with cyproheptadine for the treatment of headshaking in horses.[110]

Tryptophan

Tryptophan is the amino acid precursor of serotonin, and it has become a popular oral supplement for horses. The rationale for using tryptophan is to increase serotonin levels within the central nervous system for an additional calming effect. Because this is the least competitive amino acid for absorption, increased serotonin levels should only be expected if the normal diet is deficient.[111,112] Studies are suggesting that additional tryptophan in the diet does not affect behavior, even though plasma tryptophan levels are elevated.[113] High doses cause hemolytic anemia and bronchiolar degeneration.[114]

Caseinate

Alpha-casozepine (caseinate) is a biopeptide from milk that has anxiolytic properties.[115] It has recently become popular in horses to relieve mild anxiety. Tests in ponies suggest that the product is helpful for normal situations that may be stressful.[116] Caseinate products are available over-the-counter in powder form.

Melatonin

Melatonin is a neurohormone associated with sleep. It is produced by the pineal gland from the precursor tryptophan. Synthetic forms of melatonin are used for generalized anxiety and for correcting a day/night shift in wakefulness. In humans, melatonin levels in the blood are highest prior to bedtime, so if the drug is used, dosing is recommended in the evening.[129] In veterinary medicine, behavioral effects and an appropriate dose range have not been established under controlled conditions.[69]

Magnesium Aspartate

Magnesium is another over-the-counter product that has become popular for calming flighty horses. Doses well above values in normal daily rations will slow reaction time slightly. Levels slightly above dietary amounts show questionable changes.[117]

Pheromones

Synthetic pheromones are being used as calming agents in a variety of animal species. Commercial versions of "appeasing pheromones" are said to mimic the natural pheromone associated with the mammary gland region of the lactating female. The equine appeasing pheromone (Confidence EQ®) was originally studied as an aid for loading horses into trailers and more recently has been studied in fear-eliciting situations.[118,119] In the original trailer study, the only difference was that treated horses had a lower heart rate. The pheromone was also considered effective in a second trailer loading study. It should be noted that the research for both was conducted by the developer. A third study did not show efficacy.[120] Keeping in mind the high placebo effect in animals, experience suggests the product might be useful in mildly fearful situations, but that it is not useful in extreme situations.[121–125] The lack of extensive research trials for this product makes it difficult to know if or how well it works. No side effects have been reported, suggesting a high degree of safety.

Anabolic Steroids

A number of anabolic steroids have been used in horses, mainly in an attempt to increase muscle mass of young Quarter Horses being shown at halter. The use of these drugs is associated with undesirable side effects, particularly increased aggression and stallion-like behavior, even in mares and geldings.[126–128] Expression of these behaviors does vary slightly from normal, in that affected horses may become "fixated" on a target for aggression, or show prolonged sniffing of feces or repeated covering of other feces.

REFERENCES

1. Goodwin D. The importance of ethology in understanding the behaviour of the horse. *Equine Vet J Suppl* 1999;**28**:15–9.
2. Miller RM. The ten behavioral characteristics unique to the horse. *J Equine Vet Sci* 1995;**15** (1):13–4.
3. Hockenhull J, Creighton E. A brief note on the information-seeking behavior of UK leisure horse owners. *J Vet Behav* 2013;**8**(2):106–10.
4. Lofgren E, Voigt MA, Brady CM. Information-seeking behavior of the horse competition industry: a demographic study. *J Equine Vet Sci* 2016;**37**:58–62.
5. United States Department of Agriculture (2016): Equine 2015: baseline reference of equine health and management in the United States, 2015. https://www.aphis.usda.gov/animal_health/nahms/equine/downloads/equine15/Eq2015_Rept1.pdf, p. 180 [downloaded September 18, 2017].
6. BEVA. Survey reveals high risk of injury to equine vets. *Vet Rec* 2014;**175**:263.
7. Doherty O, McGreevy PD, Pearson G. The importance of learning theory and equitation science to the veterinarian. *Appl Anim Behav Sci* 2017;**190**:111–22.
8. Fritschi L, Day L, Shirangi A, Robertson I, Lucas M, Vizard A. Injury in Australian veterinarians. *Occup Med* 2006;**56**(3):199–203.
9. Grice AL. Getting hurt on the job. *Equi Manage* 2018;22–4. 26–28, 30.
10. Landercasper J, Cogbill TH, Strutt PJ, Landercasper BO. Trauma and the veterinarian. *J Trauma* 1988;**28**(8):1255–9.
11. Loomans JBA, van Weeren-Bitterling MS, van Weeren PR, Barneveld A. Occupational disability and job satisfaction in the equine veterinary profession: how sustainable is this 'tough job' in a changing world? *Equine Vet Educ* 2008;**20**(11):597–607.
12. Lucas M, Day L, Shirangi A, Fritschi L. Significant injuries in Australian veterinarians and use of safety precautions. *Occup Med* 2009;**59**(5):327–33.

13. Parkin TDH, Brown J, Macdonald EB. Occupational risks of working with horses: a questionnaire survey of equine veterinary surgeons. *Equine Vet Educ* 2018;**30**(4):200–5.
14. Reijula K, Räsänen K, Hämäläinen M, Juntunen K, Lindbohm M-L, Taskinen H, Bergbom B, Rinta-Jouppi M. Work environment and occupational health of Finnish veterinarians. *Am J Ind Med* 2003;**44**(1):46–57.
15. Beckstett A. The science behind horsemanship. *The Horse* 2016;**XXXIII**(1):16–25.
16. Grandin T. Horse behavior principles for safer handling. *Equine Vet* 2014;**4**:14–5.
17. Caudle AB, Pugh DG. Restraint. *Vet Clin N Am Equine Pract* 1986;**2**(3):645–51.
18. Beaver BV, Höglund DL. *Efficient livestock handling: the practical application of animal welfare and behavioral science*. New York: Academic Press; 2016. p. 213.
19. LeDoux JE. Emotion, memory and the brain. *Sci Am Spec Ed* 2002;**12**(1):62–71.
20. LeDoux JE. Emotion, memory and the brain. *Sci Am* 1994;**270**(6):50–7.
21. Jezierski T, Jaworski Z, Górecka A. Effects of handling on behavior and heart rate in Konik horses: comparison of stable and forest reared youngstock. *Appl Anim Behav Sci* 1999;**62**(1):1–11.
22. Peeters M, Verwilghen D, Serteyn D, Vandenheede M. Relationships between young stallions' temperament and their behavioral reactions during standardized veterinary examinations. *J Vet Behav* 2012;**7**(95):311–21.
23. Søndegaard E, Halekoh U. Young horses' reactions to humans in relation to handling and social environment. *Appl Anim Behav Sci* 2003;**84**(4):265–80.
24. Pearson G. Practical application of equine learning theory, part 2. *Clin Pract* 2015;**37**(6):251–4.
25. Hausberger M, Fureix C, Lesimple C. Detecting horses' sickness: in search of visible signs. *Appl Anim Behav Sci* 2016;**175**:41–9.
26. Gosling SD, John OP. Personality dimensions in nonhuman animals: a cross-species review. *Curr Dir Psychol Sci* 1999;**8**(3):69–75.
27. Saslow CA. Understanding the perceptual world of horses. *Appl Anim Behav Sci* 2002;**78**(2–4):209–24.
28. Pearson G. Practical application of equine learning theory, part 1. *Clin Pract* 2015;**37**(5):251–4.
29. Morris PH, Gale A, Howe S. The factor structure of horse personality. *Anthrozoös* 2002;**15**(4):300–22.
30. Rochais C, Henry S, Sankey C, Nassur F, Góracka-Bruzda A, Hausberger M. Visual attention, an indicator of human-animal relationships: a study of domestic horses (*Equus caballus*). *Front Psychol* 2014;**5**:1–10 [Article 108].
31. Freckleton, M. (2016): Help for a needle-shy horse. Equus 463: 88–89, 91
32. Bachmann I, Audigé L, Stauffacher M. Risk factors associated with behavioural disorders of crib-biting, weaving and box-walking in Swiss horses. *Equine Vet J* 2003;**35**(2):158–63.
33. Hockenhull J, Creighton E. The day-to-day management of UK leisure horses and the prevalence of owner-reported stable-related and handling behavior problems. *Anim Welf* 2015;**24**(1):29–36.
34. Leme DP, Parsekian ABH, Kanaan V, Hötzel MJ. Management, health, and abnormal behaviors of horses: a survey in small equestrian centers in Brazil. *J Vet Behav* 2014;**9**(3):114–8.
35. Muñoz L, Torres J, Sepúlveda O, Rehhof C, Ortiz R. Frequency of stereotyped abnormal behavior in stabled Chilean horses. *Arch Med Vet* 2009;**41**(1):73–6.
36. Parker M, Goodwin D, Redhead ES. Survey of breeders' management of horses in Europe, North America and Australia: comparison of factors associated with the development of abnormal behavior. *Appl Anim Behav Sci* 2008;**114**(1–2):206–15.

37. Tadich T, Weber C, Nicol CJ. Prevalence and factors associated with abnormal behaviors in Chilean racehorses: a direct observational study. *J Equine Vet Sci* 2013;**33**(2):95–100.
38. Waters AJ, Nicol CJ, French NP. Factors influencing the development of stereotypic and redirected behaviours in young horses: findings of a four year prospective epidemiological study. *Equine Vet J* 2002;**34**(6):572–9.
39. McGreevy P, Nicol C. Physiological and behavioral consequences associated with short-term prevention of crib-biting in horses. *Physiol Behav* 1998;**65**(1):15–23.
40. Sarrafchi A, Blokhuis HJ. Equine stereotypic behaviors: causation, occurrence, and prevention. *J Vet Behav* 2013;**8**(5):386–94.
41. Normando S, Meers L, Samuels WE, Faustini M, dberg FO. Variables affecting the prevalence of behavioural problems in horses. Can riding style and other management factors be significant? *Appl Anim Behav Sci* 2011;**133**(3–4):186–98.
42. McDonnell SM. Stall rest suggestions. *The Horse* 2012;**XXIX**(9):52–3.
43. Christie JL, Hewson CJ, Riley CB, McNiven MA, Dohoo IR, Bate LA. Management factors affecting stereotypies and body condition score in nonracing horses in Prince Edward Island. *Can Vet J* 2006;**47**(92):136–43.
44. McGreevy PD, Cripps PJ, French NP, Green LE, Nicol CJ. Management factors associated with stereotypic and redirected behavior in the Thoroughbred horse. *Equine Vet J* 1995;**27**(2):86–91.
45. Hockenhull J, Creighton E. Management practices associated with owner-reported stable-related and handling behavior problems in UK leisure horses. *Appl Anim Behav Sci* 2014;**155**:49–55.
46. McGreevy PD, French NP, Nicol CJ. The prevalence of abnormal behaviours in dressage, eventing and endurance horses in relation to stabling. *Vet Rec* 1995;**137**(2):36–7.
47. Hausberger M, Gautier E, Biquand V, Lunel C, Jégo P. Could work be a source of behavioural disorders? A study in horses. *PLoS One* 2009;**4**(10). https://doi.org/10.1371/journal.pone.0007625.
48. Nicol CJ, Badnell-Waters AJ. Suckling behavior in domestic foals and the development of abnormal oral behaviour. *Anim Behav* 2005;**70**(1):21–9.
49. Heleski CR, Shelle AC, Nielsen BD, Zanella AJ. Influence of housing on weanling horse behavior and subsequent welfare. *Appl Anim Behav Sci* 2002;**78**(2–3):291–302.
50. Mills DS, Alston RD, Rogers V, Longford NT. Factors associated with the prevalence of stereotypic behavior amongst Thoroughbred horses passing through auctioneer sales. *Appl Anim Behav Sci* 2002;**78**(2–4):115–24.
51. Hothersall B, Casey R. Undesired behavior in horses: a review of their development, prevention, management and association with welfare. *Equine Vet Educ* 2012;**24**(9):479–85.
52. Beaver BV. *Canine behavior: insights and answers*. 2nd ed. St. Louis: Saunders; 2009. p. 315.
53. Beaver BV. *Feline behavior: a guide for veterinarians*. 2nd ed. St. Louis: Saunders; 2003. p. 349.
54. Crowell-Davis SL, Houpt KA. Techniques for taking a behavioral history. *Vet Clin N Am Equine Pract* 1986;**2**(3):507–18.
55. Crowell-Davis SL. The behavioral history: Part 2. *Equine Pract* 1987;**9**(5):31–3.
56. Crowell-Davis SL. The behavioral history: Part 1. *Equine Pract* 1987;**9**(4):37–8.
57. Hockenhull J, Creighton E. The use of seven methods of preventing stable-based behavior problems in UK leisure horses and their relative effectiveness. *J Vet Behav* 2011;**6**(6):292.
58. Bulens A, Dams A, Van Beirendonck S, Van Thielen J, Driessen B. A preliminary study on the long-term interest of horses in ropes and jolly balls. *J Vet Behav* 2015;**10**(1):83–6.

59. Hart BL, Cooper LL. Integrating use of psychotropic drugs with environmental management and behavioral modification for treatment of problem behavior in animals. *J Am Vet Med Assoc* 1996;**209**(9):1549–51.
60. Crowell-Davis SL, Murray T. *Veterinary psychopharmacology*. Ames, IA: Blackwell Publishing; 2006. p. 270.
61. Nutt JG, Irwin RP. Principles of neuropharmacology. II. Synaptic transmission. In: Klawans HL, Goetz CG, Tanner CM, editors. *Textbook of clinical neuropharmacology and therapeutics*. 2nd ed. New York: Raven Press, Ltd.; 1992. p. 15–22
62. Hoyer D, Clarke DE, Fozard JR, Hartig PR, Martin GR, Mylecharane EJ, Saxena PR, Humphrey PPA. VII. International Union of Pharmacology classification of receptors for 5-hydroxytryptamine (serotonin). *Pharmacol Rev* 1994;**46**(2):157–203.
63. Levy AD, Van de Kar LD. Endocrine and receptor pharmacology of serotonergic anxiolytics, antipsychotics and antidepressants. *Life Sci* 1992;**51**(2):83–94.
64. Hollister LE. *Clinical pharmacology of psychotherapeutic drugs*. 2nd ed. New York: Churchill Livingstone; 1983. p. 258.
65. Richelson E. Pharmacology of antidepressants—characteristics of the ideal drug. *Mayo Clin Proc* 1994;**69**(11):1069–81.
66. Taylor KD, Mills DS. Headshaking syndrome. In: Robinson NE, Sprayberry KA, editors. *Current therapy in equine medicine*. 6th ed. St. Louis: Saunders; 2009. p. 103–7.
67. White MM, Neilson JC, Hart BL, Cliff KD. Effects of clomipramine hydrochloride on dominance-related aggression in dogs. *J Am Vet Med Assoc* 1999;**215**(9):128–1291.
68. Brösen K. Differences between the SSRI. *Eur Neuropsychopharmacol* 1995;**5**(3):174.
69. Simpson BS, Papich MG. Pharmacologic management in veterinary behavioral medicine. *Vet Clin N Am Small Anim Pract* 2003;**33**(2):365–404.
70. Mills DS, Simpson BS. Psychotropic agents. In: Horwitz DF, Mills DS, Heath S, editors. *BSAVA manual of canine and feline behavioural medicine*. Quedgeley, Glouscester, England: British Small Animal Veterinary Association; 2002. p. 237–48.
71. Burghardt Jr WF. Using drugs to control behavior problems in pets. *Vet Med* 1991;**86**(11):1006. p. 1068–1071, 1074–1075.
72. Baird JD, Arroyo LG, Vengust M, McGurrin MKJ, Rodriguez-Palacios A, Kenney DG, Aravagiri M, Maylin GA. Adverse extrapyramidal effects in four horse given fluphenazine decanoate. *J Am Vet Med Assoc* 2006;**229**(1):104–10.
73. Baird JD, Maylin GA. Adverse side effects in horses following the administration of fluphenazine decanoate. *Am Assoc Equine Pract Proc* 2011;**57**:57–62.
74. Marder AR. Psychotropic drugs and behavioral therapy. *Vet Clin N Am Small Anim Pract* 1991;**21**(2):329–42.
75. Shader RI, Greenblatt DJ. The pharmacotherapy of acute anxiety: a mini-update. In: Bloom FE, Kupfer DJ, editors. *Psychopharmacology: the fourth generation of progress*. New York: Raven Press Ltd; 1995. p. 1341–8.
76. Shini S. A review of diazepam and its use in the horse. *J Equine Vet Sci* 2000;**20**(7):443–9.
77. Simpson BS, Simpson DM. Behavioral pharmacotherapy. Part II. Anxiolytics and mood stabilizers. Compendium of continuing education: small animal. *Practice* 1996;**18**(11):1203–13.
78. Brown RF, Houpt KA, Schryver HF. Stimulation of food intake in horses by diazepam and promazine. *Pharmacol Biochem Behav* 1976;**5**(4):495–7.
79. Popoli M, Gennarelli M, Racagni G. Modulation of synaptic plasticity by stress and antidepressants. *Bipolar Disord* 2002;**4**:166–82.

80. Burke MJ, Preskorn SH. Short-term treatment of mood disorders with standard antidepressants. In: Bloom FE, Kupfer DJ, editors. *Psychopharmacology: the fourth generation of progress*. New York: Raven Press Ltd.; 1995. p. 1053–65
81. Overall KL. Noise phobias in dogs. In: Horwitz DF, Mills DS, Heath S, editors. *BSAVA manual of canine and feline behavioural medicine*. Quedgeley, Glouscester, England: British Small Animal Veterinary Association; 2002. p. 164–72.
82. Overall KL. The role of pharmacotherapy in treating dogs with dominance aggression. *Vet Med* 1999;**94**(12):1049–52.
83. Simpson BS, Simpson DM. Behavioral pharmacotherapy. Part I. antipsychotics and antidepressants. Compendium of continuing education: small animal. *Practice* 1996;**18** (10):1067–81.
84. King JN, Overall KL, Appleby D, Simpson BS, Beata C, Chaurand CJP, Heath SE, Ross C, Weiss AB, Muller G, Bataille BG, Paris T, Pageat P, Brovedani F, Garden C, Petit S. Results of a follow-up investigation to a clinical trial testing the efficacy of clomipramine in the treatment of separation anxiety in dogs. *Appl Anim Behav Sci* 2004;**89**(3–4):233–42.
85. Overall KL. Part 3: a rational approach: recognition, diagnosis, and management of obsessive-compulsive disorders. *Canine Pract* 1992;**17**(4):39–43.
86. Julien RM. *A primer of drug action: a concise non-technical guide to the actions, uses, and side effects of psychoactive drugs*. 7th ed. New York: W.H. Freeman and Company; 1995. p. 511.
87. Overall KL. Part 2: a rational approach: recognition, diagnosis, and management of obsessive-compulsive disorders. *Canine Pract* 1992;**17**(3):25–7.
88. Peck KE, Hines MT, Mealey KL, Mealey RH. Pharmacokinetics of imipramine in narcoleptic horses. *Am J Vet Res* 2001;**62**(5):783–6.
89. Shanley K, Overall K. Rational selection of antidepressants for behavioral conditions. *Vet Forum* 1995;30. 32–34.
90. Feinberg M. Clomipramine for obsessive-compulsive disorder. *Am Fam Physician* 1991;**43** (5):1735–8.
91. Sommi RW, Crismon ML, Bowden CL. Fluoxetine: a serotonin-specific, second-generation antidepressant. *Pharmacotherapy* 1987;**7**(1):1–15.
92. Crowell-Davis SL. Stereotypic behavior and compulsive disorder. *Compendium Equine* 2008;**3**(5):248–51.
93. Ryall RW. *Mechanisms of drug action on the nervous system*. 2nd ed. New York: Cambridge University Press; 1989. p. 232.
94. Ernst M, Zametkin A. The interface of genetics, neuroimaging, and neurochemistry in attention-deficit hyperactivity disorder. In: Bloom FE, Kupfer DJ, editors. *Psychopharmacology: the fourth generation of progress*. New York: Raven Press Ltd.; 1995. p. 1643–52
95. Schatzberg AF, Schildkraut JJ. Recent studies on norepinephrine systems in mood disorders. In: Bloom FE, Kupfer DJ, editors. *Psychopharmacology: the fourth generation of progress*. New York: Raven Press Ltd.; 1995. p. 911–31
96. Beaver BV. *Veterinarian's encyclopedia of animal behavior*. Ames: Iowa State University Press; 1994. p. 307.
97. Kelly PH. Drug-induced motor behavior. In: Iversen LL, Iversen SD, Snyder SH, editors. *Handbook of psychopharmacology: volume 8, drugs, neurotransmitters, and behavior*. New York: Plenum Press; 1977. p. 295–331.
98. Overall KL. Practical pharmacological approaches to behavior problems. In: *Purina specialty review: behavioral problems in small animals*. St. Louis, MO: Nestle Purina; 1992. p. 36–51.

99. Dodman NH, Shuster L, Court MH, Dixon R. Investigation into the use of narcotic antagonists in the treatment of a stereotypic behavior pattern (crib-biting) in the horse. *Am J Vet Res* 1987;**48**(2):311–9.
100. Schwartz J-C, Arrang J-M, Garbarg M, Traiffort E. Histamine. In: Bloom FE, Kupfer DJ, editors. *Psychopharmacology: the fourth generation of progress*. New York: Raven Press Ltd.; 1995. p. 397–405
101. Plumb DC. *Plumb's veterinary drug handbook*. 5th ed. Stockholm, WI: PharmaVet Inc; 2005. p. 1311.
102. Madigan JE, Bell SA. Owner survey of headshaking in horses. *J Am Vet Med Assoc* 2001;**219**(3):334–7.
103. Maurer BM, Dodman NH. Animal behavior case of the month. *J Am Vet Med Assoc* 2007;**231**(4):536–9.
104. Rendon RA, Shuster L, Dodman NH. The effect of the NMDA receptor blocker, dextromethorphan, on cribbing horses. Pharmacology, *Biochem Behav* 2001;**68**(1):49–51.
105. Wong DM, Alcott CJ, Davis JL, Hepworth KL, Wulf L, Coetzee JH. Use of alprazolam to facilitate mare-foal bonding in an aggressive postparturient mare. *J Vet Intern Med* 2015;**29**(1):414–6.
106. Dailey JW, Reith MEA, Steidley KR, Milbrandt JC, Jobe PC. Carbamazepine-induced release of serotonin from rat hippocampus in vitro. *Epilepsia* 1998;**39**(10):1054–63.
107. Dailey JW, Reith MEA, Yan Q-S, Li M-Y, Jobe PC. Carbamazepine increases extracellular serotonin concentration: lack of antagonism by tetrodotoxin or zero Ca^{2+}. *Eur J Pharmacol* 1997;**328**(2–3):153–62.
108. Kawata Y, Okada M, Murakami T, Kamata A, Zhu G, Kaneko S. Pharmacological discrimination between effects of carbamazepine on hippocampal basal, Ca^{2+}- and K^{+}-evoked serotonin release. *Br J Pharmacol* 2001;**133**:557–67.
109. Rogawski MA, Löscher W, Rho JM. Mechanisms of action of antiseizure drugs and the ketogenic diet. *Cold Spring Harbor Perspect Med* 2016;**6**(5):1–29. a022780.
110. Newton SA, Knottenbelt DC, Eldridge PR. Headshaking in horses: possible aetiopathogenesis suggested by the results of diagnostic tests and several treatment regimes used in 20 cases. *Equine Vet J* 2000;**32**(3):208–16.
111. Davis BP, Engle TE, Ransom JI, Grandin T. Preliminary evaluation on the effectiveness of varying doses of supplemental tryptophan as a calmative in horses. *Appl Anim Behav Sci* 2017;**188**:34–41.
112. Grimmett A, Sillence MN. Calmatives for the excitable horse: a review of L-tryptophan. *Vet J* 2005;**170**(1):24–32.
113. Hothersall B, Nicol C. Role of diet and feeding in normal and stereotypic behaviors in horses. *Vet Clin N Am Equine Pract* 2009;**25**(1):167–81.
114. Paradis MR, Breeze RG, Bayly WN, Counts DF, Laegreid WW. Acute hemolytic anemia after oral administration of L-tryptophan in ponies. *Am J Vet Res* 1991;**52**(5):742–7.
115. Beata C, Lefranc C, Desor D. Lactium™: a new anxiolytic product from milk. In: *Current issues and research in veterinary behavioral medicine*. West Lafayette, IN: Purdue University Press; 2005. p. 150–4.
116. McDonnell SM, Miller J, Vaala W. Calming benefit of short-term alpha-casozepine supplementation during acclimation to domestic environment and basic ground training of adult semi-feral ponies. *J Equine Vet Sci* 2013;**33**(2):101–6.
117. Dodd JA, Doran G, Harris P, Noble GK. Magnesium aspartate supplementation and reaction speed response in horses. *Equine Vet Sci* 2015;**35**(5):401–2.

118. Cozzi A, Lecuelle CL, Monneret P, Articlaux F, Bougrat L, Frosini CB, Pageat P. The impact of maternal equine appeasing pheromone on cardiac parameters during a cognitive test in saddle horses after transport. *J Vet Behav* 2013;**8**(2).
119. Falewee C, Gaultier E, Lafont C, Bougrat L, Pageat P. Effect of a synthetic equine maternal pheromone during a controlled fear-eliciting situation. *Appl Anim Behav Sci* 2006;**101**(1–2):144–53.
120. McDonnell SM. Sensitive and spooky horses. *The Horse* 2016;**XXXIII**(12):56–7.
121. Benedetti F. Placebo responses in animals. *Pain* 2012;**153**(10):1983–4.
122. Conzemius MG, Evans RB. Caregiver placebo effect for dogs with lameness from osteoarthritis. *J Am Vet Med Assoc* 2012;**241**(10):1314–9.
123. Drago F, Nicolosi A, Micale V, Lo Menzo G. Placebo affects the performance of rats in models of depression: is it a good control for behavioral experiments? *Eur Neuropsychopharmacol* 2001;**11**(3):209–13.
124. Nolan TA, Price DD, Caudle RM, Murphy NP, Neubert JK. Placebo-induced analgesia in an operant pain model in rats. *Pain* 2012;**153**(10):2009–16.
125. Sümegi Z, Gácsi M, Topál J. Conditioned placebo effect in dogs decreases separation related behaviours. *Appl Anim Behav Sci* 2014;**159**:90–8.
126. McDonnell SM. Anabolic steroid effects. *The Horse* 2007;**XXIV**(11):53–4.
127. Schumacher EMA, Blackshaw JK, Skelton KV. The behavioral outcomes of anabolic steroid administration to female horses. *Equine Pract* 1987;**9**(6):11–2[14–15].
128. Squires EL, Voss JL, Maher JM, Shideler RK. Fertility of young mares after long-term anabolic steroid treatment. *J Am Vet Med Assoc* 1985;**186**(6):583–7.
129. Natural Standard Patient Monograph: melatonin, http://www.mayoclinic.com/print/melatonin/NS_patient-melatonin/DSECTION+all&METOD+print [downloaded 3/19/2007].

Chapter 10

Equine Behavior Problems

Problem behaviors are often called "vices" by horsemen and women, and the synonym "misconduct" accurately describes several behaviors already covered. In theory, all behavior problems have a neurologic basis either through learning, instinct, or physiology. In a few cases, there is an associated neurological abnormality. The incidence of undesired behavior varies, depending on which behaviors are included. Although higher incidences have been reported, a more likely estimate suggests somewhere between 9% and 20% of horses have a behavior problem, with approximately half being incidental or nuisance problems and half being medically based or stereotypic/obsessive-compulsive in nature.[1–3]

Management practices have a significant impact on problem development. In the early 20th century, most horses worked on farms. Today, the number of farms in the United States has dropped by two-thirds and mechanization has become the norm. As a result, only 1.5% of households in the United States have horses. Instead of being used for work, horses are now thought of as family members by approximately 35% of the owners or as pets or companions by about 60%.[4] Modern owners are inclined to think of the horse as a big dog, not understanding horse behavior or learning. This explains why there are so many nuisance problems. In today's increasingly urban society, horses spend less time in pastures and more time in stalls, resulting in significantly higher rates of serious problems.[5,6] Stall-bound horses can "recharge" their positive attitudes if given the opportunity to spend time in a pasture with other horses.[7] This fulfills their innate need for exploration, play, social grooming, and social interaction—a reason pasturing with social peers is often part of treatment protocols.

As behavioral science progresses, problems are being catalogued more accurately by expression or cause, rather than being lumped into one large "vice" category. That said, trying to organize various problems is a lot like herding cats. Some fit nicely into a specific category. Others can be classified in several categories because they can be expressed in multiple ways. Horse owners divide problems into five broad categories (handling issues, frustration behavior, abnormal oral/ingestive behavior, aggression toward people, and locomotor stereotypies).[6] Unfortunately, this does not cover all types of potential problems, includes a judgment of motivation ("frustration"), and mixes types of problems between categories ("eat bedding" is listed as a frustration

Equine Behavioral Medicine. https://doi.org/10.1016/B978-0-12-812106-1.00010-3

behavior rather than an abnormal oral/ingestive one). The many variations of a particular behavior problem suggest different or complex causes, or at least differences in how individual horses express the same thing. Because of all these complications, the descriptions of certain problems have to be placed under one heading but could just as plausibly be put under another. Those generally considered to be a nuisance are covered in appropriate chapters. Complex problems that have a neurologic or medical component are included here.

THE BASIS OF REPETITIVE BEHAVIOR PROBLEMS

Repetitive behaviors can happen for a number of reasons. Behaviorally, stereotypies and obsessive-compulsive disorders (OCDs) fit into this category. For other behaviors, owners may inadvertently reinforce an undesired behavior by giving the horse attention. They may also attempt to withhold rewards while the horse expresses the problem behavior but eventually give in before the behavior stops. This is a strong reinforcer. Some problem behaviors, such as pawing, are learned and are only shown when the owner is present. Other repetitive behaviors are situationally dependent, such as fence walking when a social peer is gone—a form of separation anxiety.

Medical problems are associated with some repetitive behaviors. "Ticks" can be manifestations of psychomotor seizures. Some headshaking is similar to trigeminal neuralgia in humans.[8] Visual problems are blamed for some repetitive behaviors, as are occasional expressions of pain or dermatologic conditions.[9,10]

Individual horses can have more than one behavior problem. A cribbing horse might also walk off when the rider is mounting, as an example. Multiple stereotypic behaviors can occur in the same horse. Of approximately 4000 horses, 10 will stall walk and weave (0.25%), 7 will weave and crib (0.18%), and 4 will do both stall walking and cribbing (0.1%).[11]

Specific stereotypic and OCD problems will be discussed separately, but a general discussion is appropriate for a better frame of reference. As will be described under cribbing, the understanding of stereotypies is complicated because the motivation is unknown. The use of certain drugs hints at possible mechanisms. Unfortunately, no magic pill works on every case of a specific stereotypy—further suggesting their complexity. Interventions can be problematic because they do not identify or address the cause, and interrupting one component in a complex system can disrupt that system even though that specific component only has an indirect relationship.[12]

Stereotypies

Stereotypies are repetitive behaviors that gradually become ritualistic, have no obvious goal or function, and are recognizably abnormal in the context, frequency, and pattern.[2,13,14] The specific behavior may start as something the horse is motivated to do or as a substitute for what the horse is highly motivated to do.[15] The pattern solidifies over time and, once developed, the stereotypy will occupy at least 5%, and often over 25%, of the total daily time budget.[16,17] By then, the behavior is difficult to eliminate. Even if the frequency can be reduced significantly, any stress is likely to trigger the behavior again. It becomes a "default" behavior.

In humans, stereotypies begin while a child is very young. This can happen in horses, as exemplified by foals that are just a few days old circling the edges of a stall, but they generally develop within a few months after weaning. It might relate to the significant changes in the foal's nutritional management and social environment.[18] For horses that ultimately develop a stereotypy, 10.5% develop cribbing at a median age of 20 weeks, 30.3% show wood chewing by 30 weeks, 4.6% weave by 60 weeks, and 2.3% stall walk by 64 weeks.[19] Young colts are affected more than fillies, but by 3 years of age, the sex difference has disappeared.[11] Based on rodent models, it has been suggested that when an older horse develops a stereotypy, it had previously experienced significant stress while very young. This emotional "scar" remains so that when a major stress occurs to the adult, abnormal behavior surfaces.[20]

The incidence of stereotypies is highly variable due to several factors, particularly how the data was gathered. This explains the range between 5.6% and 32.5%.[3,11,19,21–27] Under poor management conditions, rates may be as high as 85%. In appropriately managed stables, the rate is closer to 20%.[2,21,28] The frequency of stereotypic behaviors will vary by breed and by the type of event the horse is trained to do.[28–30] Closely related horses show stereotypies at a rate higher than would be expected by chance, suggesting that genetics may play some role in this problem.[13] Discrepancies in data are also found when evaluating the horse's training. For example, the incidence ranges from 32.5% to 88.2% for dressage horses.[28,30] A great deal of additional data will be required to sort out the true incidence of stereotypies and assorted factors.

Today's horses live in significantly different conditions from those of their ancestors, with exercise, diet, and reproduction under human control. As a result, stereotypies are thought to develop as a coping mechanism when a horse cannot avoid stressors like social isolation or fearful stimuli, release tension or "frustration," or show species-specific behavior.[14,16,31–34] The horse is stimulated to show a particular behavior (*appetitive behavior*), typically by excitement or stress, but the environmental conditions prevent the animal from acting out that particular behavior (*consummatory behavior*) (Figure 10-1). Environmental situations are closely associated with the expression of the stereotypies. As will be mentioned for several specific problems, food plays a big role in stereotypic activity. In long barns where it takes several minutes to feed all the horses, the ones on the far end listen to grain being dumped and others eating. This could increase anticipatory stress levels. There is evidence that the personality type of affected horses is such that these individuals react more strongly to acute stressors.[35–37]

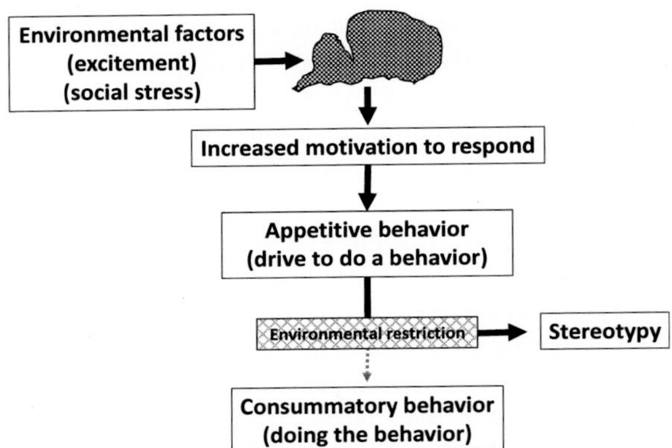

FIGURE 10-1 The interrelationship of the brain and environment in the development of stereotypic behaviors.

Physiological data gathered from horses with a variety of stereotypies and OCDs are often confusing and contradictory.[38] Test results can be higher, lower, or no different from those of controls. Both plasma and salivary cortisol levels are the same in normal horses and those with most stereotypies; however, increased levels are found in horses with oral stereotypies.[36] Timing of the samples may be important too since cortisol is significantly higher immediately before the onset of the stereotypy and greatly reduced after it is done.[39,40]

Neurologically, sensitization of the striatal region of the basal ganglia is the area thought to be primarily involved with stereotypies. Dopamine is the neurotransmitter associated with stereotypies in several species, including horses.[13,25,33,41,42]

Recently cytokines have been investigated, prompted by the known association of pain and stress and a correlation between cytokine levels and behavior changes.[43] Cytokines are nonantibody proteins released by one type of cell that then influence surrounding cells. Both of the antiinflammatory cytokines (IL-4 and IL-10) are significantly higher in cribbers than in normal horses. They were also higher in weavers but not at the same level of significance. The proinflammatory cytokine TNF-α is significantly higher in control horses that are either cribbers or weavers. These findings are opening up a new area to investigate relative to stereotypies and potentially genetic analyses.

Because stereotypic behaviors are not observed in free-ranging horses, onset must be related to how tamed horses are kept.[10,33,44–46] Certain environmental conditions increase the likelihood of stereotypic behavior and should be part of strategies used to prevent problem development. Free-ranging and pastured horses spend 60% of their time grazing. If given a choice, domestic horses will spend only 10%–20% of their time in a stall and 50%–60% in visual contact with other horses.[47] Longer duration stalling increases the risk of problems.

This is also associated with high arousal situations like waiting for food or removal of a nearby horse. Additionally, general access to food and conspecifics is limited. Other associations to stereotypy development include reductions of eating time (associated with eating cereals/concentrates rather than grazing), frequency of feeding, amount of forage, and locomotor time; conflicting rider cues; high-ranking dams; managed rather than natural weaning; confined weaning; postweaning confinement; certain types of bedding; and reduced social/visual contact with other horses.[5,13,18,19,26,28,29,33,48–55] These represent the horse's inability to control its environment.[13]

Which specific stereotypic behavior develops may relate to the type of riding the horse is used for and to the horse's perception of stress. Similar horses from the same barn have shown that repetitive licking/biting is more common if they are used for eventing, while cribbing is more common if they are used for dressage.[28] Mildly stressful events seem to be associated with cribbing. As the stress level increases, weaving and occasionally head shaking appear.[56] Because weaving is most common when the horse is relatively isolated or a social peer is removed, the response emphasizes the importance of social contact.[56]

Why does one horse develop a stereotypy when other horses in the same environment, managed the same way, do not? This suggests a neurologic/physiologic/genetic predisposition that it is easily triggered under certain environmental conditions. Stereotypic research is showing that causation is very complex. Theories suggest that stereotypic behavior releases brain endorphins to calm the stress, but it is not a linear relationship. Using the cribbing model, affected horses have resting brain β-endorphin levels lower than unaffected horses, higher than unaffected horses, and not different from unaffected horses.[35–37,57–59] Yet, the use of a narcotic antagonist like naloxone will block the expression of some stereotypies. This suggests that either plasma levels of β-endorphins do not correlate with brain levels or that opioid receptors are more sensitive in stereotypic horses.[37] Other physiological measures show thyroid stimulating hormone (TSH) and melatonin levels to be lower in weaving horses, but serotonin levels are higher.[60]

Stereotypies are very undesirable, so the horse's value and salability are greatly reduced, regardless of its ability to perform.[33,61] People go to great lengths to hide the fact that their horse has a problem. They are also concerned that nonaffected horses in the barn will learn the undesirable behavior, particularly cribbing, from affected ones. A correlation is actually very low, just 1.0%.[62] The only correlation shown is for horses at riding schools.[18,52,63] What people fail to take into account are the similarities under which all the horses are being managed and ridden, and perhaps genetic relationships as well.

Stereotypies are abnormal behaviors that have other associated costs. The cost to the horse can be physical, as happens in cribbers with abnormal wear of the teeth, loss of body condition, or colic. It has also been shown that broodmares with a stereotypy have a significantly lower conception rate than do unaffected mares.[64] Costs can also be psychological when the stereotypy becomes the preferred behavior. It interferes with the ability to learn new tasks or extinguish previously learned ones.[13,14,41,45,65] The horse is also less aware of external events, has reduced arousal, and has less distress.[14]

No single treatment is effective for all stereotypies, and treatments rarely "cure" the problem. This makes prevention important. General recommendations include low stress weaning of foals, daily pasture time, social groups, multiple meals with more roughage and less concentrates, predictability in the environment, and improved rider skills.

At least 50% of trainers and probably significantly more owners will try to deal with the stereotypy only by physically preventing it.[3,61] Another 35%–45% will try to remove factors related to the causation.[61] Only preventing expression of the stereotypy is stressful to the horse, as shown by an associated rise in cortisol.[66] Ultimately, it does not work, and as soon as possible the horse will show a rebound effect by doing the behavior longer or more frequently than it did before the prevention. The internal drive to show the behavior has not been changed, and the stress of not being able to perform it increases.

Long-term pasturing is helpful for locomotor stereotypies, but confinement at some point in the future is likely to trigger the unwanted behavior again. Oral behaviors may be less responsive to pasture management. Food dispensers as enrichment are not helpful, even for horses with orally directed behaviors.[66] Drug therapy rarely eliminates the problem either. Even if it did, treatment would probably be required for the duration of the horse's life.

Obsessive-Compulsive Disorders

OCD is a second broad category of abnormal behavior that encompasses a number of different conditions. Exactly which behavior problems can be classified as OCD is fuzzy, but there are clues. By definition an *obsession* is a haunting thought that is constantly intrusive. *Compulsion* is the uncontrollable need to act on a particular thought or obsession. OCDs are defined by their expression, their frequency and persistence, consumption of significant amounts of time, lack of an apparent function, and irrelevance to where they occur.[10] The clinical manifestations may show that the compulsion is severe enough to cause physical injury, interfere with normal activities, or become a nuisance to others.[2]

We can never know what the horse is thinking and so can never prove an "obsession," but it can be measured. As an example, a cribber typically cribs on one location in its stall. A barrier is placed between the horse and cribbing area such that the horse must push past the barrier to get to the cribbing area. By gradually increasing the difficulty of pushing open the barrier, researchers can get a quantitative measure of the significance of the behavior to the horse by how hard it is willing to work to get past the barrier. This gives some measure of the amount of "obsession."

Obsessive-compulsive behaviors are often stereotypic behaviors that have become OCDs over time. It is important to note that not all OCDs are stereotypies, and not all stereotypies are OCDs. In spite of this, some authors incorrectly equate the two. For a stereotypy to become obsessive-compulsive, it must first become emancipated from the original environment and occur as a favored behavior at any time and place. As an example, some horses will only circle their stall when confined within it—that is a stereotypy. It becomes an obsessive-compulsive problem when that horse starts to walk the same small circle when put into a large paddock or pasture too.

Research in humans and other species is providing more information about causes of compulsive behaviors. Very recent research on the genetic expression of OCDs is finding links to abnormalities in glutamate signaling and the brain's cortico-striatal circuit. Four specific genes (*NRXN1*, *HTR2A*, *CTTNBP2*, and *REEP3*) are being implicated.[67] Glutamate is the excitatory neurotransmitter that triggers postsynaptic excitation of contact neurons affecting memory and learning. When postsynaptic binding is interfered with, as is expected through these genes, excitatory levels continue to build until stereotypic/compulsive behavior results.

Dopamine is a second neurotransmitter that might be involved in stereotypies and/or compulsive behavior. It is associated with the internal reward system. By blocking this reward system with drugs like dextromethorphan, a *N*-methyl-D-aspartate (NMDA) receptor antagonist, expression of some stereotypies and compulsive behaviors is altered. This has been demonstrated in some cribbing horses. Of nine horses, four stopped cribbing for 35 to just over 60 min, four others had a reduction in their cribbing rate for a shorter period of time, and one increased its cribbing rate following dextromethorphan administration.[68]

While punishment for OCD behaviors is popular, it is ineffective at best and increases the anxiety to potentiate the problem at worst.[10] Providing free access to pastures and friendly herdmates and removing stressors are important in minimizing expressions of OCDs. Medications are usually necessary and typically include long-term use of selected serotonin reuptake inhibitors (SSRIs) and situational use of anxiolytics. "Cure" is rare, and like stereotypies, this is the default behavior that will appear in any stressful situation.

REPETITIVE ORAL AND HEAD PROBLEMS

Oral behavior problems come in many forms. While the overall incidence of these problems varies based on how the data is collected, it might be as high as 48%.[6] Some oral problems are random and situationally dependent, as discussed elsewhere. Some are repetitive stereotypies or OCDs. Studies of oral problems in general show they commonly develop around the time of weaning and are slightly more likely to happen shortly after weaning.[69] Wood chewing problems are the most common.[69] Of all serious oral problems, cribbing (cribbiting) is the one most studied.

Many oral problems are linked to the diet of the horse, particularly to foods high in digestibility but low in fiber. The fact that frequency of performance of these stereotypies is reduced with the addition of hay supports the link. There are many theories as to causes, but coexisting factors make studying the problems difficult.

Cribbing, Windsucking

The current understanding of cribbing is that it is a stereotypic OCD, is very complex, involves several body systems, and is progressive. While generally accepted that stereotypies do not exist in free-ranging horses, teeth from horses that lived 15,000–30,000 years ago show evidence of cribbing. This does not suggest they were domesticated, but more likely were wild caught and held tethered for prolonged periods for some unknown but specific purpose.[44]

Defining the Behavior

There are three closely related behavior problems: wood chewing, cribbing, and windsucking. The first is generally considered to be a nuisance behavior related to eating. It is suggested that wood eating may lead to cribbing. Cribbing begins with the horse licking the object to be grasped a couple of times. This occurs 1.7 times before each cribbing event while the horse is eating grain and 1.1 times if eating hay.[70] Next, the horse puts its upper incisors on some hard object that is about chest height, extends its neck slightly, pushes down on that object while tensing the ventral muscles of its neck, and pulls back as it puts pressure on the object (Figure 10-2). The amount of neck tension is enough to lift a 110-lb (50 kg) weight.[71] Some horses crib and suck in so hard that they produce a

FIGURE 10-2 A cribbing horse places its upper teeth on the fence. She then tenses her neck muscles as she puts downward and slightly backward pressure on the board.

characteristic, loud gasping sound. These horses are the windsuckers. The behavior has been called *aeorphagia* because it was believed that the horse is swallowing air in addition to cribbing. Fluoroscopic and endoscopic studies show that windsucking is either a variation of cribbing or a more advanced form of cribbing, not a separate stereotypy. The sound is caused by the rapid introduction of air into the proximal esophagus, expanding it to approach a diameter of 80% that of the trachea.[72,73] Tension in the ventral neck muscles is thought to create a pressure gradient in soft tissues of the throat that results in air being drawn into the esophagus.[73] Most of the air returns to the pharynx. Only a small amount moves on to the stomach, usually in association with the ingestion of food or water shortly after the air intake.[72–74] Because cribbing and windsucking are really the same, the terms are interchangeable.

A few horses show all the behaviors of cribbing without having their teeth contact a hard surface. In the past, these have been called windsuckers,[75] but now that the imaging studies have been done, it is more appropriate to call them "air cribbers" instead.

The incidence of cribbing/windsucking varies with how data was gathered and the management practices for horses being studied. Overall, the problems exist in 2.3%–10.5% of horses.[1,5,24,29,33,35,62,69,71,75] Foals can show cribbing prior to weaning. Affected foals tend to have had more early suckling terminations, show more bunting behaviors while nursing, and spend less time lying alert than other foals.[69] The likelihood of cribbing increases right after weaning occurs, perhaps as a reaction to associated stresses. Affected weanlings differ from those affected prior to weaning in that these spent more time suckling, spent approximately twice as much time nuzzling, and vocalized less than other foals.[34,69] By 20 weeks of age, there is a 10.5% incidence of cribbing in Thoroughbred and Thoroughbred-cross foals.[19,34]

The frequency of cribbing varies with the longevity of the problem, time of day, presence/absence of forage, and individual variation. Figures range from one to eight crib-bites per minute, lasting 2.5–3.5 s each. Horses typically spend 17 min/h cribbing.[68,76] This amounts to spending approximately 23% of a 24-h period cribbing.[70,76] Observations suggest how important the behavior can be. One foal spent 50% of its time cribbing.[77,78] Reports of horses cribbing for 30 s of every 5-min period for 22 h, or an average of 1603 crib-bites each 24 h, are common.[77,79,80] These examples suggest a strong motivation to perform the behavior and illustrate how extreme it can become. Tests of motivation show horses will work harder to crib (push a barrier 350 times to get to a cribbing space) than to get food (push approximately 200 times), social interaction (40–50 pushes), or released from a stall.[71] A few horses crib to the exclusion of eating.[26]

Associated Factors

Several factors contribute to the development, or at least the expression, of cribbing/windsucking. The first of these is neurological and certain brain pathways have also been implicated. Normally, the frontal cortex sends "worry" signals through the caudate nucleus, which puts a brake on the amount of signal passed on to the thalamus. The thalamus triggers excitement and sends a feedback loop to the frontal cortex. If the caudate nucleus does not reduce the input from the frontal cortex, the loop continues to escalate the stress. Secondly, cribbers show a bias toward keeping habitual response patterns, correlating with upregulation of the ventral striatum and basal ganglia dysfunction.[41,45,81] The dorsomedial striatum, an area associated with learning and exploratory behavior, is downregulated too.[41,42,81] The cribbing stereotypy is preferred above all else, suggesting a decreased output of the caudate nucleus and increased reliance on the sensorimotor putamen circuit. The result is an accelerated habit formation.[25,82]

Neurotransmitters and endogenous steroids seem to be involved as well. Cribbing is a stereotypic compulsion that probably has an associated internal reward related to opioid release. Administration of opiate drugs can induce cribbing in affected horses, and opiate antagonists reduce the incidence by 84%.[26,40] Administration of antagonists also increases resting behavior and reduces plasma β-endorphins. Cribbers commonly have significantly lower "anxiety" levels compared to nonaffected horses, supporting the theory that the behavior serves as a passive coping mechanism.[83] Like β-endorphins, baseline cortisol levels have been reported as higher, the same, or lower than in nonaffected horses. An ACTH challenge, however, shows cribbers have significantly higher cortisol levels if they do not crib during the test compared to either horses that did crib during the test or unaffected controls.[39] Even with mixed test results, evidence suggests cribbers might be more stress-susceptible.[84] This may be related to their lower vagal and higher sympathetic tone.[21] These findings alone do not make the horse a cribber; however, they probably predispose it to developing the condition if triggered by other factors.

In cribbing horses, the D1 and D2 dopamine receptor subtypes are significantly higher in number in the nucleus accumbens (ventral striatum), an area thought to be crucial in goal-directed learning by mediating the effects of the reinforcement from goal attainment.[41,81] D1 receptors are lower in the caudate nuclei.[25,33,41,85] There is a corresponding downregulation of dopamine transmission in the nigrostriatal pathways such that horses with stereotypies have difficulty changing out of habitual reinforced responses, just as do chronic amphetamine users.[42] Cribbers have a significantly lower spontaneous blink rate, indicative of dopamine receptor sensitization.[25,86] Confirmation of this is shown by administration of dopamine agonists resulting in cribbing and with attenuation of the problem by dopamine antagonists. There is also a difference in the autonomous nervous system and stress reactivity between horses that crib and those that do not.[21] Stereotypic behavior appears to reduce these.[14]

A genetic component might be the second factor related to development of cribbing. As a breed, Thoroughbreds are overrepresented in cribbing data at approximately 13%, followed by Warmbloods (7%) and Quarter Horses (5%).[48,62,71,87] While a genetic component is strongly suggested, it is difficult to prove. Within certain Thoroughbred bloodlines, the incidence of stereotypies is 13%–30%, and in Finnhorses, cribbing is considered to be 68% heritable.[88,89] To date, tests for the involvement of eight candidate genes have not revealed a connection.[90] Abnormal genetic expressions of four genes that link the cortico-striatal circuit and abnormalities in glutamate signaling have been related to OCDs in multiple species, but not yet in the horse.[67]

Physiologically, cribbers have a lower thermal threshold and heart rate while cribbing.[59] Since both are measures of stress, a reduction suggests the horse is less stressed while performing the behavior. In humans, oxidative stress has been implicated in the pathophysiology of several psychiatric disorders and pathological anxiety.[84] In horses, oxidative stress is lower in cribbers than controls and is reduced even farther while they actively crib.[84] What the significance of this is to stereotypies in general and cribbing in particular is yet to be determined.

There are a number of medical or physiological changes associated with the gastrointestinal tract of cribbers. Ulcers and gastric inflammation have been found in cribbing horses, but the reports of the number of them compared to findings in normal horses vary. Some reports indicate that these horses have a higher number of gastric ulcers than do noncribbers.[25,78] Others show there is no difference.[80] This might suggest that some animals are more sensitive to associated discomforts. A second gastric relationship is between stomach acid and management. Stomach acid is constantly being released but management systems have changed from continuous grazing to meal feeding. Acid may now be a stomach irritant. Evidence pointed to is the higher incidence of gastric ulcers and inflammation in some studies.[20,34,75,78] In addition, there is less neutralizing saliva produced and swallowed because of less chewing or biting.[77] The theory that cribbing behavior is done to increase saliva production to act as a natural antacid has been shown to be incorrect.[77,91,92] Cribbing produces 1–2 mL of saliva during 20 crib bites. Eating grain produces 15–30 times more from the same horse.[91] Some foals seem to be more sensitive to the stresses of weaning, using abnormal oral activities to cope with abdominal discomfort.[34,77,78] Antacids reduce the frequency of cribbing.[34,93]

Three hormones have been implicated in cribbing. Gastrin is secreted from G cells in the stomach and causes the release of gastric acid. After eating grain, serum levels of gastrin are higher in horses that crib than in normal horses.[80] Ghrelin is a gastroprotective hormone produced in the stomach X-A-like cells following stimulation by gastrin. The two work synergistically to stimulate gastric acid production in anticipation of food.[94] Some studies find that levels of ghrelin are higher in cribbers, but other studies do not.[58,95] Leptin is the third hormone related to eating. It influences the hypothalamic regulation of appetite and modulates the reward of eating in association with body mass. Leptin concentrations are lower in cribbers compared to nonaffected horses.[95] If prevented from cribbing, horses will increase ingestive behavior. If that is blocked too, there is a relative stasis of foregut motility, suggesting these are important for normal gut function.[35]

The lower intestinal tract is part of the complexity of cribbing. There is conflicting information about whether cribbers are more likely to colic, but once they have colicked, there is a high rate of additional colic bouts.[75,96–98] There is an 85.7% likelihood of a repeat colic, compared to 34.9% in noncribbers that previously colicked.[98] An environmental effect for this has not been ruled out, however. Between 47% and 68% of horses affected with epiploic foramen entrapment are cribbers, and cribbers that colic are eight times more likely to colic due to epiploic foramen entrapment than other types of colic.[96,99–101] There is a dramatic increase in intraabdominal pressure that begins when cribbing starts, increases throughout the duration of the cribbing bout, and continues for at least 30 min after it has stopped.[102] It is likely this is a cause and effect relationship.

Even near the terminal end of the intestines, there are differences in cribbers from nonaffected horses. Cribbing behavior tends to peak about the time that ingesta reaches the cecum, and fecal pH is lower in affected horses as well.[75,78] The transit time for food is altered too. While there is no difference in the time it takes food to go from the mouth to the cecum, it takes much longer to get from the cecum to the anus.[103] What the connection of these findings is to cribbing has yet to be determined.

Sleep patterns are altered. Cribbers spend significantly less time resting than do normal horses.[103] It is not known if the problem behavior reduces the desire to sleep, or if the lack of sleep causes the horse to start cribbing.

Stress and cribbing are closely associated, with stressful environments more likely to be associated with high cribbing rates. Temperaments of certain individuals suggest they are prone to stress responses. These horses are more likely to develop aggression toward neighboring horses, as an example. For horses that have no visual or physical contact with other horses, the incidence of cribbing is twice as high as for those with contact, and it is higher in horses stabled next to an aggressive neighbor.[104] Horses with only visual contact with other horses crib more commonly than do those having physical contact, suggesting another type of stress.[29]

Diet is a cribbing associative factor. Feeding programs that are high in concentrates and low in forage are related to the increased development and performance of cribbing.[19,34,49,51,57,105] Sweet feed diets increase the likelihood that cribbing will start after weaning fourfold.[75] An adult cribber getting sweet feed will spend 33% of the day cribbing, but this is reduced to 17% if fed oats instead.[70,106] Studies suggest it is the taste of the food that increases cribbing, perhaps through opioid and dopaminergic systems.[105] Horses are most likely to crib in the 2- to 8-h period after eating concentrated feed.[79] Those fed twice a day, when compared to three times a day, are also more likely to crib.[29] Serum selenium levels are significantly lower in cribbers than in noncribbers, especially while showing the problem behavior.[107]

While horse owners frequently comment that their horse "learned" to crib by watching other horses, studies suggest that observational learning is rare in horses. Learning to crib has only been correlated to an incidence of 1%.[62]

The type of show events and the amount of exercise a horse gets affect which stereotypy the animal will show, even when other things are constant. Cribbing is related to dressage and high school horses.[28] Horses exercised for 20 min a day spend 30.6% of the day cribbing, while those not exercised spend 25.3%.[70]

Treating the Cribber

A number of different treatments have been tried to get cribbers to stop. Of the various treatments, physically preventing the behavior is the most common, and several ways have been tried. Physical limitations were the first used. Muzzles will reduce the behavior by approximately one-half.[108,109] Of the various physical methods, the cribbing strap remains one of the most popular. When a cribbing strap is used alone, the frequency of cribbing is reduced by half to two-thirds. Use of a cribbing strap significantly elevates cortisol levels.[40] Also supporting this evidence that physically preventing cribbing is stressful, research shows there is a highly significant increase in the cribbing rate rebound when the strap is removed.[75,110] Another study contradicts this because it found no differences in cortisol levels and no rebound effect.[16] Use of a cribbing strap and other preventative measures is accompanied by an elevation in plasma beta-endorphin levels, at least in some studies.[35]

Another treatment suggestion is that surfaces where a horse could crib can be covered in ways that discourage the behavior.[108,109] Spike strips and rounded metal surfaces are examples.

Punishing the behavior with electric shock has also been tried. This extreme measure first used electric fencing inside the stall. When dog shock collars came out, they were tried on cribbers under the guise of being positive punishment. Equine versions of electric shock collars now come in styles that are triggered remotely by the owner or by neck tension. The remote-controlled versions depend on a trigger person being present 24/7 to punish each and every attempt at cribbing—not very realistic. In addition, horses quickly equate the presence of person with "do not crib." The behavior will resume as soon as the person leaves. If the person hides to activate the shock, a dummy collar might fool the horse when no one was around.[109] Only shock collars that do not depend on a person being present are helpful, because the unwanted behavior must be punished every time for the technique to work.

Crib rings are "C" shaped metal wires placed between the upper incisor teeth. They are then mechanically pinched into the gums to form almost a complete ring. When used, cribbing is significantly reduced from 6.75 h/day before insertion to 0.5 h/day afterwards.[16] The initial day of surgical implantation is associated with a rise in cortisol levels, but not thereafter. The rings are not a long-term solution, however, because they fall out in a week or less. The associated pain, even when eating, has significant welfare considerations too.

Several surgical techniques have been tried to stop cribbing. One of the first techniques created a permanent fistula from the buccal cavity to the outside, making it difficult for the horse to keep its mouth airtight.[109] In subsequent years, the myectomy of the sternomandibular, omohyoid, and sternothyrohyoid muscles has shown success in 80%–88% of cribbing cases.[111,112] Another procedure involves the bilateral neurectomy of the ventral branch of the spinal accessory nerve, which innervates the sternomandibular muscle. The success rate for this surgery approximates 60%.[111] Combinations of the myectomy and neurectomy procedures have success rates falling somewhere in-between. The results of surgery are reasonably good for eliminating the expression of the behavior, reducing the number of colics, promoting weight gain, and improving performance.[112] On the negative side, surgical treatments do not address the initial cause of the behavior nor the ongoing stress experienced from not being able to show the stereotypy.[113] The outcomes are disfiguring, sometimes ineffective, and may have secondary complications.[113] The rebound effect is the reason the behavior reappears in some horses. Most try to crib postsurgery, but only some can physically manage to do so.

Drug therapy can be tried to stop cribbing. Antacids reduce the frequency of cribbing, improve the condition of the stomach, and make the horse with ulcers or inflammation more comfortable.[25,34,75,78,93] Coconut oil has been suggested to have the same soothing quality. Narcotic antagonists like naloxone work to block the β-endorphin internal reward that cribbing produces. Unfortunately, narcotic antagonists are expensive, require intravenous administration, only work for a short time, and do not work on every cribber.[26,114] Other drug choices target various neurotransmitters. SSRIs and tricyclic antidepressants (TCAs) have become popular for stereotypic conditions. The associated increase in interneuron serotonin helps reduce anxiety levels from environmental factors. On the down side, SSRIs and TCAs do not work on all cribbers and they are expensive. Dextromethorphan, an NMDA receptor antagonist, reduces cribbing in some horses.[68] It works by blocking receptors where dopamine would attach and thus blocking dopamine's reinforcement of the behavior. Dopamine antagonists like acepromazine reduce the behavior, but the amount of accompanying sedation is considered too great.[26] One of the newer products tried on cribbers is the equine appeasing pheromone. Controlled studies are yet to be done, but there is some anecdotal evidence that it reduces cribbing.

Welfare concerns of only preventing the behavior must be raised because the internal drive remains. As soon as possible, the behavior rebounds, often at a higher rate, at least for a while. A more reasonable treatment goal should be to reduce the behavior rather than to eliminate it. This requires incorporating several strategies into a treatment plan, including some of those described previously.

Diet is an important consideration in any cribbing/windsucking treatment protocol. Frequent or continuous feeding of forage, ideally in a pasture setting, is appropriate. While it is best to eliminate concentrated feeds, if they must be fed, doing so multiple times a day can help reduce cribbing frequency.[23,34]

Since cribbing is rarely stopped, it is helpful to provide an appropriate surface that will reduce the amount of wear on the teeth. A steel bar with a dense rubber coating is one suggestion. Other techniques attempt to redirect the oral behavior from cribbing to an alternative like using the horse's muzzle or tongue instead. Licking a roller that dispenses a sweet flavored liquid does not significantly change the behavior[27]; however, a device that requires the horse to push a wheel to dispense small portions of its food does.[115]

Acupuncture has been tried as a cribbing treatment. In one study, 64% of horses with gastrointestinal symptoms improved or stopped cribbing when treatment used seven acupuncture points, three of which are specific for gastrointestinal disorders.[75]

Behavior modification alone is not particularly successful. There are certain precautions that should be taken to prevent accidentally reinforcing the behavior. The horse should not be fed while it is actively cribbing, as an example. Punishing the behavior is also inappropriate.[26]

Different types of environmental enrichment have anecdotally shown evidence of successfully reducing cribbing.[109] Pasture access not only modifies the diet but also allows social interaction with other horses. At the very least, horses that spend 12 or more hours a day in pasture are half as likely to colic.

Prevention

Preventing the onset of cribbing is difficult because there are numerous contributing factors, although certain management techniques can help reduce the likelihood it will develop. Multiple meals throughout the day are best, especially if concentrated feeds must be given. Even then, oats are better than sweet feed. Keeping horses on pasture with herdmates is desirable, but if 24-h pasture access is not practical, the longer it can be made available, the better. When stalled, horses on straw bedding are less likely to crib, probably because straw provides additional fiber when hay and grass are not available. Reducing stress is important in preventing cribbing. Foals of submissive mares are best protected from stereotypy development if there are no bossy mares in the herd and if gradual weaning is used instead of forced weaning. Training techniques emphasizing positive reinforcement are considered less stressful as well. A final recommendation is to avoid neighbor horses that crib. This is suggested more for the owner's comfort than the horse's likelihood of learning from its neighbor.

Excessive Licking

All horses will occasionally lick, but some show extreme repetitive licking. The behavior is commonly directed toward the inside of feed boxes or stall walls, but other objects can be targeted, including human skin. Excessive licking is a problem in approximately 14% of horses.[5] Excessive licking is more common in eventing and jumping horses and is significantly more frequent in reining horses than in horses used for general pleasure riding.[28,116] Individually stalled weanlings are another subset of affected horses.[50]

Specific causes of excessive licking have not been identified, suggesting multifactorial causation. In dogs, excessive licking of objects is often associated with some problem in the gastrointestinal tract. Since horses are commonly affected with ulcers and gastric inflammation, a similar connection seems likely. Treatment depends on the results of a medical workup. A second concern is associated with horses that are deficient in minerals, particularly salt. There are also horses that lick people excessively. Some do so when nervous, as at a horse show, while others seem to "like" licking people but were never discouraged. Treatment for them is a matter of stepping back from the horse. If that is insufficient, a firm "No" accompanied by a firm thump on the forehead with the flat of a hand usually works.

In general, excessive licking does not harm facilities or the horse, so it draws less attention than cribbing. Yet it is an abnormal behavior and suggests that the horse might be experiencing stress.

Head Nodding (Tossing)

Head nodding (tossing) represents a different problem than headshaking. Here the nose is repetitively moved up and down. Head nodding is a normal response to insect pests, but when there is no obvious physical stimulus, the behavior is worrisome at best and serious if it makes the horse unstable. The vertical up and down nodding behavior is often blamed on rider error, resistance to tucked head positions, or ill-fitting tack, but sometimes it is a stereotypy. Affected horses are more likely to develop the problem gradually and be relatively young. The nodding is most common when the horse is socially isolated, decreasing significantly once the barnmate returns.[75] An aluminum mirror or life-size picture of another horse's face will reduce the incidence of nodding if a social peer is not possible.[117,118]

Head nodding or bobbing is also used to describe the head motion associated with lameness. This is unrelated to the stereotypic expression.

Headshaking

Headshaking is a behavior problem characterized by recurrent, sudden, and severe shaking of the head up and down and/or side to side. The three most common clinical presentations described by owners are flipping of the head, acting as if an insect was flying up the nostril, and rubbing the muzzle on objects. Approximately 55% of affected horses show the behavior at rest and when exercising. Only 4% headshake only at rest.[8] The motion can be severe enough to throw the horse and rider off balance, making the horse dangerous to ride.[75] Owners frequently comment that when not headshaking the horse was very reliable.[8] Headshaking occurs in approximately 1.4% of horses.[2]

When the onset occurs in an older horse or begins abruptly in the spring or early summer, medical conditions must be considered. In 64% of cases, headshaking is seasonal. Light seems to play a role in the seasonal onset, because the condition is significantly exaggerated in spring and early summer. It is described as worse in bright light and improves on rainy days in 57.8% of affected horses, at night (74.4%), and when indoors (76.9%). In a fourth of these horses, the duration of each shaking bout tends to increase in subsequent years.[8,75,119] It is thought the behavior is a response to pain or irritation, or that it is an extreme form of nodding. Unfortunately, extensive medical workups seldom reveal a specific cause. In one study of 78 horses, only 4 horses had a specific diagnosis.[8] Thoroughbreds are three times more likely to be affected than other breeds, and geldings are twice as likely to be affected than nongeldings.

A number of treatments have been tried with minimal success. The ones that work best include percutaneous electrical stimulation and medicating with cyproheptadine and/or carbamazepine. While helpful, they do not completely relieve symptoms in most horses.[8,120–122] It is speculated that affected horses have a condition similar to trigeminal neuralgia in humans. In the case of horses, headshaking may involve the ophthalmic branch of the trigeminal as well as the maxillary portion.[8]

Head Shy

Horses that are head shy are not showing a repetitive behavior, but this problem can resemble nodding or shaking. Head shy horses are usually just difficult to bridle or halter, because they quickly move their head away. Some are reactive to any hand movement near the head. Causes usually relate to vision or learning. People forget that a horse's vision is compromised around its head. There are several blind areas, such as under the jaw and directly in front of the forehead. Even for areas not blinded, the ability to focus on items within a few feet is diminished. Items brought toward the face may seem to disappear and then startle the animal when it is suddenly touched. Specific spoken cues can be helpful to avoid startling the horse so it learns to anticipate that something is about to touch it.

Avoidance of certain situations can be related to shying. As a lazy animal, a horse does not want to work, and for them, riding is work. They learn that being caught in pasture, having a halter put on, or being bridled ultimately leads to work. If the person attempting to do these things stops before catching the horse, avoidance is rewarded, and the horse continues the behavior. Avoidance can also be part of an interactive "game." Horses purposefully trying to avoid having a halter or bridle put on by moving their head exasperate the owner. Prolonging the horse-human battle can be rewarding to the animal. A quick, firm "No" coupled with a firm thump on the forehead with the flat of a hand is a reminder to hold still. It also has a surprise value that helps reinforce the lesson. Avoidance of halter or bridle must be differentiated from the head shy behavior resulting from fear of closeness of a hand or other object. A horse that has been ear twitched, beaten, or abused in other ways can be head shy or show avoidance as a protective reaction. Desensitization is necessary to regain the horse's trust.

Lip/Tongue Flapping

Some horses will flap their lower lip as they rapidly move their head up and down. Others stick their tongue out the side of their mouth where it may just hang or flap as the head goes up and down. These behaviors are usually associated with stressful situations, representing a coping mechanism. Young horses that roll their tongue out the side often show the behavior when asked to begin learning something new.[123] This suggests that uncertainty of what is now expected is stressful. As the task is mastered, the frequency of the tongue protrusions decreases.

Anecdotally, tryptophan has helped some of these animals, suggesting dietary tryptophan was insufficient for normal serotonin levels.[124] When flapping behaviors occur in stalled horses, enriching the environment with toys, a licking ball, or additional hay can provide things to occupy the horse. Lip and tongue flapping become ingrained over time. When a horse is ridden, head nodding becomes problematic, while a droopy lower lip or protruding tongue are more unsightly than worrisome to the rider.

Teeth Grating

Teeth grating involves the horse rubbing its teeth against an object, such as bars on a stall, a board, or the wall to produce a sound reminiscent of chalk screeching on a blackboard. This vice results in abnormal wear on the incisors, so it is important to investigate possible causes. Medical problems such as ulcers or oral problems should be ruled out. Once eliminated, other causes are considered. Some horses grate their teeth when they are anxious or upset by a nearby new horse. For others, the behavior is triggered by anticipation, particularly for food. Horses that lack environmental stimuli can show it too. If the behavior is immediately followed by attention or feed, the horse may learn to do it for the associated reward.

Treatments for teeth grating not due to medical problems relate to associated events. The need for social contact with other horses and pasture access are important. If occurrences only happen in the presence of a particular person who increases interactions while the behavior is occurring, that specific person needs to be part of the solution. As soon as the behavior starts, the horse should be given no attention, not even eye contact, until the behavior stops. Human attention is then given immediately when appropriate behavior occurs.

As the horse learns it no longer gets attention for making the noise, the time between its stopping and the human's attention reward can gradually increase. Similarly, no food is given unless the behavior is appropriate. With consistency, the teeth grating will be extinguished.

Teeth Grinding

Teeth grinding, also called *bruxism*, occurs when the teeth in the upper and lower arcades are rubbed against each other. The sound produced here is an "I-know-it-when-I-hear-it" type of noise that is almost always associated with pain or discomfort. The behavior is common with gastrointestinal (GI) pains in particular, occurring in approximately 1% of horses.[1]

After an appropriate medical workup for possible GI issues, treatment should begin with pain and other appropriate medication. Consideration should be given to long-term ulcer treatment because of the high frequency of ulcers in horses. If teeth grinding continues or if stress is expected to be a contributing factor, antianxiety medications, including SSRIs, may be added to the protocol.

REPETITIVE LOCOMOTOR PROBLEMS

Many common behavior problems like bucking or rearing result from a specific stimulus such as an inappropriate cue or an attempt to avoid having to do something. There are, however, movement problems, other than lameness, that have a medical or neurologic component. Additionally, modern management practices may relate to the cause. Free-ranging horses spend a significant portion of their day in motion: stalled horses do not. As a result, the drive to move and the high-energy diets most horses eat provide motivation to do something. This partially accounts for the high incidence of locomotor stereotypies like stall walking or weaving in stabled horses.[125]

Circling/Stall Walking

The stereotypic behavior of walking in circles is a relatively common locomotor problem. The problem is usually called stall walking (*box-walking*) because that is where it is usually seen. As the stereotypy becomes an OCD, horses will walk in circles of the same size even though they are in a large paddock or pasture. It is suggested that rapid circling is related to separation anxiety or a desire to get out of the stall, and slower circling is stereotypic.[71]

The incidence of stall walking ranges from 0.7% of horses to 8%, with one outlier of 19% reported.[1,2,5,24,29,33,34,36,75,87] The behavior is more common in mares and young horses, with the average age of onset being 64 weeks.[19,29,75] Just over 2% of foals show stall walking by 16 months of age,[19] with some circling when only a few days of age. The frequency and duration of stall walking bouts increase over time too.[87] Arabians are the breed most affected, followed by Thoroughbreds and then Warmbloods.[48,75,87] Of locomotor problems, this is more common than weaving in endurance horses compared to eventing or dressage horses.[30,75]

As with most stereotypies, this one can be associated with management issues, particularly minimal social contact with other horses. Another association is the need for exercise. Horses turned out more often and for long periods in warm seasons but confined in winter are more likely to be affected.[75,87]

Treatment and prevention are similar in nature. Social contact between horses; significant time in paddocks or pastures; forage, particularly grass, rather than concentrates; multiple small meals spread throughout the day; and gradual weaning with a group of foals are helpful in preventing the development of and managing circling, just as they are for other stereotypies.[19,34,48]

Fence Walking

Fence walking is the behavior named for the repetitive back-and-forth walking along a fence. Characteristically, the horse walks a certain distance, turns toward the fence, and returns on the same path. As the behavior becomes more ingrained in its expression, the horse will flip its nose up as it turns at each end of the path. As fence walking continues, a well-worn path develops (Figure 10-3). While showing the behavior, the horse is aware of things going on around it. As an example, it might kick out at a dog that comes up from behind. The most common trigger is the departure of another horse to which the fence walker is closely attached. It can also be triggered when access is prevented from other nearby horses.

FIGURE 10-3 This arena is used to exercise the Lipizzaner breeding stallions for the Spanish Riding School of Vienna at their Piber, Austria farm. The path along the fence is limited to this area of the arena as the result of stereotypic fence walking by this Lipizzaner stallion. He directs his attention toward the mare barn on the other side. Notice that the stallions have also chewed the fence, but only in the same section of the arena.

Pawing

Pawing is recognized as a problem in 0.2% to 3.6% of horses.[1,2] Origins of this repetitive behavior are probably numerous. Pawing is reinforced with the presentation of food, attention, or release. Horses quickly make the connection between pawing and reward, even though the owner does not. Once the horse equates pawing with reinforcement, breaking that connection is difficult. In one controlled study to reduce pawing, positive reinforcement with a food treat was given only when the horse stood with all 4 ft. on the ground. It took between 25 and 40 training sessions to reduce pawing to less than 14% of the original amount.[126] Pawing in anticipation of being fed usually stops as soon as the feed is put in the stall, but not always. While chewing, the horse may continue to paw the ground or paw the air. Continued pawing is indicative of a stereotypy. Besides anticipation, pawing can be related to impatience, discomfort, and learning. Individually stalled weanlings are also commonly affected.[50]

Treatment for pawing begins with a medical examination, especially knowing the high incidence of ulcers in horses. While not all discomfort is related to ulcers and not all ulcers are related to discomfort, any detectable problem should be ruled out as a cause or contributing factor. For horses that paw until the feed is put into the feedbox, the owner should stop moving toward the stall, and even take a step back, as soon as the pawing begins. If pawing continues, the person takes another step back, and then another. Putting the food in the stall rewards the undesired behavior, but stopping or backing away does not. Then when the pawing stops, the person begins to walk forward again until the pawing resumes. No food is put into the stall if the horse is pawing. By giving small portions each time instead of a full meal, the "no paw" sessions can be done multiple times a day to help speed learning. For the horse that paws while eating, lowering the level of the feeder or feeding on the ground makes it harder to paw (Figure 10-4).

Young horses, in particular, tend to paw when they are tied in one location. Lessons of patience are important for standing tied to trailers or other objects when necessary. Until this is learned, the area must be safe for them, including softer ground, the rope tied at head level or above, and minimal slack in the rope to avoid the chance the horse could get its foot caught. For safety reasons, someone should be nearby with a sharp pocket knife just in case the horse gets into trouble and the rope has to be cut. Many people will begin by tying the horse to a stout rubber inner tube that is securely attached to an object.[127] This permits some give if the horse fights the restraint but snaps back to ensure the horse does not win the battle.

Weaving

Weaving is a relatively common stereotypy in which a horse rhythmically shifts its weight between right and left forelimbs while swinging its head back and

FIGURE 10-4 This gelding paws when anticipating food as well as while chewing it, to the point of wearing a hole near the feeder. When he is fed out of a tub on the ground, he will try to paw but stops after a few low-level attempts.

forth. Careful observation shows that the rear feet also move in cadence of walking. This problem is not well researched, but the estimated incidence is 0.6%–9.5%.[1,2,5,24,29,33,34,36,75,87] In foals, 4.6% show the weaving behavior by 15 months of age.[19,118] Then, the incidence of weaving increases with age.[87] Stallions are affected significantly more often than mares or geldings, and Thoroughbreds have the highest incidence of the various breeds, followed by Warmbloods.[48,87,128]

In the long term, physical problems can develop as a result of weaving behavior. Affected horses can develop strained ligaments and often lose condition. They also are poor performers.[33]

Environmental events are most commonly identified as triggers for weaving. The behavior is anticipatory in stables where management is highly predictable, being most common during periods of activity, like feed preparation, and in the hour before daily turnout.[25,60,79,129] Horses with minimal social contact and those that can only see another horse across an aisle have a higher incidence of weaving.[25,29,51,54,128] Horses having visual closeness, and perhaps with the ability to touch horses in adjoining stalls, show less weaving (Figure 5.3).[129] This suggests the behavior relates to the horse's social needs and the resulting "frustration" of not being able to get that contact.[34,118] The concept of "frustration" or stress is supported by the fact that weaving behavior is not reduced with naloxone, indicating there is no accompanying internal reward.[40] The association of feeding protocols with weaving is questionable. Twice a day meals make weaving more likely than do three times a day meals.[29] On the other hand, increasing the number of feedings of concentrated food results in a higher frequency of weaving, particularly in the time preceding the food presentation.[23] Forage intake of less than 6.8 kg/day reduces the risk of weaving, while nonstraw bedding increases the risk for abnormal behaviors in general and weaving in particular.[51]

Neurological studies suggest that weaving, like cribbing, may be complex and involve multiple neurotransmitters. Serotonin levels are higher in weavers than in nonaffected horses. They also have reduced levels of blood magnesium, TSH, melatonin, and adrenocorticotropic hormone (ACTH).[60] SSRIs can reduce symptoms by as much as 95%, where 43.5 weaving motions per minute dropped to less than 1.[130] Even with a stressor such as social separation, the number of motions remained low, at approximately 19 motions/min. Dopamine antagonists like acepromazine reduce the behavior by approximately 40% (24 motions/min), although there is also sedation. Opioid antagonists reduce it by 30% (32 motions/min).[130]

Medically, weavers are almost four times more likely to have a repeat colic than are nonweavers.[131] While this is not as high as the incidence associated with cribbing, it is significant.

Treatment often incorporates the use of antiweave panels in the stall door (Figure 10-5). Used as a lone treatment, these panels are associated with a significant increase in plasma cortisol, and thus should be considered stressful.[40] To address this stress, horses often learn to back away from the door and resume the weaving motion inside the stall.[33,40]

The goal of therapy for weaving horses should be directed at reducing the frequency of the behavior by minimizing its triggers, not at stopping the behavior altogether. Since feed anticipation is common, concentrates should be eliminated from the diet or fed multiple times, perhaps by a dispenser on a timer. The alternative is ad libitum forage. Lack of social peers is another causative factor, so paddock/pasture time with other horses is important. When this is not possible, than some type of stablemate, such as a dog, goat, or pony, might make an acceptable substitute. If adjacent stall contact is not feasible, an aluminum mirror or life-size picture of another horse's face can significantly reduce the incidence of weaving.[117,118] The addition of exercise as pasture or riding time reduces the incidence of weaving. Free-ranging horses take an estimated 10,000 strides daily as part of their normal feeding pattern, an amount far more than what a stabled horse would take.[25] Drug therapy using SSRIs or dopamine antagonists can be used for severe cases.

NEUROLOGICAL AND MEDICALLY RELATED BEHAVIOR PROBLEMS

Physiological causes of behavior problems are the ones that veterinarians are most comfortable dealing with because they are the closest to traditional illnesses. Medical conditions can be reported as a "behavior problem," but a

FIGURE 10-5 Many barns have antiweave panels on the stall doors, even for horses that are not weavers. It allows the horse to put its head outside the stall to watch barn activities and see other horses more easily. For weavers, these panels severely limit the amount of left-to-right head movement, although most learn to step back to weave.

diagnostic workup shows the physiological origin. As science untangles the mysteries of stereotypies and obsessive-compulsive behaviors, it is likely they will be included within this broad category in the future.

Blindness

Blindness in one or both eyes occurs in approximately 2% of the horse population, presenting some unique considerations for the animal's long-term care.[132] With the loss of one sense, horses begin to substitute other senses. In humans, vision is the primary sense, but in horses, several senses are of approximately equal importance. Hearing will become more important for the blind horse, so lots of soft talking is helpful to it to let it know where the person is and even what they want. The typical head tilt will develop over time.

At first, blind horses are cautious in their paddocks and stalls, but over time, they develop a mental map of their surroundings. They can also do well in pastures that do not have significant obstacles and when ridden by a trusted person who signals obstacles. Precautions include being sure the horse can identify water and feeding locations. Care must be taken to eliminate low-hanging and sharp objects, including barbed wire, to avoid injuries. Gravel can be used in less safe areas of a pasture to signal a warning. When no changes occur in these areas, blind horses do fine. Cases are reported where another horse, or even a dog or goat, becomes a leader for the blind animal, allowing it to go into less familiar pasture areas.

Depression

Clinical depression is a controversial diagnosis in animals, even dogs. There are studies that suggest horses can experience this condition. Support for this diagnosis includes the horse standing, eyes fixed and open, with the neck flat and the same height as the withers. In addition, there is the absence of ear and head movements, reduced response to tactile stimulation, indifference to the sudden approach of a human, higher reactivity to novel objects in familiar locations, and lower than normal cortisol levels following exercise.[133,134] Unfortunately, the number of animals studied has been small, so additional data is needed before this can be identified as a true condition in the horse.

Narcolepsy

Narcolepsy is a sleep disorder that can be confused with sleep deprivation. The condition of narcolepsy is a brain disorder characterized by excessive daytime sleepiness, rapid eye movement (REM) sleep, cataplexy, and collapse. The condition can be genetically transmitted. In Suffolks, Shetland ponies, and American Miniature horses, the condition typically shows up in foals by the age of 6 months.[135–137] Narcolepsy can also occur spontaneously in adult horses of varying breeds.

The specific pathophysiology has not been identified in horses, but in humans and dogs, there is a suggested relationship to a deficiency in the hypothalamic hypocretin (orexin) system. A mutation in the *hypocretin receptor-2* gene affects postsynaptic hypocretin neurotransmission. The acquired form of narcolepsy in dogs is thought to relate to low hypocretin-1 levels in the central nervous system.[135,137,138] Similarly, low hypocretin-1 levels have been reported in narcoleptic horses.[135] Dogs also have a hypersensitive state in the overall muscarinic cholinergic system. The upregulated muscarinic receptors in the pontine reticular formation impair dopamine release, which then affects the emotional state.[139] This is the relationship between pleasurable things triggering narcoleptic attacks.

In humans and dogs, excitement or anticipation are commonly associated with the onset of a narcoleptic episode. While food anticipation can trigger an equine episode, other reported triggers include being tied in a wash rack, saddling, and standing in a stall or pasture.[137]

Two variations of narcolepsy have been reported in horses. Foals tend to show a sudden buckling of the knees and then fall into recumbency with a flaccid paralysis.[139] They appear to be in rapid eye movement (REM) sleep and all spinal reflexes are lost during the attack. These attacks are often triggered by specific stimuli, particularly restraint. The second variation begins when the horse is at least 2 years old.[139] The head gradually lowers and the horse's front end begins to lower and the weight is shifted caudally. The fetlocks may flex, resulting in lesions on the dorsal aspects of fetlocks and carpi. Most horses recover quickly without falling all the way to the ground. Triggering events may be specific, such as when being groomed or saddled.

Because resolution of an episode may occur with noise or touch, it is hard to differentiate narcolepsy or atypical narcolepsy from sleep deprivation. The diagnosis can be confirmed by drugs that stimulate cholinergic activity in the brain, such as with the slow intravenous injection of the anticholinesterase drug physostigmine salicylate.[139,140] A narcoleptic attack occurs in an affected horse within a few minutes, but not in normal animals. Atropine can reverse the effects.

Treatments are not always successful, and controlled studies regarding effectiveness of various treatment protocols have not been done. Drugs that stimulate the monoamine systems in the brain are used to suppress or minimize narcolepsy in several species.[140] Imipramine is the drug of choice because it simultaneously stimulates the aminergic activity of norepinephrine, dopamine, and serotonin related neurons by blocking reuptake and inhibiting cholinergic activity.[140] This is a normal part of keeping animals awake. In some cases, particularly Miniature Horse foals and Thoroughbreds, narcolepsy completely resolves spontaneously.[136,139]

Seizures

While seizures are a neurologic disorder, their manifestation can be confused with behavior problems. Therefore, they need to be included in many lists of possible differential diagnoses. Seizure disorders occur in horses, with the grand mal seizure being the easiest to recognize as a neurologically important medical condition. Fortunately, grand mal seizures are not as common in horses as they are in humans and dogs.[141] Focal (partial) seizures are more likely to be confused with behavior problems because of their limited scope of expression. Depending on what part of the brain is affected, focal seizures can result in a sensation or a motor sign in an otherwise alert and conscious horse that could be confused with a behavior problem. Headshaking and self-mutilation in stallions are conditions where focal seizures should be included as a differential diagnosis. Complex focal seizures affect mentation and would need to be considered as a differential for depression and chronic pain.

Sleep Deprivation

Horses can become sleep deprived and show behaviors similar to those of narcoleptic animals. As the standing horse rapidly slips into deep sleep, weight is shifted caudally and the front end lowers. Individuals affected over a prolonged period of time will show wear lesions on the front of the fetlocks, and/or they will just lie down a lot. Six categories of causes for sleep deprivation have been described.[142] The first category relates to horses that have pain or physical discomfort. Severe arthritis, myopathies, gastric ulcers, and even late-term pregnancy can be associated with this category. A second category relates to the environment—not adjusting to a new stall, too much light, or bad weather. Long-term standing without mental stimulation (also called "*monotony-associated*") is a third category of sleep deprivation causes.[142] A fourth category relates to very aggressive horses, particularly geldings, that are continuously aggressive to one or all horses in a group, missing sleep in order to keep watch. These horses can be helped with the addition of an alpha mare. Lyme disease, even without evidence of joint pain, is responsible for the fifth type of sleep deprivation. Lastly, the most recently identified category resembles sleep terror in humans and is diagnosed using videography.[142]

Equine Self-Mutilation Syndrome

Several years ago, a variation of excessive grooming was identified that had other behavior components. The sequence of behaviors begins with the horse glancing at its flank or occasionally at the chest area.[143] This progresses to biting of the area, resulting in skin lesions (Figure 10-6). There is a squeal, followed by the horse kicking out with a hind leg, bucking, spinning, or rolling. Random muscle twitching of the head or neck might also occur.[132] The equine self-mutilation syndrome (ESMS) primarily occurs in stallions and has a reported incidence of 1.9% in stallions, 0.7% in geldings, and rare in mares.[143,144] Symptoms may begin before sexual maturity but can start in adult stallions too.[132,143,145] As with other stereotypies, stress or anticipation are associated with the expression of this behavior, particularly the anticipation of food. A genetic component is suggested, at least in some of the horses.[132,143,144,146] There may be a relationship with medical problems including pain and lameness.[147] Management protocols are sometimes implicated when the horse has been stalled for a prolonged healing time while being maintained on a high-energy diet. Compounding this is social isolation for the stallion.[145]

A few horses show the behavior almost constantly and are totally dissociated from anything else in the environment. At this point, the behavior is probably an OCD.[144] These horses are usually euthanized because of quality of life issues.

Treatment should be directed at reducing the frequency of the behavior, with the knowledge that it cannot be eliminated completely in most horses. Castration of affected stallions results in improvement in about 70% of them.[143]

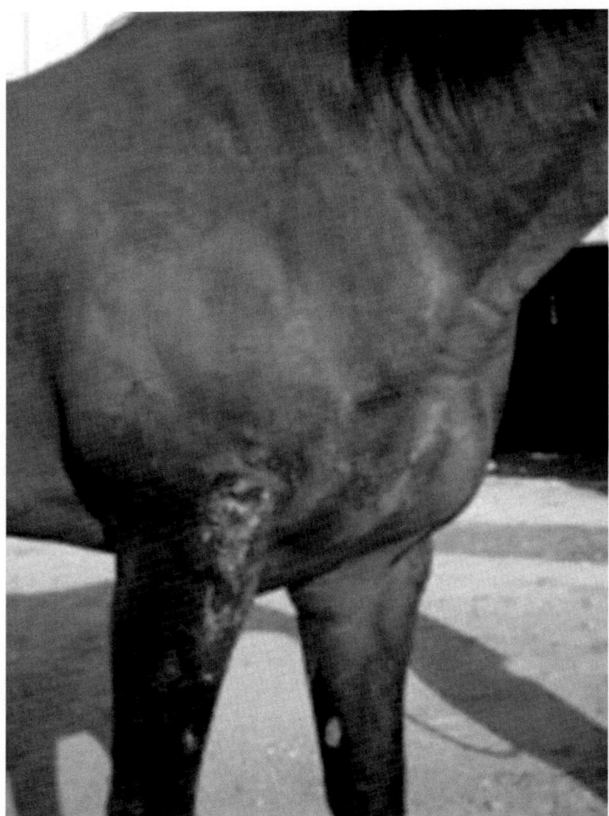

FIGURE 10-6 Horses with the self-mutilation syndrome may bite themselves in the chest, thorax, forelimb, or flank region, resulting in skin lesions. This stallion directed his bites to the chest and forearm.

Changes in social stabling helps 83%.[143] While diet changes alone do not make a difference, increased roughage and minimal grain is recommended to reduce energy intake. Increased exercise such as hand walking and increased social interactions are important too. Even if social contact with other horses is not possible, visual contact or the addition of a nonequine companion can be helpful.[145,148] Several medications have been tried for therapeutic management of the self-mutilation syndrome. Nalmefene, a narcotic antagonist, reduces the behavior's frequency in proportion to the dose used.[146] Dopamine antagonists like acepromazine and fluphenazine are also helpful, as expected with stereotypies in general.[143,149] Alpha-2 antagonists like detomidine and serotonin agonists like buspirone and fluoxetine are somewhat helpful.[144,149]

There is a similar condition that is speculated to have a different etiology. The horse will look at its flank, circle, kick, and perhaps squeal. The self-biting is absent. Visual observation strongly suggests the horse is "feeling" something. There is a condition in humans (inclusion body myositis) and in cats (feline hyperesthesia syndrome) where affected individuals show spontaneous electromyography (EMG) activity in the lumbar epaxial muscles. While EMGs have not been done on affected horses, the possibility of some type of myositis/myopathy needs to be considered in extremely affected animals. Current treatment recommendations in cats include not touching the areas because of hypersensitivity, use of pain medications, and doses of an SSRI to reduce anxiety. Treatment is palliative rather than curative.

REFERENCES

1. Borstel UKV, Erdmann C, Maier M, Garlipp F. Relationships between owner-reported behavior problems and husbandry; use and management of horses. *J Vet Behav* 2016;**15**:92–3.
2. Dallaire A. Stress and behavior in domestic animals: temperament as a predisposing factor to stereotypies. *Ann N Y Acad Sci* 1993;**697**:269–74.
3. Tadich T, Weber C, Nicol CJ. Prevalence and factors associated with abnormal behaviors in Chilean racehorses: a direct observational study. *J Equine Vet Sci* 2013;**33**(2):95–100.
4. American Veterinary Medical Association. *U.S. Pet Ownership & Demographics Sourcebook*. Schaumburg, IL: American Veterinary Medical Association; 2012.
5. Hockenhull J, Creighton E. The day-to-day management of UK leisure horses and the prevalence of owner-reported stable-related and handling behaviour problems. *Anim Welf* 2015;**24**(1):29–36.
6. Hockenhull J, Creighton E. Management practices associated with owner-reported stable-related and handling behavior problems in UK leisure horses. *Appl Anim Behav Sci* 2014;**155**:49–55.
7. Löckener S, Reese S, Erhard M, Wöhr A-C. Pasturing in herds after housing in horseboxes induces a positive cognitive bias in horses. *J Vet Behav* 2016;**11**:50–5.
8. Madigan JE, Bell SA. Owner survey of headshaking in horses. *J Am Vet Med Assoc* 2001;**219**(3):334–7.
9. Christoffersen M, Lehn-Jensen H, Bøgh IB. Referred vaginal pain: cause of hypersensitivity and performance problems in Mares? A clinical case study. *J Equine Vet Sci* 2007;**27**(1):32–6.
10. Crowell-Davis SL. Stereotypic behavior and compulsive disorder. *Compendium Equine* 2008;**3**(5):248–50251.
11. Mills DS, Alston RD, Rogers V, Longford NT. Factors associated with the prevalence of stereotypic behavior amongst thoroughbred horses passing through auctioneer sales. *Appl Anim Behav Sci* 2002;**78**(2–4):115–24.
12. Low M. Stereotypies and behavioural medicine: confusions in current thinking. *Aust Vet J* 2003;**81**(4):192–8.
13. Broom DM, Kennedy MJ. Stereotypies in horses: their relevance to welfare and causation. *Equine Vet Educ* 1993;**5**(3):151–4.
14. Mason GJ. Stereotypies: a critical review. *Anim Behav* 1991;**41**(6):1015–37.
15. Latham NR, Mason GJ. Maternal deprivation and the development of stereotypic behavior. *Appl Anim Behav Sci* 2008;**110**(10):84–108.
16. Albright JD, Witte TH, Rohrbach BW, Reed A, Houpt KA. Efficacy and effects of various anti-crib devices on behaviour and physiology of crib-biting horses. *Equine Vet J* 2016;**48**(6):727–31.
17. Flannigan G, Stookey JM. Day-time time budgets of pregnant mares housed in tie stalls: a comparison of draft versus light mares. *Appl Anim Behav Sci* 2002;**78**(2–4):125–43.

18. Nicol C. Understanding equine stereotypies. *Equine Vet J Suppl* 1999;**28**:20–5.
19. Waters AJ, Nicol CJ, French NP. Factors influencing the development of stereotypic and redirected behaviours in young horses: findings of a four year prospective epidemiological study. *Equine Vet J* 2002;**34**(6):572–9.
20. Nicol, C.J. (2000): Equine stereotypies. In: Recent advances in companion animal behavior problems, Houpt, K.A., ed., International Veterinary Information Service, Ithaca, NY (www.ivis.org) [downloaded October 27,2017].
21. Bachmann I, Bernasconi P, Herrmann R, Weishaupt MA, Stauffacher M. Behavioural and physiological responses to an acute stressor in crib-biting and control horses. *Appl Anim Behav Sci* 2003;**82**(4):297–311.
22. Christie JL, Hewson CJ, Riley CB, McNiven MA, Dohoo IR, Bate LA. Management factors affecting stereotypies and body condition score in nonracing horses in Prince Edward Island. *Can Vet J* 2006;**47**(2):136–43.
23. Cooper JJ, Mcall N, Johnson S, Davidson HPB. The short-term effects of increasing meal frequency on stereotypic behaviour of stabled horses. *Appl Anim Behav Sci* 2005;**90**(3–4):351–64.
24. Muñoz L, Torres J, Sepúlveda O, Rehhof C, Ortiz R. Frequency of stereotyped abnormal behaviour in stabled Chilean horses. *Arch Med Vet* 2009;**41**(1):73–6.
25. Roberts K, Hemmings AJ, McBride SD, Parker MO. Causal factors of oral versus locomotor stereotypy in the horse. *J Vet Behav* 2017;**20**:37–43.
26. Simpson BS. Behavior problems in horses: cribbing and wood chewing. *Vet Med* 1998;**93**(11):999–1004.
27. Stanley SO, Cant JP, Osborne VR. A pilot study to determine whether a tongue-activated liquid dispenser would mitigate abnormal behavior in pasture-restricted horses. *J Equine Vet Sci* 2015;**35**(11 – 12):973–6.
28. Hausberger M, Gautier E, Biquand V, Lunel C, Jégo P. Could work be a source of behavioural disorders? A study in horses. *PLoS One* 2009;**4**(10). https://doi.org/10.1371/journal.pone.0007625.
29. Dezfouli MM, Tavanaeimanesh H, Naghadeh BD, Bokaei S, Corley K. Factors associated with stereotypic behavior in Iranian stabled horses. *Comp Clin Pathol* 2014;**23**(5):1651–7.
30. McGreevy PD, French NP, Nicol CJ. The prevalence of abnormal behaviours in dressage, eventing and endurance horses in relation to stabling. *Vet Rec* 1995;**137**(2):36–7.
31. Cooper JJ, Albentosa MJ. Behavioural adaptation in the domestic horse: potential role of apparently abnormal responses including stereotypic behaviour. *Livest Prod Sci* 2005;**92**(2):177–82.
32. Hothersall B, Casey R. Undesired behavior in horses: A review of their development, prevention, management and association with welfare. *Equine Vet Educ* 2012;**24**(9):479–85.
33. McBride S, Hemmings A. A neurologic perspective of equine stereotypy. *J Equine Vet Sci* 2009;**29**(1):10–6.
34. Waran NK, Clarke N, Farnworth M. The effects of weaning on the domestic horse (*Equus caballus*). *Appl Anim Behav Sci* 2008;**110**(1–2):42–57.
35. McGreevy P, Nicol C. Physiological and behavioral consequences associated with short-term prevention of crib-biting in horses. *Physiol Behav* 1998;**65**(1):15–23.
36. Pell SM, McGreevy PD. A study of cortisol and beta-endorphin levels in stereotypic and normal Thoroughbreds. *Appl Anim Behav Sci* 1999;**64**(2):81–90.
37. Wickens CL, Heleski CR. Crib-biting behavior in horses: a review. *Appl Anim Behav Sci* 2010;**128**(1–4):1–9.
38. Fureix C, Benhajali H, Henry S, Bruchet A, Prunier A, Ezzaouia M, Coste C, Hausberger M, Palme R, Jego P. Plasma cortisol and faecal cortisol metabolites concentrations in stereotypic and non-stereotypic horses: do stereotypic horses cope better with poor environmental conditions? *BMC Vet Res* 2013;**9**:3.

39. Briefer Freymond S, Bardou D, Briefer EF, Bruckmaier R, Fouché N, Fleury J, Maigrot A-L, Ramseyer A, Zuberbühler K, Bachmann I. The physiological consequences of crib-biting in horses in response to an ACTH challenge test. *Physiol Behav* 2015;**151**:121–8.
40. McBride SD, Cuddleford D. The putative welfare-reducing effects of preventing equine stereotypic behaviour. *Anim Welf* 2001;**10**(2):173–89.
41. Parker M, Redhead ES, Goodwin D, McBride SD. Impaired instrumental choice in crib-biting horses (*Equus caballus*). *Behav Brain Res* 2008;**191**(1):137–40.
42. Parker MO. *Behavioural correlates of the equine stereotypy phenotype*. https://eprints.soton.ac.uk/67410/1/Matt_Parker_PhD_FINAL_pdf; 2008 [downloaded September 7,2017].
43. Alberghina D, De Pasquale A, Piccione G, Vitale F, Panzera M. Gene expression profile of cytokines in leukocytes from stereotypic horses. *J Vet Behav* 2015;**10**(6):556–60.
44. Bahn PG. Crib-biting: tethered horses in the paleolithic? *World Archeol* 1980;**12**(2):212–7.
45. Hemmings A, McBride SD, Hale CE. Perseverative responding and the aetiology of equine oral stereotypy. *Appl Anim Behav Sci* 2007;**104**(1–2):143–50.
46. McDonnell SM. Important lessons from free-running equids. *Equine Vet J* 1998;**30**(S27):58–9.
47. Houpt KA. Equine behavior problems in relation to humane care. *Anthrozoös* 1987;**1**(3):184–7.
48. Bachmann I, Audigé L, Stauffacher M. Risk factors associated with behavioural disorders of crib-biting, weaving and box-walking in Swiss horses. *Equine Vet J* 2003;**35**(2):158–63.
49. Cooper JJ, Mason GJ. The identification of abnormal behavior and behavioural problems in stabled horses and their relationship to horse welfare: a comparative review. *Equine Vet J* 1998;**30**(S27):5–9.
50. Heleski CR, Shelle AC, Nielsen BD, Zanella AJ. Influence of housing on weanling horse behavior and subsequent welfare. *Appl Anim Behav Sci* 2002;**78**(2–3):291–302.
51. McGreevy PD, Cripps PJ, French NP, Green LE, Nicol CJ. Management factors associated with stereotypic and redirected behavior in the Thoroughbred horse. *Equine Vet J* 1995;**27**(2):86–91.
52. Ninomiya S. Social leaning and stereotypy in horses. *Behav Process* 2007;**76**(1):22–3.
53. Parker M, Goodwin D, Redhead ES. Survey of breeders' management of horses in Europe, North America and Australia: comparison of factors associated with the development of abnormal behaviour. *Appl Anim Behav Sci* 2008;**114**(1–2):206–15.
54. Redbo I, Redbo-Torstensson P, Odberg FO, Hedendahl A, Holm J. Factors affecting behavioural disturbances in race-horses. *Anim Sci* 1998;**66**(2):475–81.
55. Sarrafchi A, Blokhuis HJ. Equine stereotypic behaviors: causation, occurrence, and prevention. *J Vet Behav* 2013;**8**(5):386–94.
56. Young T, Creighton E, Smith T, Hosie C. A novel scale of behavioural indicators of stress for use with domestic horses. *Appl Anim Behav Sci* 2012;**140**(1–2):33–43.
57. Gillham SB, Dodman NH, Shuster L, Kream R, Rand W. The effect of diet on cribbing behavior and plasma β-endorphin in horses. *Appl Anim Behav Sci* 1994;**41**(3–4):147–53.
58. Hemmann K, Radkallio M, Kanerva K, Hänninen L, Pastell M, Palviainen M, Vainio O. Circadian variation in ghrelin and certain stress hormones in crib-biting horses. *Vet J* 2012;**193**(1):97–102.
59. Lebelt D, Zanella AJ, Unshelm J. Physiological correlates associated with cribbing behavior in horses: changes in thermal threshold, heart rate, plasma β-endorphin and serotonin. *Equine Vet J* 1998;**30**(S27):21–7.
60. Binev R. Weaving horses. Etiological, clinical and paraclinical investigation. *Int J Adv Res* 2015;**3**(3):629–36.
61. McBride SD, Long L. Management of horses showing stereotypic behavior, owner perception and the implications for welfare. *Vet Rec* 2001;**148**(26):799–802.

62. Albright JD, Mohammed HO, Heleski CR, Wickens CL, Houpt KA. Crib-biting in US horses: breed predispositions and owner perceptions of aetiology. *Equine Vet J* 2009;**41** (5):455–8.
63. Nagy K, Schrott A, Kabai P. Possible influence of neighbours on stereotypic behaviour in horses. *Appl Anim Behav Sci* 2008;**111**(3–4):321–8.
64. Benhajali H, Ezzaouia M, Lunel C, Charfi F, Hausberger M. Stereotypic behaviours and mating success in domestic mares. *Appl Anim Behav Sci* 2014;**153**:36–42.
65. Hausberger M. Lower learning abilities in stereotypic horses. *Appl Anim Behav Sci* 2007;**107** (3–4):299–306.
66. Henderson JV, Waran NK. Reducing equine stereotypies using an Equiball™. *Anim Welf* 2001;**10**(1):73–80.
67. Noh HJ, Tang R, Flannick J, O'Dushlaine C, Swofford R, Howrigan D, Genereux DP, Johnson J, van Grootheest G, Grünblatt E, Andersson E, Djurfeldt DR, Patel PD, Koltookian M, Hultman CM, Pato MT, Pato CN, Rasmussen SA, Jenike MA, Hanna GL, Stewart SE, Knowles JA, Ruhrmann S, Grabe H-J, Wagner M, Rück C, Mathews CA, Walitza S, Cath DC, Feng G, Karlsson EK, Lindblad-Toh K. Integrating evolutionary and regulatory information with multispecies approach implicates genes and pathways in obsessive-compulsive disorder. *Nat Commun* 2017;**8**:https://doi.org/10.1038/s41467-017-00831-x[Article Number 774].
68. Rendon RA, Shuster L, Dodman NH. The effect of the NMDA receptor blocker, dextromethorphan, on cribbing in horses. *Pharmacol Biochem Behav* 2001;**68**(1):49–51.
69. Nicol CJ, Badnell-Waters AJ. Suckling behaviour in domestic foals and the development of abnormal oral behaviour. *Anim Behav* 2005;**70**(1):21–9.
70. Whisher L, Raum M, Pina L, Pérez L, Erb H, Houpt C, Houpt K. Effects of environmental factors on cribbing activity by horses. *Appl Anim Behav Sci* 2011;**135**(1–2):63–9.
71. Houpt KA. Behavior in horses, In: *Notes from talk given at the southwest veterinary symposium, Ft. Worth, TX*; 2010.
72. Lane JG. Recent studies on crib-biting horses. *Equine Vet J* 1998;**30**(S27):59–61.
73. McGreevy PD, Richardson JD, Nicol CJ, Lane JG. Radiographic and endoscopic study of horses performing an oral based stereotypy. *Equine Vet J* 1995;**27**(2):92–5.
74. Houpt KA. New perspectives on equine stereotypic behavior. *Equine Vet J* 1995;**27**(2):82–3.
75. Mills DS, Taylor KD, Cooper JJ. Weaving, headshaking, cribbing, and other stereotypies. *Am Assoc Equine Pract Proc* 2005;**51**:221–30.
76. Ellis AD, Redgate S, Zinchenko S, Owen H, Barfoot C, Harris P. The effect of presenting forage in multi-layered haynets and at multiple sites on night time budgets of stables horses. *Appl Anim Behav Sci* 2015;**171**:108–16.
77. Hothersall B, Nicol C. Role of diet and feeding in normal and stereotypic behaviors in horses. *Vet Clin N Am Equine Pract* 2009;**25**(1):167–81.
78. Nicol CJ, Davidson HPD, Harris PA, Waters AJ, Wilson AD. Study of crib-biting and gastric inflammation and ulceration in young horses. *Vet Rec* 2002;**151**(22):658–62.
79. Clegg HA, Buckley P, Friend MA, McGreevy PD. The ethological and physiological characteristics of cribbing and weaving horses. *Appl Anim Behav Sci* 2008;**109**(1):68–76.
80. Wickens CL, McCall CA, Bursian S, Hanson R, Heleski CR, Liesman JS, McElhenney WH, Trottier NL. Assessment of gastric ulceration and gastrin response in horses with history of crib-biting. *J Equine Vet Sci* 2013;**33**(9):739–45.
81. Roberts K, Hemmings A, Moore-Colyer M, Hale C. Cognitive differences in horses performing locomotor versus oral stereotypic behavior. *Appl Anim Behav Sci* 2015;**168**:37–44.
82. Parker M, McBride SD, Redhead ES, Goodwin D. Differential place and response learning in horses displaying an oral stereotypy. *Behav Brain Res* 2009;**200**(1):100–5.

83. Nagy K, Bodó G, Bárdos G, Bánszky N, Kabai P. Differences in temperament traits between crib-biting and control horses. *Appl Anim Behav Sci* 2010;**122**(1):41–7.
84. Omidi A, Vakili S, Nazifi S, Parker MO. Acute-phase proteins, oxidative stress, and antioxidant defense in crib-biting horses. *J Vet Behav* 2017;**20**:31–6.
85. McBride SD, Hemmings A. Altered mesoaccumbens and nigro-striatal dopamine physiology is associated with stereotypy development in a non-rodent species. *Behav Brain Res* 2005;**159**(1):113–8.
86. Karson CN. Spontaneous eye-blink rates and dopaminergic systems. *Brain* 1983;**106**(3):643–53.
87. Luescher UA, McKeown DB, Dean H. A cross-sectional study on compulsive behavior (stable vices) in horses. *Equine Vet J* 1998;**30**(S27):14–8.
88. Hemmann K, Raekallio M, Vainio O, Juga J. Crib-biting and its heritability in Finnhorses. *Appl Anim Behav Sci* 2014;**156**:37–43.
89. Vecchiotti GG, Galanti R. Evidence of heredity of cribbing, weaving and stall-walking in Thoroughbred horses. *Livest Prod Sci* 1986;**14**(1):91–5.
90. Hemmann K, Ahonen S, Raekallio M, Vainio O, Lohi H. Exploration of known stereotypic behaviour-related candidate genes in equine crib-biting. *Animal* 2014;**8**(3):347–53.
91. Houpt KA. A preliminary answer to the question of whether cribbing causes salivary secretion. *J Vet Behav* 2012;**7**(5):322–4.
92. Moeller BA, McCall CA, Silverman SJ, McElhenney WH. Estimation of saliva production in crib-biting and normal horses. *J Equine Vet Sci* 2008;**28**(2):85–90.
93. Mills DS, Macleod CA. The response of crib-biting and windsucking in horses to dietary supplementation with an antacid mixture. *Ippologia* 2002;**13**(2):33–41.
94. Fukumoto K, Katayama K, Katayma T, Miyazatao M, Kangawa K, Murakami N. Synergistic action of gastrin and ghrelin on gastric acid secretion in rats. *Biochem Biophys Res Commun* 2008;**374**(1):60–3.
95. Hemmann KE, Koho NM, Vainio OM, Raekallio MR. Effects of feed on plasma leptin and ghrelin concentrations in crib-biting horses. *Vet J* 2013;**198**(1):122–6.
96. Archer DC, Pinchbeck GL, French NP, Proudman CJ. Risk factors for epiploic foramen entrapment colic: an international study. *Equine Vet J* 2008;**40**(3):224–30.
97. Malamed R, Berger J, Bain MJ, Kass P, Spier SJ. Retrospective evaluation of crib-biting and windsucking behaviours and owner-perceived behavioural traits as risk factors for colic in horses. *Equine Vet J* 2010;**42**(8):686–92.
98. Scantlebury CE, Archer DC, Proudman CJ, Pinchbeck GL. Recurrent colic in the horse: incidence and risk factors for recurrence in the general practice population. *Equine Vet J* 2011;**43**(Suppl. 39):81–8.
99. Archer DC, Freeman DE, Doyle AJ, Proudman CJ, Edwards GB. Association between cribbing and entrapment of the small intestine in the epiploic foramen in horses: 68 cases (1991–2002). *J Am Vet Med Assoc* 2004;**224**(4):562–4.
100. Archer DC, Pinchbeck GL, French NP, Proudman CJ. Risk factors for epiploic foramen entrapment colic in a UK horse population: a prospective case-control study. *Equine Vet J* 2008;**40**(4):405–10.
101. Archer DC, Proudman CJ, Pinchbeck G, Smith JE, French NP, Edwards GB. Entrapment of the small intestine in the epiploic foramen in horses: a retrospective analysis of 71 cases recorded between 1991 and 2001. *Vet Rec* 2004;**155**(25):793–7.
102. Albanese V, Munsterman AS, DeGraves FJ, Hanson RR. Evaluation of intra-abdominal pressure in horses that crib. *Vet Surg* 2013;**42**(6):658–62.

103. McGreevy PD, Webster AJF, Nicol CJ. Study of the behavior, digestive efficiency and gut transit times of crib-biting horses. *Vet Rec* 2001;**148**(19):592–6.
104. Houpt, K.A. (1994): Personal communication.
105. Albright J, Sun X, Houpt K. Does cribbing behavior in horses vary with dietary taste or direct gastric stimuli? *Appl Anim Behav Sci* 2017;**189**:36–40.
106. Houpt KA. The effect of diet on cribbing, In: *Proceedings of the American College of Veterinary Behaviorists Scientific paper session*; 2003. p. 76.
107. Omidi A, Jafari R, Nazifi S, Parker MO. Potential roles of selenium and zinc in the pathophysiology of crib-biting behavior in horses. *J Vet Behav* 2018;**23**:10–4.
108. Kennedy MJ, Schwabe AE, Broom DM. Crib-biting and wind-sucking stereotypies in the horse. *Equine Vet Educ* 1993;**5**(3):142–7.
109. McGreevy PD, Nicol CJ. Prevention of crib-biting: a review. *Equine Vet J* 1998;-**30**:35–8Supplement 27.
110. McGreevy PD, Nicol CJ. The effect of short term prevention on the subsequent rate of crib-biting in Thoroughbred horses. *Equine Vet J* 1998;**30**:30–4. Supplement 27.
111. Fjeldborg J. Evaluation of two different surgical techniques in the treatment of cribbing. *Equine Vet J* 1998;**30**(S27):61–2.
112. Ritzberger-Matter G, Kaegi B. Retrospective analysis of the success rate of surgical treatment of aerophagia in horses at the veterinary surgical clinic, University of Zurich. *Equine Vet J* 1998;**30**(S27):62.
113. Marcella KL. Common behavior problems in horses. *Equine Pract* 1988;**10**(6):22–6.
114. Dodman NH, Shuster L, Court MH, Dixon R. Investigation into the use of narcotic antagonists in the treatment of a stereotypic behavior pattern (crib-biting) in the horse. *Am J Vet Res* 1987;**48**(2):311–9.
115. Mazzola, S., Palestrini, C., Cannas, S., Fè, E., Bagnato, G.L., Vigo, D., Frank, D. and Minero, M. (2016): Efficacy of a feed dispenser for horses in decreasing cribbing behaviour. Vet Med Int 2016: ID 4698602 [downloaded February 13,2017].
116. Leme DP, Parsekian ABH, Kanaan V, Hötzel MJ. Management, health, and abnormal behaviors of horses: a survey in small equestrian centers in Brazil. *J Vet Behav* 2014;**9**(3):114–8.
117. McAfee LM, Mills DS, Cooper JJ. The use of mirrors for the control of stereotypic weaving behavior in the stabled horse. *Appl Anim Behav Sci* 2002;**78**(2–4):159–73.
118. Mills DS, Riezebos M. The role of the image of a conspecific in the regulation of stereotypic head movements in the horse. *Appl Anim Behav Sci* 2005;**91**(1–2):155–65.
119. Mills DS, Cook S, Taylor K, Jones B. Analysis of the variations in clinical signs shown by 254 cases of equine headshaking. *Vet Rec* 2002;**150**(8):236–40.
120. Newton SA, Knottenbelt DC, Eldridge PR. Headshaking in horses: possible aetiopathogenesis suggested by the results of diagnostic tests and several treatment regimes used in 20 cases. *Equine Vet J* 2000;**32**(3):208–16.
121. Roberts VLH, Patel NK, Tremaine WH. Neuromodulation using percutaneous electrical nerve stimulation for the management of trigeminal-mediated headshaking: a safe procedure resulting in medium-term remission in five of seven horses. *Equine Vet J* 2016;**48**(2):201–4.
122. Wilkins, P.A. (1997): Cyproheptadine: Medical treatment for photic headshakers. Compend Contin Educ Pract Vet: Equine, 98–99,111.
123. McDonnell SM. Tongue trouble. *The Horse* 2009;**XXVI**(9):54–5.
124. McDonnell SM. Young horse habits. *The Horse* 2010;**XXVII**(4):64–5.
125. Normando S, Meers L, Samuels WE, Faustini M, Ödberg FO. Variables affecting the prevalence of behavioural problems in horses. Can riding style and other management factors be significant? *Appl Anim Behav Sci* 2011;**133**(3–4):186–98.

126. Fox AE, Belding DL. Reducing pawing in horses using positive reinforcement. *J Appl Behav Anal* 2015;**48**(4):1–5.
127. Miller RM. Common misbehaviors (vices in hand and under saddle). *J Equine Vet Sci* 1997;**17**(2):67–9.
128. Ninomiya S, Sato S, Sugawara K. Weaving in stabled horses and its relationship to other behavioural traits. *Appl Anim Behav Sci* 2007;**106**(1–3):134–43.
129. Cooper JJ, McDonald L, Mills DS. The effect of increasing visual horizons on stereotypic weaving: implications for the social housing of stabled horses. *Appl Anim Behav Sci* 2000;**69**(1):67–83.
130. Nurnberg HG, Keith SJ, Paxton DM. Consideration of the relevance of ethological animal models for human repetitive behavioral spectrum disorders. *Biol Psychiatry* 1997;**41**(2):226–9.
131. Scantlebury CE, Archer DC, Proudman CJ, Pinchbeck GL. Management and horse-level risk factors for recurrent colic in the UK general equine practice population. *Equine Vet J* 2015;**47**(2):202–6.
132. Marcella KL. *Poor socialization can stem from a variety of circumstances.* DVM Newsmagazine; 2006. p. 2E–5E.
133. Fureix C, Jego P, Henry S, Lansade L, Hausberger M. Towards an ethological animal model of depression? A study on horses. *PLoS One* 2012;**7**(6). https://doi.org/10.1371/journal.pone.0039280.
134. Rochais C, Henry S, Fureix C, Hausberger M. Investigating attentional processes in depressive-like domestic horses (*Equus caballus*). *Behav Process* 2016;**124**:93–6.
135. Aleman M, Williams DC, Holliday T. Sleep and sleep disorders in horses. *Am Assoc Equine Pract Proc* 2008;**54**:180–5.
136. Geering RR. Clinical aspects of equine narcolepsy. *Equine Vet J* 1998;**30**(S27):50.
137. Hines MT. *Narcolepsy: more common than you think?* In: *Proceedings of the North American veterinary conference*; 2005. p. 189–90. www.ivis.org/ [downloaded 4/6/2017].
138. Reed SM, Andrews FM. Disorders of the neurologic system. In: Reed SM, Bayley WM, Sellon DC, editors. *Equine internal medicine*. St. Louis: Saunders; 2010. p. 545–681.
139. Moore LA, Johnson PJ. Narcolepsy in horses. *Compend Contin Educ Pract Vet* 2000;**21**(1):86–90.
140. Peck KE, Hines MT, Mealey KL, Mealey RH. Pharmacokinetics of imipramine in narcoleptic horses. *Am J Vet Res* 2001;**62**(5):783–6.
141. Lacombe VA. Seizures in horses: diagnosis and classification. *Vet Med Res Rep* 2015;**6**:301–8.
142. Bertone JJ. 6 types of sleep deprivation in horses. *The Horse* 2017;**XXXIV**(12):14.
143. Dodman NH, Normile JA, Shuster L, Rand W. Equine self-mutilation syndrome. *J Am Vet Med Assoc* 1994;**204**(8):1219–23.
144. Luescher UA. More on self-mutilative behavior in horses. *J Am Vet Med Assoc* 1993;**203**(9):1252–3.
145. McClure SR, Chaffin KM, Beaver BV. Nonpharmacologic management of stereotypic self-mutilative behavior in a stallion. *J Am Vet Med Assoc* 1992;**200**(12):1975–7.
146. Dodman NH, Shuster L, Court MH, Patel J. Use of a narcotic antagonist (nalmefene) to suppress self-mutilative behavior in a stallion. *J Am Vet Med Assoc* 1988;**192**(11):1585–6.
147. Bedford SJ, McDonnell SM, Tulleners E, King D, Habecker P. Squamous cell carcinoma of the urethral process in a horse with hemospermia and self-mutilation behavior. *J Am Vet Med Assoc* 2000;**216**(4):551–3.
148. Houpt KA. Self-directed aggression: a stallion behavior problem. *Equine Pract* 1983;**5**(2):6–8.
149. Dodman NH, Shuster L, Patronek GJ, Kinney L. Pharmacologic treatment of equine self-mutilation syndrome. *Int J Appl Res Vet Med* 2004;**2**(2):90–8.

Chapter 11

Equine Welfare

Animal welfare is a topic of great concern to horsemen and horsewomen. It is also a big concern to the general public. The ideal "natural horse" is the free-ranging mustang, even though the public lacks a personal, firsthand perspective of what the horse's day-to-day life actually involves, either in the West or in the barn. Cruelty to horses upsets everyone, or at least it should. Perceptions of what constitutes good welfare differ based on an individual's background, sex, experience, and ethical views. These differences make it particularly important to look at a broad view of what constitutes equine welfare and how it relates to behavior.

Ideas about what constitutes good welfare change over time, particularly in developed countries, and several things drive this change. Women now dominate the horse industry in both the United States and Europe.[1–3] They tend to be the biggest proponents of "persuasive rather than coercive techniques."[1] Urbanization of society means people depend on several sources for their information, rather than learning horse-handling techniques handed down through the generations. Because this information comes later in life, urban and suburban dwellers are more likely to question the humaneness of techniques and listen to alternative sources than in the past.[1,4] The public's expression of empathy concerns approximately 60% of horse show exhibitors, with concerns that it might be extreme. There is also a significant belief that behavior should be part of any welfare assessment.[2] The third significant change is the information explosion. This has allowed easy access to thousands of articles and books about different horse-handling techniques.[1] As significant as the previous three changes are, ready access to electronic media is perhaps even more important. Both good and bad occurrences can go viral because of phone videos. Reputations of trainers, riders, farriers, and veterinarians can be destroyed in minutes, and not always for justifiable reasons. The smart trainer rides as if a video of what s/he does could show up on YouTube at any time. For those who show horses, role models play an important part of how their opinions are influenced (Table 11-1).

Equine Behavioral Medicine. https://doi.org/10.1016/B978-0-12-812106-1.00011-5

TABLE 11-1 Factors Influencing an Exhibitor's Decision About Their Show Horse[2]

Type of Influence	% Rated as Very or Extremely Important
Association's handbook or rules	58.6%
Hired trainer's opinion	48.1%
Hired riding instructor's opinion	46.7%
Judge's opinion	37.1%
Judge's placing of individuals	27.2%
Perceived social acceptance	21.4%
Superior competitor's opinion	20.2%
Seeing a superior competitor doing it	18.3%
Close friend's opinion	18.2%
Family's opinion	18.2%

Other possibilities like veterinarians, farriers, and the Internet were not part of the survey.

DEFINING ANIMAL WELFARE

Welfare is defined in dictionaries as involving the physical and emotional wellbeing of the animal. To be a little more specific, the World Organization for Animal Health (OIE) defines animal welfare as "how an animal is coping with the conditions in which it lives." A slightly different version defines welfare as "the physical and psychological state of an animal in regards to its attempt to cope with its environment."[5] Thus, welfare includes physical traits that can be measured, and psychological components that must be assessed behaviorally. The ability to scientifically evaluate various components associated with welfare differentiates it from animal rights, which are individually held philosophical beliefs.

Because all aspects of the animal's state of being should be subjected to evaluation, outcomes can range from very negative to very positive (Figure 11-1). Using several assessment tools and collectively compiling the results ensures a better picture of welfare status compared to using a single parameter. Broadly, there are three major inputs to a horse's welfare. The first is the animal itself—its genetics, personality, past experiences, and physical state. The environment in which the animal lives is a second important aspect of welfare, and the third is people that interact with the horse, including trainers, riders, stable help, and others. How they interact is important in how the horse and environment interface.

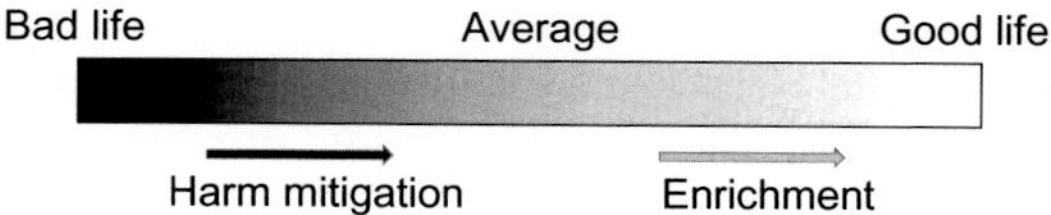

FIGURE 11-1 Welfare is a spectrum of conditions ranging from very poor to very good. At some point in between, harm is mitigated and life becomes worth living.

From the inputs come the outputs, which can also be combined into three broad categories. Health and production measures are the first of the three. This category is the focus for veterinarians and animal scientists—disease treatment, disease prevention, and production statistics. Included would be the evaluation of the body condition, physical appearance, soundness, morbidity, and mortality. Physiological measures constitute the second assessment category. Things like heart rate, cortisol levels, and blood counts provide relatively objective data. Behavior is the third output. Within this would be responses to being handled and abnormal behaviors, including those of discomfort or pain, fear or anxiety, and stress.[6] Some people argue that domesticated animals should be free to express all natural behaviors. In other words, all horses should have a lifestyle approaching that of mustangs, and all mustangs should be free-ranging. Two factors are not considered with such an argument. The first is that free-ranging animals can have poor welfare such as that associated with overgrazing, drought, aggression, insects, and freezing temperatures. The other is that many behaviors are stimulus driven and not internally generated.[7] Without the stimulus, there is no need for the animal to do the behavior. Another group of concerned individuals argue that horses should not be used in competitive sports and cite figures of breakdowns and deaths in support of their ethical argument. They further argue that human competitors can choose whether to compete or not, but horses have no say.[8] The result of such an argument is that horses would be worked at a job like plowing a field or pulling a buggy, used for pleasure riding, or become "pasture ornaments." Horses relegated to these categories could be viewed in less favorable terms by owners and have poorer welfare than competition horses. A final group opposes the use of the horse, and other domestic animals, for any purpose.[8] In that case, domestic animals would no longer have a purpose and could just as well go extinct. After all, domestication happened because there was a human need.

RELATIONSHIPS BETWEEN BEHAVIOR AND WELFARE

"It is an essential condition for keeping horses that handlers, riders, trainers, farriers and veterinarians have proper knowledge of the behavior of the horse in order to fulfil their natural needs and guarantee their welfare."[9]

The thoughts of a horse can never actually be known, so psychological evaluation of welfare is dependent on behaviors shown by the animal. One perspective of importance is to measure the horse's preference. Preference tests are commonly used to study taste likes and dislikes.[10, 11] The same concept can be used to look at other preferences, such as the type of bedding, the timing of lights, and amount of exercise.[12, 13] The results provide some idea of how the animal would prefer to be kept.

How hard a horse is willing to work to achieve a goal is one measure of preference. The horse is taught to push a panel to reach a specific goal and measure of how much force is used or how many times they will repetitively push to reach that goal indicates the relative importance. While a horse will push a panel approximately 200 times to get food, it will do so approximately 50 times to get to a friendly horse and 40 times to get out into a paddock.[13]

Preference tests have been useful relative to stall environments. Fifty-two percent of horses stand where they can see other horses, but if they cannot see another horse, they move around more.[13] Fifty-five percent of horses prefer shavings to straw, and more prefer straw-like substances to sawdust.[13, 14] In a choice of shavings to concrete, 65% prefer the shavings and of those that prefer concrete, they do not lie down. Deeper bedding is preferred to slight amounts.[13] If taught how to turn lights or heaters on and off, horses show an 18-h maximum for lights being on and 2h for a heater in a New England winter.[13] Similarly, horses can be trained to indicate when they want a blanket on or off by touching specific locations on a panel.[15]

Exercise preferences show similarities to humans—horses really do not want to work; 9 of 10 horses will go to their stall rather than go to a treadmill.[16] If allowed to choose a stall, paddock access by themselves, or paddock access with other horses, the latter is chosen. Even then, the amount of time spent in the paddock is approximately a half hour before returning to a stall.[16] Food remains preferable to exercise or social interaction.[17]

Preference tests do have limitations. Just because a horse prefers one thing over another, that choice is not always in the best interest of the animal. A person might prefer chocolate and potato chips to broccoli and carrots, but the vegetables are better for them.

Comparative studies of horses kept in enriched environments or standard ones are useful at looking at whether environments can make subtle differences in welfare.[18–20] Control group weanlings were kept in individual stalls bedded with shavings, fed three times a day, and turned out into individual paddocks three times a week. Those in the enriched environment group were kept in large, individual, straw-bedded stalls, fed three times a day, exposed to various objects, odors, and music, and turned into pasture 17h a day. Enriched group weanlings learned better, had fewer behavioral indicators of stress, and had an induced expression of genes involved in cell growth and proliferation.[19] Other studies have shown that broodmares spend less time moving, standing resting, standing alert, and aggressing and more time with positive social interactions and eating, when hay is provided in paddocks.[18] Access to herdmates improves the general attitude of horses that otherwise spend long periods alone, and retaining the same pasturemates is better for overall attitude too.[20, 21]

Using science to answer questions about what is best for the horse can provide meaningful information for the animal's welfare.[22] Examples include varied subjects such as how to measure rein tension, how much rein tension is too much, how to measure pain, and how to more accurately determine which weanling will excel in a specific performance event.[23]

MEASURING ANIMAL WELFARE

For a long time, welfare was measured by "I know it if I see it." This is still a significant component for assessing recreational horse welfare. However, that philosophy does not always hold up to scrutiny because it is based on the person's previous experience. As an example, to a person raised on a farm where old, rusty farm equipment is scattered throughout the pasture and where that pasture is surrounded by barbed wire in disrepair, a modern Thoroughbred stallion barn in Kentucky or show barn in Texas would look spectacular (Figure 11-2). At the other extreme, that person would see little wrong with a lean-to shelter that also has a lot of old twine and equipment or a paddock made from barbed wire nailed to trees. Wellbeing is correlated with having the most accurate assessment of stable conditions. On the other hand,

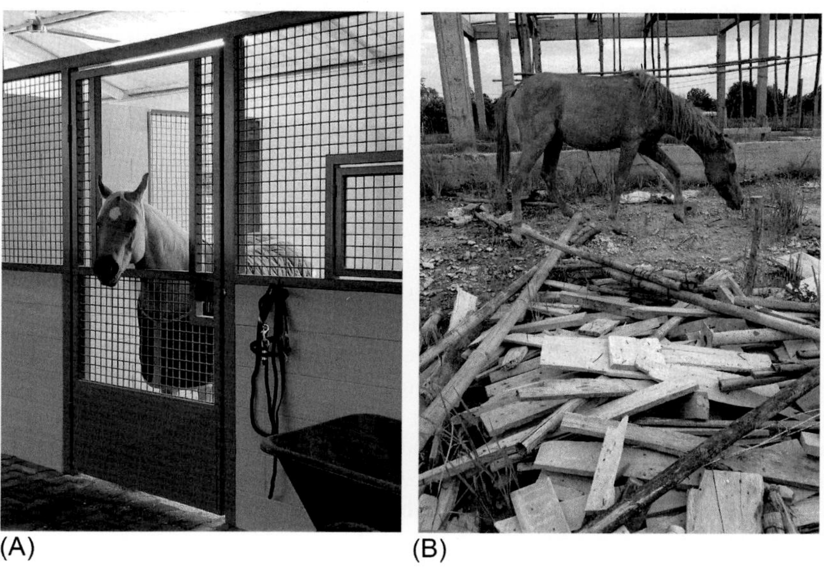

(A) (B)

FIGURE 11-2 Horses are kept in a number of types of environments, from pristine barns to those that contain equipment and discarded trash that could injure the animal. (A) The stall doors in this barn allow horses to easily see other horses, metal openings between stalls and to the outside that give the horse multiple views, and the stall is free of objects that might injure the mare. (B) The barnyard where this horse stays is littered with potentially hurtful objects including boards having protruding nails. *(Photo B: Courtesy of Carla Carleton and used with permission.)*

"overexposure" to the animals is correlated with underreporting of abnormal behaviors, indicating that constant exposure may dampen or change one's perceptions.[24] This is a common happening in the dairy industry, where producers only recognize one of every three or four lame cows.[25]

Animal welfare is multifactorial.[26] That suggests that assessments should evaluate a number of different measures or standards, scoring each outcome.[27,28] The best welfare assessments use a combination of criteria to provide several views for evaluation. A condition may be evaluated as present or absent, or it may be scored on a scale of 1–5 or 1–10. At the end of the assessment, the scores in each measure/standard provide a summary overview of the animal's welfare status, and areas needing improvement can be singled out descriptively. Several assessment standards follow as suggested measures to be combined in different ways for determining the level of welfare.

Performance Standards

Performance standards assess the animal state relative to what is considered acceptable for similar animals. This category can also be subdivided.

Animal-Based Measures

Animals are compared to others of their own type, such as weanlings compared to other weanlings. For a foal intended to show in a halter futurity, it is desirable for the growth rate to be ahead of the curve compared to one that will not be shown. Behavior can also be included in this assessment.[29] Stereotypic stall-walking foals would receive poor scores, and foals interacting in play with conspecifics would receive good ones.

Outcome Criteria

Productivity, health, number of breedings to achieve pregnancy, and performance success are examples of outcomes that can be measured for performance standards. These criteria are commonly used in production animals—pounds of milk, numbers of piglets born and weaned, and calving intervals are examples. For horses, health assessments should include criteria like the body condition score, hydration status, skin chafing or injuries, scars, whip or spur lesions, and coat and hoof condition.[27] Behavioral assessments would look for abnormal problems in addition to normal activity.[9] An important consideration is that prey species are good at hiding illness until they are very compromised, so superficial evaluations may not be adequate.

Prohibited Practices

Certain practices are not acceptable at any time. It does not mean these things do not occur: it does mean that owners, trainers, and the public are in general agreement that they should not happen at all. Practices considered abusive, painful, fear-inducing, or cruel by the majority of people involved in the horse industry would fall into this category. Although stacked wedge pads and chains do not seem to be stressful to Tennessee Walking Horses,[30] other things are. Soring is an extreme example of a prohibited practice that is distasteful to the public and within the general horse industry. There may be trainers raised in the Walking Horse industry who have always used soring with horses to get the "big lick." They might not consider how abusive the practices are. However, the techniques used are common enough and considered bad enough to have resulted in federal legislation (the Horse Protection Act of 1970) to get them stopped.[31, 32] Self-regulation by the Walking Horse industry has not been successful, and budgetary shortfalls within the United States Department of Agriculture limit the number of shows they can inspect.[33, 34]

There are other prohibited practices of concern within the horse industry. Horse tripping is one.[35] While this is part of Mexican-style rodeos (*charreadas*) and part of the Mexican culture, the events are considered to be inhumane in the United States. Also of concern are several lesser known "training" abuses like tying the horse's head up or around to its tail for long periods.[4] Beating, doping, and using hidden equipment to amplify bit movements are but a few others. When prize money and egos are involved, a small group of people will try anything to win.

Certain practices may not be considered abusive by industry standards but may be opposed by some segment of the public. Because public opinion is getting a louder voice, these opinions are starting to be forced on the industry. Society tends to view animal welfare as a moral rather than science-based issue.[36] This is most true in the cattle and swine industry, but it is increasingly true in the horse industry as well. Public demands around the unwanted horse and processing of horsemeat are vivid examples.

A number of horse organizations are becoming proactive in policing their own, rather than waiting until things get so bad that governmental oversight is mandated. Drug testing has been part of the industry for many years. For smaller organizations with limited budgets, all drug use at their sanctioned shows is prohibited. Larger groups with more resources for reporting and testing prohibit only performance altering medications and practices, recognizing that some medications are appropriate for equine athletes.[37] More recently, show stewards and judges are being educated about and charged to respond to problem incidents. While total elimination of abuse is ideal, it is not likely to happen because of the high value reward associated with winning. Appropriate rules do provide a method to punish those caught abusing the horse, and they help deter people who are tempted to try something considered abusive.

Input-Based Standards

Input-based standards are usually engineering based. Some European countries have recommendations for the size of box stalls,[9] but most countries do not.

Other environmental features are appropriate to consider as part of this standard. The assessment of physical locations should include noise level, environmental temperature, appropriate ventilation, ceiling height, type and quality of bedding, paddock and stall size, relative humidity, safety from predators, presence of waterers, and feeder levels.[3,6,9,28]

Because input-based standards are not well defined in the horse industry, subjective evaluations are used for assessments. The problem is perception and what it is based upon. There certainly are differences of opinion about what is an appropriate size for a box stall or a paddock, as examples. Even set standards can be problematic. The sizes of box stalls appropriate for a Miniature Horse and a Percheron are quite different. Other concerns have been expressed, although not researched. Concern had been voiced about the prolonged stay of pregnant mares in tie stalls associated with the collection of urine used to produce estrogen. Studies, however, did not find significant problems in the mares' welfare.[38] Paddock size and type of fencing have been studied. Horses use the border areas less when in a small paddock or one fenced with electric wires, but salivary cortisol and heart rate variability do not indicate stress reactions.[39] It is just a behavioral preference.

Of particular concern relative to physical standards is their relevance to facilities that take in rescued and retired horses. Aged geldings and horses with physical problems are the most difficult of these horses to place in new homes, compounding the rescue problem.[37] Unfortunately, alternatives are scarce and often poorly funded. There is no inspection system for horse retirement or rescue facilities. Similarly, mustangs that are removed from the range are difficult to place in new homes, and governmental funding for their long-term care is shrinking.

The Five Freedoms

In 1992, the Farm Animal Welfare Council in the United Kingdom created The Five Freedoms as guidelines for the production of farm animals (Figure 11-3). Since that time, these have come to be viewed as a mandate, and their application expanded to nonproduction animals as well.

As can be read into each of the Freedoms, 100% compliance is unlikely. Vaccinations cause some pain but are necessary to protect against devastating diseases. Normal behavior would include interhorse aggression. Escape from predators is a normal behavior but is also associated with fear and distress. An additional issue is who is to decide which behaviors if not all and what constitutes discomfort, distress, and fear. The Freedoms represent goals to be aspired to, not endpoints. Each has limitations.[40]

Five Domains

About the same time that the Five Freedoms were developed, Dr. D.J. Mellor at Massey University in New Zealand developed the Five Domains. This was done

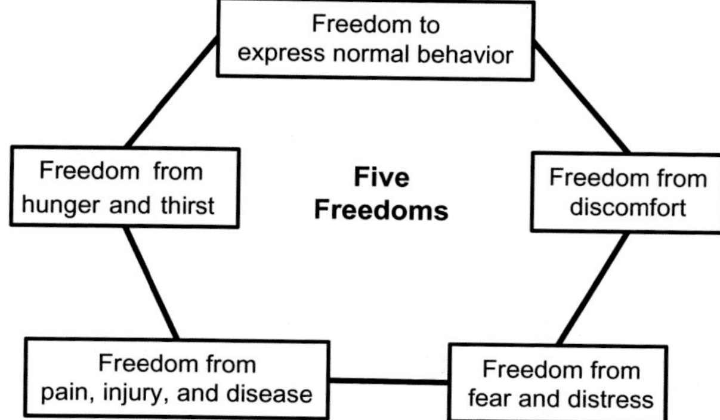

FIGURE 11-3 The Five Freedoms were developed as guidelines for livestock producers in the United Kingdom but have since been applied to many other species.

in recognition that welfare should be based on "quality of life" rather than "free from" concepts and should apply to livestock and nonlivestock species. The Domains have recently gained popularity, perhaps because of recognition of animals as sentient. The first four domains are broad in scope (nutrition, environment, health, and behavior) (Figure 11-4). The fifth domain, mental,

Physical/functional domains							
Survival-related factors						**Situational factors**	
1. Nutrition		**2. Environment**		**3. Health**		**4. Behavior**	
Restrictions	Opportunities	Unavoidable	Available	Present	Absent	Impediments	Facilitation
H_2O intake Food intake 1° Food type	Drink enough Eat enough Roughage	Too hot Dirt stall floor	Stall fans Shavings	Lameness Coughing Colic	Sound Clear lungs Normal GI	No exercise No horse contact Limited sleep	Daily riding Stall window Enough sleep

Affective experience domain							
5. Mental status							
Negative	Positive	Negative	Positive	Negative	Positive	Negative	Positive
Thirst Hunger Malnutrition GI pain	H_2O available Food present Eat til full GI comfort	Discomforts: Thermal Physical Visual	Comforts: Neutral No pain No glare	Illness Pain Exhaustion	No illness Pain free Physically fit	Loneliness Anxiety Stressed	Social contacts Secure Relaxed

Welfare status

FIGURE 11-4 The Five Domains model, including examples of negative/positive variations shown, is useful for evaluating a horse's welfare if put into a species-appropriate context.[41] The Five Domains can be used alone or as part of a broader type of evaluation process.

involves the sensory inputs from the other four domains and ultimately is used to determine the animal's mental state.

Record-Based Standards

Record-based standards rely on physiological data such as blood counts, cortisol levels, heart rate, blood chemistry, and the health record. Because several require invasive procedures, they are not data that would be obtained frequently. Radiographic and thermographic imaging and testing for nervous sensation are important additional tools within this standard for assessing a horse's welfare state.

Subjective Evaluations

Subjective evaluations are the "I know it if I see it" responses. Some things are obvious enough that general agreement would support the conclusion, like emaciation or no water sources present. As previously pointed out though, an evaluator's background and beliefs can strongly bias interpretations and can add ethical considerations to the assessment. Inclusion of ethical views is becoming an increasingly important component of this evaluation, particularly as the public joins the discussions.[4,42] Efforts are ongoing to identify ways to add quantification techniques to subjective evaluations. Examples include the use of pain scores or behavioral scales to indicate stress.[43–47] Visual observations can include quantifiable information about feeding intervals and being sure fresh, clean water is available.

GENERAL CONCEPTS OF WELFARE IN HORSES

Worldwide, the majority of horses are used in agriculture and for transportation. It is in developed countries where horses are used primarily for sport and recreation. Issues about equine welfare vary due to cultural differences, economic issues, environmental extremes, and interpretation of the assessor. The concern is determining where the line of acceptability begins and where use becomes abuse.[8] Studies show that environmental enrichment with social contacts, sensory stimulation, and multiple feedings decreases aberrant behavior, prolonged alertness, and pinned back ears, while increasing the amount of lateral (deep) sleep.[48] Such practices can be useful for evaluating or improving a horse's "emotional" state, or at least its welfare.[29] The discussion which follows is not intended to "solve" the various issues, but rather to highlight some of the more common welfare concerns. Because of human ingenuity, the scope of issues never can be completely identified. As one problem is handled, another will surface.

Working Horses

People in undeveloped and developing countries remain heavily dependent on the use of horses and donkeys for subsistence. In these environments, the welfare is considered poor when judged by standards of developed countries. Issues of concern are numerous but typically relate to appropriate health and foot care, living environments, and reasonable workloads and expectations.[3,6] In most cases, poor welfare results from a lack of knowledge rather than intentional abuse. The loss of a horse has severe consequences for the family, and the economic burden of replacement is great. Owners care about their animals and are receptive to education about basic health care and nutrition.

Even in developed countries, horses have a number of jobs. For urban carriage horses and police horses, concerns center around the amount of time they work, living conditions, exposure to exhaust pollution, and concussive forces of working on asphalt. The welfare status for these animals should include availability of adequate water and shade, protection from temperature extremes, safety from traffic, correct fitting of the harness, appropriate training of the driver or rider, and appropriate weights of the carriage loads.[6,49–51] Working cattle ranch and dude ranch horses also have jobs, and each has unique concerns that relate to animal welfare. Rodeo bucking stock are generally well cared for, but travel can present challenges. Horses used to work fighting bulls have the potential of being gored.[3]

Show and Performance Horses

Show and exhibition horses are those most likely to be individually stalled throughout the year, and that can negatively impact social contact. Training, showing, traveling, and living conditions are other areas that impact the welfare of these horses.

Training

Welfare issues tend to center around abusive or questionable training techniques. Positive and negative reinforcement are useful for teaching horses, but when punishment is used, it can easily become abusive. The goal of training is to increase a horse's motivation to do something and then perfect the lesson over time. Even though starting young horses under saddle results in a stress response, they will return to normal physiological patterns within a few weeks.[52] If the horse experiences significant stress or encounters very low or high arousal stimuli, learning becomes difficult and training less effective.[53, 54]

Welfare and abuse concerns vary by breed and type of training. As an example, the position of the horse's head can be an issue. In dressage and reining, there is a tendency to want the neck hyperflexed (*Rollkür*, *Low-Deep-and-Round*) such that the plane of the nose is tilted toward the chest instead of perpendicular to the ground.[3,22] In hyperflexion, the horse's vision and air passages are compromised, resulting in conflict and stress responses.[22, 55–58] To achieve this posture, some trainers use prolonged bit pressure so compliance results from pain avoidance rather than learning. In some classes like western pleasure, the neck is lower than horizontal, and gaits are unnaturally slow. Extreme animation in the gaits are emphasized in some Tennessee Walking Horse, Saddlebred, and Hackney classes.

Training equipment can also raise welfare concerns, particularly the bits and how they might be used. Snaffle bits are commonly used on young horses. Initially the horse shows more chewing and mouth opening than without the bit, but this decreases as familiarity increases. Some bits are severe, potentially injuring the gums or the roof of the mouth.[22] During foundation training, horses that wear bitless bridles do as well as those wearing bits and show less physiological stress.[59] Tie-downs restrict upward head movements and can also restrict vision. When nosebands are used to prevent opening the mouth, the damage can be worse if they are excessively tight.

Showing

Just as with training, concern associated with showing depends on event and breed. One of the universal influences on welfare is stress.[53] There are multitudes of internal and external factors that influence the stress response, almost to the point of wondering if anything can go right. Small amounts of stress can speed learning, but major stress limits it by triggering the survival mode.

Internal stressors are those associated with the animal itself, including conformation, movement, temperament, genetics, health, and physiology.[53] The "off day" is part of this picture. Stress of illness or lameness, coupled with the horse's temperament and physiology, will dramatically impact performance, at least in most horses. The lack of sleep, as happens during the first 2 or 3 days at a new show ground or following a long trailer ride, compounds other intrinsic issues. Another often-neglected question is whether the horse is suitable for the job it is asked to do. Not every animal has the right conformation, emotional makeup, or ability to perform successfully in a particular event. Expected compliance regardless of ability is when abusive training becomes more likely. Not every Quarter Horse is right for western pleasure or reining, nor every Warmblood suited for dressage or 3-day eventing.

External stressors are also numerous, and the primary one is the rider. While longtime professional trainers are the most consistent between the training pen and show ring, amateurs are more likely to vary how they ride because of show ring jitters. This less experienced group is more likely to compromise the show horse's welfare.[60] The horse gets confused because signals are unconsciously changed. How long and hard the rider warms up or rides a horse may be inappropriate, even to the point of exhaustion. This is a particular concern for long-distance endurance horses, ones out of shape, and those warming up for extremely physical classes. The use and abuse of equipment can relate to external stress. While an abusive bit may be used at home for training, hidden appliances in the mouth may cause pain when hit by the bit—even a bit that passes preclass inspection. In other cases, an ill-fitting saddle, inadequate padding, or excessively tight cinch can be problematic.

Drug testing has been a part of the horse show and racing worlds for several years. Two welfare concerns are relative. The first is that certain drugs alter performance, and that is not acceptable.[61] The second is that horses with a true medical need should be able to receive appropriate treatment.[61] On one side of this medical need discussion is the argument that if the horse needs medication, it should be too ill to perform. Since illness is a stressor, perhaps the horse should not show. On the other side is the fact that for some futurities, exhibitors/owners have paid a lot of money, invested a lot of time, and do not want to lose the opportunity for their moment of fame.[60] If the condition can be appropriately treated without medically altering performance, welfare might be minimally affected. Without this provision, appropriate treatment may be delayed so the horse can show anyway.

Body-altering procedures have opened up a completely new area of concern. At one extreme is the soring done on some Tennessee Walking Horses, but examples of other extremes would include the alcohol blocking of tails to prevent their movement.[34, 62] In some breeds where animation is desired in the halter horses, atropine and dark stalls dilate the pupils to heighten the effect of bright arena lights on the horse's arousal level. Tail docking is primarily done in some draft horse breeds for cosmetic reasons, and it leaves an animal unable to fight off insect pests.[62]

Conflict behaviors that indicate stress vary with the way a horse is shown, so one size (or in this case, behavior) does not fit all. This is true even at the elite levels. In some horses, tail swishing indicates the excessive use of spurs or overtraining. However, it is common in dressage horses, even in those never experiencing poor training. Pulling the reins out of the rider's hands, potentially a conflict behavior, happens more with show jumpers.[63]

Traveling

Horses usually travel to and from the show by trailer or van. Physical and mental stress accompany these travels, regardless of how or how often it is done.[9,64–66] The distance traveled vs. stress relationship is difficult to study and is associated with some mixed findings. The first issue regarding travel is to be sure the horse is comfortable with being loaded. Training the horse to load by setting small but progressive goals with positive reinforcement minimizes stress and avoidance behaviors.[67] Then, short trips from home will acclimate the horse for upcoming longer distance travel. For performance horses, travel for short distances is unlikely to affect performance.[65] On the other hand, it might, if the level of competition is close and/or the horse has less traveling experience.[65] The farther the trip, the greater is the need for recovery from stress. A recovery period of 2 h for a 3-h trailer ride is appropriate to allow muscle enzyme concentrations to return to normal.[65, 68] Long-distance travel is associated with dehydration and an increased risk of respiratory problems, particularly if the horse is not able to lower its head or has shown behavioral signs of stress .[69, 70] Depending on the duration of travel, periodic breaks are recommended so water can be offered, and the horse can lower its head. Travel over 8 h requires at least a full day for the immune system to recover.

Home Environment

The home environment for a show horse is likely to be more constricting and problematic than that for horses used only for recreational riding. Many show horses do not get pasture time or a chance to interact with other horses. Futurity-bound foals are stalled early in life too. Early stalling, concentrated diets rather than roughage, minimal contact with conspecifics, and the inability to see other horses are all associated with problem behaviors, warranting special precautions.[3,71–73] Stereotypies are indicators of negative welfare and may be adaptive behaviors to relieve stress.[9,71,73,74] They indicate something has gone wrong in the past. Because they dramatically reduce the value of the horse, there is a temptation to block the stereotypy any way possible. However, without addressing underlying causes, doing so compromises welfare.[73, 74]

Horses stabled alone for 13–16 h a day are more likely to be associated with aggression and have handling issues.[72, 75] Those that spend 22–24 h a day in a stall have abnormal oral and ingestive behaviors.[72] Spending significant amounts of time in stalls just before and after weaning can negatively impact a foal's behavior and welfare.[76]

Environmental enrichment is important for improving welfare. Weanlings housed in a paddock or pasture with others of their same age spend the same amount of time in social interactions as do free-ranging horses.[76] This would be ideal for adult horses too, but adult groups are less stable. Mixing new horses can be associated with aggression. Fortunately, horses quickly solve differences without major fights, and any remaining aggressions are low in frequency and form.[77] When pasture/paddock time is restricted, even a few hours of outdoor time with a compatible horse provides opportunities for social interactions.[78] Stall enrichments are necessary to minimize the negative welfare implications of long hours spent confined, and seeing other horses is even more important to the horse than is its stall size.[79] When visual contact is not possible, aluminum mirrors can be used. Open windows through which horses can see outdoors, even without seeing other horses, are also beneficial.[80]

Race Horses

Horse racing has a number of welfare issues for consideration. On-track euthanasia and breakdowns are top concerns for public and horse enthusiasts.[79] These usually involve torn ligaments, tendon tears, or bone fractures. When the full weight of a horse comes down on a single limb, the concussive forces are tremendous, and injuries will happen, especially with deviations in conformation. It has been suggested that Thoroughbreds might have a genetic predisposition to injury, whether racing or not.[37] Racecourse surfaces vary by location, and their condition plays some role in injuries.[37] Soft surfaces, particularly dry sand, help absorb some of the concussion.[81] As the surface is dampened, the foot flight pattern changes to a shorter stride length and higher stride frequency.[81] Footing that is too hard, too soft, or sloppy is a welfare issue.

A concern in racing Thoroughbreds is the use of drugs. Some enhance racing performance; others prevent the horse from racing at its full potential.[37] The industry is working with veterinarians to minimize drug use, establish rules for which drugs are permitted and banned, and perform drug tests as a deterrent for abuse.[34] The goal is to minimize illegal drug use, realizing it will not stop completely, particularly at smaller tracks. The racing industry is also working on injury reporting and prevention, safety, and care for retired racehorses.[32]

Thoroughbreds are not the only racehorses. Standardbreds run as trotters and pacers. Quarter Horses sprint up to 770 yards (704 m), although 440 yards (402 m) is the most common race distance. Other horses participate in timed endurance rides, barrel racing events, pole bending classes, wild horse races, and chuckwagon races, to name some of the higher profile events. Each race has a unique set of concerns for the safety and welfare of the horse. When the concept is to "go fast" and money or egos are involved, other types of abuses will be tried, from bats and spurs to electronic shocking devices and hidden sharp objects. As with performance horses, long-term stalling issues, diet, bedding, little contact with conspecifics, training methods, development of stereotypies, and transportation are problems that occur within the racehorse population too.[9,82]

When racing is no longer an option, concern shifts to the fate of the excess animals. It takes a large number of foals to get a few top racehorses, and the surplus must find a suitable home. Retirement from racing means a no longer useful horse must be placed somewhere else. Many of these animals find a second career as show horses, but as show events become so highly specialized, that second career may become harder to find. Riding horses and the hunter/-jumper industry take much of this surplus. Old geldings and animals with physical problems are the hardest to place.[37] Some horses have personalities that make them difficult for an average owner to handle and so end up as unwanted.

Horses Used for Recreation

The majority of horses in developed countries are used for recreation, and concern for their wellbeing has prompted a high number of welfare investigations in recent years.[83] These animals live in a wide variety of conditions, from being maintained on the owner/rider's property, to staying in rented stalls or being a rent horse (sometimes called a hack horse) for a variety of riders. As a result, facilities in which the horses are kept range from very nice to totally inadequate, introducing a plethora of potential welfare issues.

The quality of care is ultimately the responsibility of the owner, and therein lies the biggest welfare concern.[3,83] Cost and lack of owner knowledge are major factors that drive issues relating to the size and quality of the facility, schedule of care, quantity and quality of hay or pasture, type of feed, availability and cleanliness of the water, presence of salt, and health care.[9,26,84] Appropriate health and foot care require routine attention from qualified professionals.[3] Neglect usually happens because of the lack of knowledge about proper care, rather than from not caring. It is easy for a novice to fail to appreciate the level of responsibility and associated costs to properly care for such a large animal. The cheapest part of owning a horse is often the purchase price.

Other things affect welfare in addition to the environment. Behavior problems are common in horses used for recreation, representing strategies of coping with stressful environments and improper training.[9] Owners who do not understand the implications of a problem behavior may actually complicate it by reinforcement. Health problems and attitude changes can be difficult for a novice owner to recognize.[29,] Clearer, well-defined behavioral signs would be helpful, but still need to be developed.[85] Tack may not be appropriately sized and the competence of the person trimming the feet questionable. The pet horse often receives poor quality training, which is further complicated by inconsistent or conflicting rider cues. The amount of exercise it gets is another welfare consideration. Light exercise is less stressful compared to no exercise, even when riders are clumsy and inexperienced.[86] At the other extreme is overuse. This tends to occur because of sporadic riding where the horse is ridden for long times, but only on Saturday. Poor conditioning programs are not just problems for the "weekend warriors" but also for owners who decide to enter local endurance competitions for fun.[87]

In suburban and rural areas, horses are more likely to be kept in a pasture, at least part of the day. This also increases the likelihood that there are multiple animals, but these are animals of the owner's choosing and not necessarily ones that get along.[88] The group also does not mirror a natural harem group because it is comprised of mares and geldings. While groups allow for the social interaction, the presence of an overly aggressive individual, or "bully," can cause chronic stress for some individuals.[88] As a result, stressed horses need to be identified and separated from the problem horse.

Free-Ranging Horses

Mustangs in the western regions of the United States and brumbies in Australia are the free-ranging horses that face a number of welfare issues. The idealized mustang is running free on the North American plains, at least in the public's mind. As "wild" animals, their quality of welfare is primarily dependent on nature, but when that environment no longer supports the number of animals grazing it, humans must step in to affect welfare. In the United States, the Bureau of Land Management is charged with mustang welfare, and this is a hot button issue for equine enthusiasts.[34] The sheer number of free-ranging horses and a relatively successful reproductive rate have resulted in overgrazing, soil compaction, and other related environmental issues. Concerns about exploding populations and land stewardship have led to roundups, short and long-term drylot pens, adoption programs, temporary sterilization of mares, and for a short while, the possibility of slaughter. These programs are costly and excessive numbers still run.

Associated welfare issues during the roundups and while confined in drylots include injuries, mare/foal separation, the mixing of horses from different harem groups and herds, and intermingling of stallions resulting in significant aggression. Serious efforts go into the adoption program, but that introduces transportation problems. These horses are not taught how to load, travel with strangers in tight spaces, or unload. Not all mustangs make good adoptees, and available horses and burros outnumber potential new homes. Excess captives are dependent on humans for their long-term care and welfare. The unadopted horses become a unique subset of the unwanted horse population.

Unwanted Horses

For many years, horse slaughter was legal in the United States. It remains so in several countries around the world, representing a significant economic impact wherever it exists.[65] With the closing of the last three U.S.-based horse processing plants in 2007, a pressing problem surfaced—what to do with all the unwanted horses. To understand the growth of the unwanted horse population, it is necessary to understand what issues result in them not being wanted in the first place. Financial limitations, the horse's illness or injury, the horse's behavior, and family issues are commonly cited reasons the animal is no longer wanted.[89]

As the number of unwanted horses began to climb, acceptable options had to change. Some horses are shipped to Canada or Mexico, but the long trips, often in double-deck trailers are problematic in multiple ways. These trailers are not suited for horses because of narrow doorways and low ceilings.[90] More injuries occur during transport on double-deckers (29.2%) compared to single level, straight-deck trailers (8.0%).[91] Some injuries are associated with aggression because horses are hauled in mixed groups. This may include stallions and animals not familiar with one another.[70] Weight loss of 4% is common on extended 30-h trips if the horses have access to water and as much as 12.8% if not.[79, 91, 92] Stress indicators, muscle fatigue, and dehydration become problematic.[91] There can also be trailer ventilation and design issues since the equipment was originally designed to haul cattle.[90]

Rescue and retirement facilities provide a possible option for unwanted horses, but two significant problems remain with this option. The first is the lack of enough facilities with an appropriate capacity to take in unadoptable unwanted horses. The second is the expense of long-term horse care. Lack of funding has a dramatic impact on the ability of animal control centers and horse rescue groups to take in and maintain significant numbers of horses for any length of time.[93]

Euthanasia of an unwanted horse is another option, although not a desirable one to most people. The size of the carcass makes this an expensive option. There is a fee for euthanasia and a disposal fee, whether for a backhoe to bury the body, landfill disposal, use of an incinerator/biodigester, or pick up for rendering.

Some people simply neglect their unwanted horse. This is extremely distasteful to the public, and rescue attempts to bring the animal back to health are expensive. Rather than "neglect" the horse, others turn the animal loose or leave them to fend for themselves or for others to find, neither of which is realistic or good for the horse.

End-of-Life Issues

One of the hardest decisions a person faces in horse care is when and how to say "goodbye" in a humane way. Abandonment is not an option and every effort should be made to avoid prolonged and painful deaths.[6] Euthanasia means "humane death" by definition. Unfortunately, when done to a large animal, it still looks traumatic. After death, there remains the problem of how to dispose of the body. Associated costs amount to several hundred dollars and disposal options can be significantly limited in certain areas. Since novice horse owners are most often making these decisions and usually have not thought about end-of-life welfare issues, people who work in the equine industry should have a knowledge of various options so they can guide less informed people through the process.[3] In addition to information about options for disposal, it is helpful to know professionals who can assist an owner through the grief process if they have difficulty dealing with the horse's death and their decisions relating to it.

REFERENCES

1. Miller RM. The revolution in horsemanship. *J Am Vet Med Assoc* 2000;**216**(8):1232–3.
2. Voigt M, Hiney K, Richardson JC, Waite K, Borron A, Brady CM. Show horse welfare: horse show competitors' understanding, awareness, and perceptions of equine welfare. *J Appl Anim Welf Sci* 2016;**19**(4):335–52.
3. World Horse Welfare and Eurogroup for Animals. *Removing the blinkers: The health and welfare of European Equidae in 2015*. http://www.eurogroupforanimals.org/wp-content/uploads/EU-Equine-Report-Removing-the-Blinkers.pdf; 2015 [downloaded July 2, 2015].
4. Rollin BE. Equine welfare and emerging social ethics. *J Am Vet Med Assoc* 2000;**216**(8):1234–7.

5. Broom D. Indicators of poor welfare. *Br Vet J* 1986;**142**(6):524–6.
6. World Organisation for Animal Health. Welfare of working equids. In: *Terrestrial animal health code*. Paris: World Organisation for Animal Health; 2017. p. 10 [Chapter 7.12].
7. Veasey JS, Waran NK, Young RJ. On comparing the behavior of zoo housed animals with wild conspecifics as a welfare indicator. *Anim Welf* 1996;**5**(1):13–24.
8. Campbell MLH. When does use become abuse in equestrian sport? *Equine Vet Edu* 2013;**25** (10):489–92.
9. Minero M, Canali E. Welfare issues of horses: an overview and practical recommendations. *Ital J Anim Sci* 2009;**8**:219–30. Suppl. 1.
10. Goodwin D, Davidson HPB, Harris P. Selection and acceptance of flavours in concentrate diets for stabled horses. *Appl Anim Behav Sci* 2005;**95**(3–4):223–32.
11. Randall RP, Schurg WA, Church DC. Response of horses to sweet, salty, sour and bitter solutions. *J Anim Sci* 1978;**47**(1):51–5.
12. Dawkins MS. Behavioural deprivation: a central problem in animal welfare. *Appl Anim Behav Sci* 1988;**20**(3–44):209–25.
13. Houpt KA. Equine welfare, In: *Presentation at southwest veterinary symposium, Ft. Worth, TX*; 2010.
14. Ninomiya S, Aoyama M, Ujiie Y, Kusunose R, Kuwano A. Effects of bedding material on the lying behavior in stabled horses. *J Equine Sci* 2008;**19**(3):53–6.
15. Mejdell CM, Buvik T, Jørgensen GHM, Bøe KE. Horses can learn to use symbols to communicate their preferences. *Appl Anim Behav Sci* 2016;**184**:66–73.
16. Lee J, Floyd T, Erb H, Houpt K. Preference and demand for exercise in stabled horses. *Appl Anim Behav Sci* 2011;**130**(3–4):91–100.
17. von Borstel UK, Keil J. Horses' behavior and heart rate in a preference test for shorter and longer riding bouts. *J Vet Behav* 2012;**7**(6):362–74.
18. Benhajali H, Richard-Yris M-A, Ezzaouia M, Charfi F, Hausberger M. Foraging opportunity: a crucial criterion for horse welfare? *Animal* 2009;**3**(9):1308–12.
19. Lansade L, Valenchon M, Foury A, Neveux C, Cole SW, Yayé S, Cardinaud B, Lévy F, Moisan M-P. Behavioral and transcriptomic fingerprints of an enriched environment in horses (*Equus caballus*). *PLoS One* 2014;**9**(12). https://doi.org/10.1371/journal.pone.0114384.
20. Löckener S, Reese S, Erhard M, Wöhr A-C. Pasturing in herds after housing in horseboxes induces a positive cognitive bias in horses. *J Vet Behav* 2016;**11**:50–5.
21. Henry S, Fureix C, Rowberry R, Bateson M, Hausberger M. Do horses with poor welfare show 'pessimistic' cognitive biases? Science of. *Nature* 2007;**104**(1–2).
22. McGreevy PD. The advent of equitation science. *Vet J* 2007;**174**(3):492–500.
23. Visser EK, Van Reenen CG, Engel B, Schilder MBH, Barneveld A, Blokhuis HJ. The association between performance in show-jumping and personality traits earlier in life. *Appl Anim Behav Sci* 2003;**82**(4):279–95.
24. Lesimple C, Hausberger M. How accurate are we at assessing other's well-being? The example of welfare assessment in horses. *Front Psychol* 2014;**5**:.
25. Flower FC, Weary DM. Gait assessment in dairy cattle. *Animal* 2009;**3**(1):87–95.
26. Lesimple C, Poissonnet A, Hausberger M. How to keep your horse safe? An epidemiological study about management practices. *Appl Anim Behav Sci* 2016;**181**:105–14.
27. Ali ABA, El Sayed MA, Matoock MY, Fouad MA. A welfare assessment scoring system for working equids—a method for identifying at risk populations and for monitoring progress of welfare enhancement strategies (trialed in Egypt). *Appl Anim Behav Sci* 2016;**176**:52–62.
28. Viksten SM, Visser EK, Nyman S, Blokhuis HJ. Developing a horse welfare assessment protocol. *Anim Welf* 2017;**26**(1):59–65.

29. Stratton R, Cogger N, Beausoleil N, Waran N, Stafford K, Stewart M. *Indicators of good welfare in horses*. Wellington, New Zealand: Ministry for Primary Industries; 2014. p. 48.
30. Everett JB, Schumacher J, Doherty TJ, Black RA, Amelse LL, Krawzel P, Coetzee JF, Whitlock BK. Effects of stacked wedge pads and chains applied to the forefeet of Tennessee walking horses for a five-day period on behavioral and biochemical indicators of pain, stress, and inflammation. *Am J Vet Res* 2018;**79**(1):21–32.
31. DeHaven WR. The horse protection act—a case study in industry self-regulation. *J Am Vet Med Assoc* 2000;**216**(8):1250–8.
32. Larson E. Equine welfare. *Horse* 2011;**XXVIII**(3):14–6.
33. American Veterinary Medical Association. *Soring: Unethical and illegal*. https://www.avma.org/KB/Resources/Reference/AnimalWelfare/Documents/soring_in_horses_factsheet.pdf; 2013. Accessed 15 December 2017.
34. Leitch M. Equine welfare issues: an overview. *Am Assoc Equine Pract Proc* 2010;**56**:357–9.
35. Larkin M. Horse tripping can fly under the radar. *J Am Vet Med Assoc* 2016;**249**(8):852.
36. Larson E. Keynote: unwanted horses. *Horse* 2011;**XXVIII**(3):11–3.
37. Munday GD. Racing. *J Am Vet Med Assoc* 2000;**216**(8):1243–6.
38. Flannigan G, Stookey JM. Day-time time budgets of pregnant mares housed in tie stalls: a comparison of draft versus light mares. *Appl Anim Behav Sci* 2002;**8**(2–4):125–43.
39. Glauser A, Berger D, van Dorland HA, Gygax L, Bachmann I, Howald M, Bruckmaier RM. No increase stress response in horses on small and electrically fenced paddocks. *Appl Anim Behav Sci* 2015;**167**:27–34.
40. Hockenhull J, Whay HR. A review of approaches to assessing equine welfare. *Equine Vet Edu* 2014;**26**(3):159–66.
41. Mellor DJ, Beausoleil NJ. Extending the 'Five Domains' model for animal welfare assessment to incorporate positive welfare states. *Anim Welf* 2015;**24**(3):241–53.
42. Heleski CR, Anthony R. Science alone is not always enough: the importance of ethical assessment for a more comprehensive view of equine welfare. *J Vet Behav* 2012;**7**(3):169–78.
43. Bussières G, Jacques C, Lainay O, Beauchamp G, Leblond A, Cadoré J-L, Desmaizières L-M, Cuvelliez SG, Troncy E. Development of a composite orthopaedic pain scale in horses. *Res Vet Sci* 2008;**85**(2):294–306.
44. Dalla Costa E, Minero M, Lebelt D, Stucke D, Canali E, Leach MC. Development of the horse grimace scale (HGS) as a pain assessment tool in horses undergoing routine castration. *PLoS One* 2014;**9**(3). https://doi.org/10.1371/journal.pone.0092281.
45. Gleerup KB, Forkman B, Lindegaard C, Andersen PH. An equine pain face. *Vet Anaesth Analg* 2015;**42**(1):103–14.
46. Van Loon JPAM, Back W, Hellebrekers LJ, van Weeren PR. Application of a composite pain scale to objectively monitor horses with somatic and visceral pain under hospital conditions. *J Equine Vet Sci* 2010;**30**(11):641–9.
47. Young T, Creighton E, Smith T, Hosie C. A novel scale of behavioural indicators of stress for use with domestic horses. *Appl Anim Behav Sci* 2012;**140**(1–2):33–43.
48. Valenchon M, Lévy F, Neveux C, Lansade L. Horses under an enrichement program showed better welfare, stronger relationships with humans and less fear. *J Vet Behav* 2012;**7**(6).
49. American Veterinary Medical Association. *Use of horses in urban environments*. https://www.avma.org/KB/Policies/Pages/Use-of-Horses-in-Urban-Environments.aspx; 2017. Accessed 15 December 2017.
50. Mason C. Welfare issues of urban work horses, In: *Proceeding of the American Veterinary Medical Convention*; 2015.

51. Merriam JG. Urban carriage horses 1999—status and concerns. *J Am Vet Med Assoc* 2000;**216**(8):1261–2.
52. Fazio E, Medica P, Aveni F, Ferlazzo A. The potential role of training sessions on the temporal and spatial physiological patterns in young Friesian horses. *J Equine Vet Sci* 2016;**47**:84–91.
53. Bartolomé E, Cockram MS. Potential effects of stress on the performance of sport horses. *J Equine Vet Sci* 2016;**40**:84–93.
54. Olczak K, Nowicki J, Klocek C. Motivation, stress and learning—critical characteristics that influence the horses' value and training method—a review. *Ann Anim Sci* 2016;**16**(3):641–52.
55. Christensen JW, Beekmans M, van Dalum M, VanDierendonck M. Effects of hyperflexion on acute stress responses in ridden dressage horses. *Physiol Behav* 2014;**128**:39–45.
56. Lenapfel K, Link Y, Borstel UKV. Prevalence of different head-neck positions in horses shown at dressage competitions and their relation to conflict behavior and performance marks. *PLoS One* 2014;**9**(8). https://doi.org/10.10.1371/journal.pone.0103140.
57. Smiet E, Van Dierendonck MC, Sleutjens J, Menheere PPCA, van Breda E, de Boer D, Back W, Wijnberg ID, van der Kolk JH. Effect of different head and neck positions on behavior, heart rate variability and cortisol levels in lunged Royal Dutch Sport horses. *Vet J* 2014;**202**(1):26–32.
58. von Borstel UU, Duncan IJH, Shoveller AK, Merkies K, Keeling LJ, Millman ST. Impact of riding in a coercively obtained Rollkur posture on welfare and fear of performance horses. *Appl Anim Behav Sci* 2009;**116**(2–4):228–36.
59. Quick JS, Warren-Smith AK. Preliminary investigations of horses' (*Equus caballus*) responses to different bridles during foundation training. *J Vet Behav* 2009;**4**(4):169–76.
60. Voigt M, Hiney K, Croney C, Waite K, Borron A, Brady C. Show horse welfare: the viewpoints of judges, stewards, and show managers. *J Appl Anim Welf Sci* 2016;**19**(2):183–97.
61. Miller MW. Use and abuse of medications at horse shows—an emerging welfare issue? *Am Assoc Equine Pract Proc* 2010;**56**:360–4.
62. American Veterinary Medical Association. *Tail alteration in horses*. https://www.avma.org/KB/Policies/Pages/Tail-Alteration-in-Horses.aspx; 2017. Accessed 15 December 2017.
63. Górecka-Bruzda A, Kosińska I, Jaworski Z, Jezierski T, Murphy J. Conflict behavior in elite show jumping and dressage horses. *J Vet Behav* 2015;**10**(2):137–46.
64. Cross N, van Doorn F, Versnel C, Cawdell-Smith J, Phillips C. Effects of lighting conditions on the welfare of horses being loaded for transportation. *J Vet Behav* 2008;**3**(1):20–4.
65. Padalino B. Effects of the different transport phases on equine health status, behavior, and welfare: a review. *J Vet Behav* 2015;**10**(3):272–82.
66. Waran NK, Cuddeford D. Effects of loading and transport on the heart rate and behavior of horses. *Appl Anim Behav Sci* 1995;**43**(2):71–81.
67. Slater C, Dymond S. Using differential reinforcement to improve equine welfare: shaping appropriate truck loading and feet handling. *Behav Process* 2011;**86**(3):329–39.
68. Tateo A, Padalino B, Boccaccio M, Maggiolino A, Centoducati P. Transport stress in horses: effects of two different distances. *J Vet Behav* 2012;**7**(1):33–42.
69. Padalino B, Raidal SL, Knight P, Celi P, Jeffcott L, Muscatello G. Behaviour during transportation predicts stress response and lower airway contamination in horses. *PLoS One* 2018;**13**(3).
70. Weeks CA, McGreevy P, Waran NK. Welfare issues related to transport and handling of both trained and unhandled horses and ponies. *Equine Vet Edu* 2012;**24**(8):423–30.
71. Cooper JJ, Albentosa MJ. Behavioural adaptation in the domestic horse: potential role of apparently abnormal responses including stereotypic behavior. *Livest Prod Sci* 2005;**92**(2):177–82.

72. Hockenhull J, Creighton E. Management practices associated with owner-reported stable-related and handling behavior problems in UK leisure horses. *Appl Anim Behav Sci* 2014;**155**:49–55.
73. Sarrafchi A, Blokhuis HJ. Equine stereotypic behaviors: causation, occurrence, and prevention. *J Vet Behav* 2013;**8**(5):386–94.
74. Hothersall B, Casey R. Undesired behavior in horses: a review of their development, prevention, management and association with welfare. *Equine Vet Edu* 2012;**24**(9):49–485.
75. Yarnell K, Hall C, Royle C, Walker SL. Domesticated horses differ in their behavioural and physiological responses to isolated and group housing. *Physiol Behav* 2015;**143**:51–7.
76. Heleski CR, Shelle AC, Nielsen BD, Zanella AJ. Influence of housing on weanling horse behavior and subsequent welfare. *Appl Anim Behav Sci* 2002;**78**(2–4):291–302.
77. Fureix C, Bourjade M, Henry S, Sankey C, Hausberger M. Exploring aggression regulation in managed groups of horses *Equus caballus*. *Appl Anim Behav Sci* 2012;**138**(3/4):216–28.
78. Werhahn H, Hessel EF, Van den Weghe HFA. Competition horses housed in single stalls (I): behavior and activity patterns during free exercise according to its configuration. *J Equine Vet Sci* 2012;**32**(1):45–52.
79. Houpt, K.A. (2012): Equine welfare. In: Recent advances in companion animal behavior problems, Houpt, K.A., ed., International Veterinary Information Service (www.ivis.org) [downloaded October 27, 2017].
80. Ninomiya S, Kusunose R, Obara Y, Sato S. Effect of an open window and conspecifics within view on the welfare of stabled horses, estimated on the basis of positive and negative behavioural indicators. *Anim Welf* 2008;**17**(4):351–4.
81. Chateau H, Holden L, Robin D, Falala S, Pourcelot P, Estoup P, Denoix J-M, Crevier-Denoix N. Biomechanical analysis of hoof landing and stride parameters in harness trotter horses running on different tracks of a sand beach (from wet to dry) and on an asphalt road. *Equine Vet J* 2010;**42**(Suppl. 38):488–95.
82. McGreevy PD, Cripps PJ, French NP, Green LE, Nicol CJ. Management factors associated with stereotypic and redirected behavior in the thoroughbred horse. *Appl Anim Behav Sci* 1995;**44**(2–4):270–1.
83. Hemsworth LM, Jongman E, Coleman GJ. Recreational horse welfare: the relationships between recreational horse owner attributes and recreational horse welfare. *Appl Anim Behav Sci* 2015;**165**:1–16.
84. McNeill LR, Bott RC, Mastellar SL, Djira G, Carroll HK. Perceptions of equid well being well-being in South Dakota. *J Appl Anim Welf Sci* 2018;**21**(1):40–68.
85. Hall C, Huws N, White C, Taylor E, Owen H, McGreevy P. Assessment of ridden horse behavior. *J Vet Behav* 2013;**8**(2):62–73.
86. Kang O-D, Lee W-S. Changes in salivary cortisol concentration in horses during different types of exercise. *Asian Australas J Anim Sci* 2016;**29**(5):747–52.
87. Frazier DL. Who speaks for the horse—the sport of endurance ridi
ng and equine welfare. *J Am Vet Med Assoc* 2000;**216**(8):1258–61.
88. VanDierendonck MC, Spruijt BM. Coping in groups of domestic horses—review from a social and neurobiological perspective. *Appl Anim Behav Sci* 2012;**138**(3–4):194–202.
89. Unwanted Horse Coalition (2009): 2009 Unwanted horses survey: creating advocates for responsible ownership. http://www.unwantedhorsecoalition.org/wp-content/uploads/2015/09/unwanted-horse-survey.pdf [downloaded December 21, 2017].
90. American Veterinary Medical Association. *Humane transport of equines*. https://www.avma.org/KB/Policies/Pages/Humane-Transport-of-Equines.aspx; 2017. Accessed 15 December 2017.

91. Stull CL. Responses of horses to trailer design, duration, and floor area during commercial transportation to slaughter. *J Anim Sci* 1999;**77**(11):2925–33.
92. Reece VP, Friend TH, Stull CH, Grandin T, Cordes T. Equine slaughter transport—update on research and regulations. *J Am Vet Med Assoc* 2000;**216**(8):12.
93. Animal Welfare Council. *A survey of animal control centers and the unwanted horse*. http://animalwelfarecouncil.com/wp-content/uploads/2013/05/unwantedhorsessurvey.pdf; 2009. Accessed 21 December 2017.

Index

Note: Page numbers followed by *f* indicate figures and *t* indicate tables.

D

E

F

G

H

I

N

R

S

T

U

V

W

Printed in the United States
By Bookmasters